R Stark County District Library AUG -- 2005
Main Library
715 Market Ave. N
Canton, OH 44702
330-452-0665
www.starklibrary.org

D1716033

SCRIBNER *TURNING POINTS* LIBRARY

Tobacco
in History and Culture

AN ENCYCLOPEDIA

Tobacco in History and Culture
AN ENCYCLOPEDIA

Editorial Board

Editor in Chief

JORDAN GOODMAN
Honorary Research Fellow
Wellcome Trust Centre
University College, London

Board Members

MARCY NORTON
Assistant Professor of History
George Washington University

MARK PARASCANDOLA
Cancer Prevention Fellow
National Cancer Institute

SCRIBNER *TURNING POINTS* LIBRARY

Tobacco
in History and Culture

AN ENCYCLOPEDIA

JORDAN GOODMAN
Editor in Chief

VOLUME
2
Native Americans–
Zimbabwe

Index

CHARLES SCRIBNER'S SONS
An imprint of Thomson Gale, a part of The Thomson Corporation

Detroit • New York • San Francisco • San Diego • New Haven, Conn. • Waterville, Maine • London • Munich

Tobacco in History and Culture: An Encyclopedia

Jordan Goodman

© 2005 by Thomson Gale,
a part of the Thomson Corporation.

Thomson and Star Logo are trademarks and Gale and Charles Scribner's Sons are registered trademarks used herein under license.

For more information, contact
Charles Scribner's Sons
27500 Drake Road
Farmington Hills, MI 48331-3535
Or visit our Internet site at
http://www.gale.com/scribners

ALL RIGHTS RESERVED
No part of this work covered by the copyright hereon may be reproduced or used in any form or by any means—graphic, electronic, or mechanical, including photocopying, recording, taping, Web distribution, or information storage retrieval systems—without the written permission of the publisher.

For permission to use material from this product, submit your request via Web at http://www.gale-edit.com/permissions, or you may download our Permissions Request form and submit your request by fax or mail to:

Permissions Department
The Gale Group, Inc.
27500 Drake Rd.
Farmington Hills, MI 48331-3535
Permissions Hotline:
248-699-8006 or 800-877-4253, ext. 8006
Fax: 248-699-8074 or 800-762-4058

LIBRARY OF CONGRESS CATALOGING-IN-PUBLICATION DATA

Tobacco in history and culture : an encyclopedia / Jordan Goodman, editor in chief.
  2 v. cm. – (Scribner turning points library)
  Includes bibliographical references and index.
  Contents: v. 1. Addiction–Music, popular–v. 2. Native Americans–Zimbabwe.
  ISBN 0-684-31405-3 (set hardcover : alk. paper) – ISBN 0-684-31406-1 (volume 1) – ISBN 0-684-31407-X (volume 2) – ISBN 0-684-31453-3 (e-book)
  1. Tobacco–History–Dictionaries. 2. Tobacco–Social aspects–Dictionaries. 3. Smoking–History–Dictionaries. I. Goodman, Jordan. II. Series.
  GT3020.T594 2004
  394.1'4–dc22   2004007109

This title is also available as an e-book.
ISBN 0-684-31453-3 (set)
Contact your Gale sales representative for ordering information.

Printed in the United States of America
10 9 8 7 6 5 4 3 2 1

Tobacco in History and Culture
AN ENCYCLOPEDIA

Contents

Color Plates
EIGHT PAGES OF COLOR PLATES APPEAR NEAR THE CENTER OF EACH VOLUME

Volume 1
Preface ix
Timeline xiii

A
Addiction 1
Additives 6
Advertising 11
Advertising Restrictions 19
Africa 23
Age 31
Air Travel 33
Alcohol, Tobacco, and Other Drugs 35
American Tobacco Company 42
Antismoking Movement Before 1950 45
Antismoking Movement From 1950 51
Antismoking Movement in France 59
Appetite 61
Archaeology 66
Architecture 71
Arents Collection 73

B
Bad Habits in America 77
Black Patch War 80
Body 81
Botany (History) 85
Brazil 88
British American Tobacco 92
British Empire 96

C
Calumets 103
Camel 105
Caribbean 107
Chemistry of Tobacco and Tobacco Smoke 111
Chesapeake Region 117
Chewing Tobacco 125
China 129
Christianity 135
Cigarettes 144
Cigars 150
Class 153
Connoisseurship 155
Consumption (Demographics) 167
Cuba 186

D
Developing Countries 193
Disease and Mortality 199
Doctors 214
Documents 216
Dutch Empire 219

E
English Renaissance Tobacco 223
Ethnicity 226

F
Film 231
Fire Safety 238
French Empire 240

G
Genetic Modification 245
Gitanes/Gauloises 248
Globalization 251

H
Hallucinogens 259

I
Industrialization and Technology 261
Insurance 267
Intellectuals 270
Iranian Tobacco Protest Movement 272
Islam 274

J
Japan 277
Judaism 281

K
Kentucky 285
Kretek 287

L
Labor 291
"Light" and Filtered Cigarettes 298
Literature 302

v

CONTENTS

Litigation 307
Lobbying 314
Lucky Strike 318
Lung Cancer 320

M
Marketing 327
Marlboro 337
Mayas 339
Medical Evidence (Cause and Effect) 343
Menthol Cigarettes 348
Mexico 350
Middle East 355
Missionaries 360
Music, Classical 364
Music, Popular 369

Volume 2

Timeline vii

N
Native Americans 375
Nazi Germany 382
New Deal 385
Nicotine 387

O
Oceania 393
Opium 396
Origin and Diffusion 397
Ostracism 403

P
Philip Morris 407
Philippines 410
Pipes 414

Plantations 423
Politics 428
Portuguese Empire 440
Processing 447
Product Design 450
Prohibitions 456
Psychology and Smoking Behavior 467
Public Relations 473

Q
Quitting 481
Quitting Medications 488

R
Regulation of Tobacco Products in the United States 491
Retailing 496

S
"Safer" Cigarettes 505
Sailors 509
Secondhand Smoke 512
Shamanism 517
Sharecroppers 522
Slavery and Slave Trade 525
Smoking Clubs and Rooms 533
Smoking Restrictions 535
Smuggling and Contraband 542
Snuff 547
Social and Cultural Uses 551
Soldiers 568
South and Central America 570
South Asia 575
South East Asia 579
Spanish Empire 585
Sponsorship 593

Sports 597
State Tobacco Monopolies 599

T
Taxation 603
Therapeutic Uses 609
Tobacco as an Ornamental Plant 616
Tobacco Control in Australia 619
Tobacco Control in the United Kingdom 622
Tobacco Industry Science 625
Tobacco Mosaic Virus 628
Toxins 630
Trade 634

U
United States Agriculture 653

V
Virginia Slims 663
Visual Arts 665

W
Warning Labels 675
Women 679

Y
Youth Marketing 689
Youth Tobacco Use 694

Z
Zimbabwe 699

Contributors List 703
Index 707

Tobacco in History and Culture
AN ENCYCLOPEDIA

Timeline

c. 50,000 B.C.E.: Australia populated. Humans there may have begun chewing tobacco species: *Nicotiana. gossei, N. ingulba, N. simulans, N. benthamiana, N. cavicola, N. excelsior, N. velutina,* and *N. megalosiphon.*

c. 15,000–10,000 B.C.E.: Americas south of the Arctic populated. Humans there may have begun to pick and use wild tobacco species.

c. 5000 B.C.E.: Maize-based agriculture develops in central Mexico, probable beginnings of tobacco cultivation as well.

c. 1400–1000 B.C.E.: Remains of cultivated and wild tobacco dating from this period have been found in High Rolls Cave in New Mexico. Dates established by radiocarbon methods.

1492: Columbus sees Taíno (Indians of Greater Antilles) with leaves that are probably tobacco. Two men among Columbus's crew explore the interior of Cuba and see people smoking.

1518: Juan de Grijalva, leader of expedition to Yucatan and Gulf of Mexico, accepts offerings of cigars or pipes.

1535: Publication of Gonzalo Fernández de Oviedo's *Historia general de las Indias,* which has first published reference to tobacco. It condemns it as a "vile vice" but also notes that the habit spread to "Christians" and black slaves as well.

1535: Jacques Cartier encounters natives using tobacco on the island of Montreal.

1555: Franciscan Friar André Thevet of Angoulême (France) witnesses Brazil's Tupinamba Indians smoking tobacco; following year sows tobacco seeds in France.

1560: Jean Nicot, France's ambassador to Portugal, writes of tobacco's medicinal properties, describing it as a panacea. Nicot sends *rustica* plants to French court.

1561: Nicot sends snuff to Catherine de Medici, the Queen Mother of France, to treat her son Francis II's migraine headaches.

1565: Sir John Hawkins's expedition observes Florida natives using tobacco.

1571: Publication of Nicolas Monardes's *Segunda parte del libro, de las cosas que se traen de nuestras Indias Occidentales, que sirven al uso de medicina* [The second part of the book of the things brought from our Occidental Indies which are used as medicine], which has the most extensive and positive description of tobacco to that date.

1583: Council of Lima declares that priests cannot consume tobacco in any form before saying mass, under threat of excommunication.

1585: Francis Drake expedition trades for tobacco with Island Caribs of Dominica.

1587: Gilles Everard's *De herba panacea* (Antwerp) is first publication devoted entirely to tobacco.

1588: Thomas Hariot publishes *A Brief and True Report of the New Found Land of Virginia,* in which he describes Virginia native people smoking tobacco.

1595: Anthony Chute publishes *Tabacco,* the first book in the English language devoted to the subject of tobacco.

1600: Franciscan missionary presents tobacco seeds and tobacco tincture to Tokugawa Ieyasu, who will become Shogun of Japan in 1603.

1603: Spanish colonies of Cumaná and Caracas (Venezuela) produce 30,000 pounds of tobacco.

1604: King James I publishes *A Counterblaste to Tobacco,* in which he condemns tobacco smoking as unhealthy, dirty, and immoral.

1606: King of Spain prohibits the cultivation of tobacco in Caribbean and South America to thwart contraband trade between Spanish settlers and English and

vii

TIMELINE

Dutch traders. Edict rescinded in 1612.

1607: Inhabitants of Sierra Leone seen sowing tobacco.

1607: Jamestown, the first permanent English colony in the Americas, is founded.

1612: John Rolfe raises Virginia's first commercial crop of "tall tobacco."

1617: Mughal Shah Jahangir (reigned 1605–1627) bans smoking because tobacco consumption creates "disturbance in most temperaments."

1624: Texts by Chinese physicians Zhang Jiebin (1563–1640) and Ni Zhumo (c. 1600) mention tobacco in section on pharmacopoeia.

1627: Tobacco cultivation in Ottoman territory is banned.

1636: First state tobacco monopoly established in Castille (Spain).

1642: Papal Bull forbids clerics in Seville from using tobacco in church and other holy places.

1674: Tobacco monopoly established in France.

1682: Virginia colonists rebel when the government fails to decree a cessation in tobacco crops after bumper crops lead to low prices. Disgruntled planters destroy thousands of tobacco plants; six ringleaders are executed.

1698: In Russia, Peter the Great agrees to a monopoly of the tobacco trade with the English, against church wishes.

1724: Pope Benedict XIII learns to smoke and use snuff, and repeals papal bulls against clerical smoking.

1753: Linnaeus names the plant genus *nicotiana*. and describes two species, *nicotiana rustica*. and *nicotiana tabacum*.

1760: Pierre Lorillard establishes a "manufactory" in New York City for processing pipe tobacco, cigars, and snuff. P. Lorillard is the oldest tobacco company in the United States.

1794: U.S. Congress passes the first federal excise tax on snuff, leaving chewing and smoking tobacco unaffected.

1827: First friction match invented.

1828: Isolation of nicotine from tobacco by Wilhelm Posselt and Karl Reimann.

1832: Paper-rolled cigarette is invented in Turkey by an Egyptian artilleryman.

1839: Discovery that flue-curing turns tobacco leaf a bright brilliant yellow and orange color. The bright-leaf industry is born.

1843: French tobacco monopoly begins to manufacture cigarettes.

1847: In London, Philip Morris opens a shop that sells hand-rolled Turkish cigarettes.

1849: J. E. Liggett and Brother is established in St. Louis, Missouri, by John Edmund Liggett.

1854: Philip Morris begins making his own cigarettes. Old Bond Street soon becomes the center of the retail tobacco trade.

1868: British Parliament passes the Railway Bill of 1868, which mandates smoke-free cars to prevent injury to nonsmokers.

1880: James Bonsack is granted a patent for his cigarette-making machine.

1881: James Buchanan (Buck) Duke starts to manufacture cigarettes in Durham, North Carolina.

1889: Five leading cigarette firms, including W. Duke Sons & Company, unite. "Buck" Duke becomes president of the new American Tobacco Company.

1890: *My Lady Nicotine*, by Sir James Barrie, is published in London.

1890–1892: Popular revolts against imposition of British-controlled monopoly on sale of tobacco take place in Iran.

1899: Lucy Payne Gaston founds the Chicago Anti-Cigarette League, which grows by 1911 to the Anti-Cigarette League of America, and by 1919 to the Anti-Cigarette League of the World.

1902: Imperial Tobacco (U.K.) and American Tobacco Co. (U.S.) agree to market cigarettes in their respective countries exclusively, and to form a joint venture, the British American Tobacco Company (BAT), to sell both companies' brands abroad.

1907: The U.S. Justice Department files anti-trust charges against American Tobacco.

1908: The U.K. Children Act prohibits the sale of tobacco to children under 16, based on the belief that smoking stunts children's growth.

1910: Gitanes and Gauloises cigarette brands are introduced in France.

1911: U.S. Supreme Court dissolves Duke's trust as a monopoly, in violation of the Sherman Anti-Trust Act (1890). The major companies to emerge are American Tobacco Co., R.J. Reynolds, Liggett & Myers Tobacco Company (Durham, N.C.), Lorillard, and British American Tobacco (BAT).

1913: R.J. Reynolds introduces the Camel brand of cigarettes.

1913: China has its first harvest of Bright leaf tobacco, grown from imported American seeds and using American growing methods.

1916: Henry Ford publishes an anti-cigarette pamphlet titled *The Case against the Little White Slaver*.

1924: Philip Morris introduces Marlboro, a women's cigarette that is "Mild as May."

TIMELINE

1927: Long Island Railroad grants full rights to women in smoking cars.

1933: United States Agricultural Adjustment Act of 1933 compels tobacco farmers to cut back on output by reducing acreage devoted to tobacco production, in return for price supports. They are saved from economic ruin.

1938: Dr. Raymond Pearl of Johns Hopkins University reports to the New York Academy of Medicine that smokers do not live as long as nonsmokers.

1950: Five important epidemiological studies show that lung cancer patients are more likely to be smokers than are other hospital patients.

1954: Results from two prospective epidemiological studies show that smokers have higher lung cancer mortality rates than nonsmokers. The studies were conducted by E. Cuyler Hammond and Daniel Horn in the U.S. and Richard Doll and Austin Bradford Hill in the U.K.

1957: First Japanese-made filter cigarette, Hope, is put on the market.

1964: *Smoking and Health: Report of the Advisory Committee to the Surgeon General*, the first comprehensive governmental report on smoking and health, is released at a highly anticipated press conference. It concludes that smoking is a cause of lung cancer, laryngeal cancer, and chronic bronchitis and "is a health hazard of sufficient importance in the United States to warrant appropriate remedial action."

1965: U.S. Congress passes the Federal Cigarette Labeling and Advertising Act, requiring health warnings on all cigarette packages stating "Caution—cigarette smoking may be hazardous to your health."

1970: U.S. Congress enacts the Public Health Cigarette Smoking Act of 1969. Cigarette advertising is banned on television and radio.

1970: World Health Organization (WHO) takes a public position against cigarette smoking.

1972: First report of the surgeon general to identify involuntary (secondhand) smoking as a health risk.

1977: American Cancer Society (ACS) sponsors the first national "Great American Smokeout," a grassroots campaign to help smokers to quit.

1986: Congress enacts the Comprehensive Smokeless Tobacco Health Education Act, requiring health warnings on smokeless (spit) tobacco packages and advertisements and banning smokeless tobacco advertising on radio and television.

1988: Liggett Group (L&M, Chesterfield) ordered to pay Antonio Cipollone $400,000 in compensatory damages for its contribution to his wife Rose Cipollone's death (she died in 1984). First-ever financial award in a liability suit against a tobacco company. However, the verdict was later overturned on appeal, and the lawsuit was dropped when the family could no longer afford to continue.

1988: Publication of *The Health Consequences of Smoking: Nicotine Addiction*, the first surgeon general's report to deal exclusively with nicotine and its effects.

1990: Airline smoking ban goes into effect, banning smoking on all scheduled domestic flights of six hours or less.

1991: U.S. Food and Drug Administration (FDA) approves a nicotine patch as a prescription drug.

1992: World Bank establishes a formal policy on tobacco, including discontinuing loans or investments for tobacco agriculture in developing countries.

1994: Six major domestic cigarette manufacturers testify before the U.S. House Subcommittee on Health and the Environment that nicotine is not addicting and that they do not manipulate nicotine in cigarettes.

1995: *Journal of the American Medical Association (JAMA)* publishes a series of articles describing the contents of secret documents from the Brown & Williamson Tobacco Corporation indicating that the industry knew early on about the harmful effects of tobacco use and the addictive nature of nicotine.

1996: President Bill Clinton announces the nation's first comprehensive program to prevent children and adolescents from smoking cigarettes or using smokeless tobacco. Under the plan, the Food and Drug Administration would regulate cigarettes as drug-delivery devices for nicotine.

1998: California becomes the first state in the nation to ban smoking in bars.

1999: U.S. Department of Justice sues the tobacco industry to recover billions of dollars spent on smoking-related health care, accusing cigarette makers of a "coordinated campaign of fraud and deceit."

1999: Attorneys general of 46 states and 5 territories sign a $206 billion Master Settlement Agreement with major tobacco companies to settle Medicaid lawsuits.

2000: In Canada, Health Minister Allan Rock unveils new health labels that include color pictures.

2000: U.S. Supreme Court issues a 5–4 ruling that existing law does not provide the Food and Drug Administration with authority over tobacco or tobacco marketing, thus invalidating the 1996 Clinton Administration's regulations.

2001: BAT breaks into Vietnam market, announces that it has been granted a license for a $40 million joint venture with

TIMELINE

Vintaba to build a processing plant in Vietnam.

2003: First stage of the Tobacco Advertising and Promotion Act 2002 bans new tobacco sponsorship agreements, advertising on billboards and in the press, and free distributions. The ban also covers direct mail, Internet advertising, and new promotions.

2003: New York City's smoking ban goes into effect, forbidding smoking in all restaurants and bars, except for a few cigar lounges.

2004: Complete public smoking ban goes into effect in Ireland.

Native Americans

Many Native Americans throughout North and South America believe that tobacco is so powerful that it was involved in the very act of creating the world. In the Pima or O'odham origin story, for example, Blue Gopher lit a huge cigarette made out of Coyote's tobacco wrapped in a cornhusk. He puffed toward the east in a great white cloud that cast a shadow over the land. A carpet of grass grew in the shadow. Blue Gopher scattered the seeds of other plants across the grassy area, thereby causing corn to grow.

In one version of the Navajo creation story, Sky Father and Earth Mother smoked tobacco, before creation began, to help them plan the awesome task that lay ahead. Morning Star—a Crow Indian deity—turned into the first tobacco plant after he fell from the sky. The first tobacco grew from the head of Earth Mother, one of the Haudenosaunee (Iroquois) creator spirits, while the Cahuilla creator Mukat drew the first tobacco and pipe from his heart, then made the sun to light them. After he was killed, tobacco grew from his heart. The Kickapoo creator Kitzihiat also used a piece of his heart to make the first tobacco. Pulekukewerek, one of the Yurok creator *woges*, grew from a tobacco plant; then tobacco continued to grow from the palms of his hands, so that he never ran out.

The Huichol in the mountains of western Mexico have similar beliefs, as do the Shipibo along the upper Amazon in Peru and the Haida and Tlingit in southern Alaska and many native peoples in between. In one version of the Huichol creation story, the first tobacco grew from the semen of Deer Person, one of their most powerful deities, who turned into corn and peyote and whose blood is still used to nourish corn and bless babies. Huichol tobacco belongs to Grandfather Fire—the most powerful deity of all—tobacco was once a hawk and even today it is the spiritual essence of the gods. Huichol tobacco (*makutsi*) is also the most powerful tobacco on earth, almost as powerful as peyote and able to cause visions, with up to 18 percent nicotine.

The belief that tobacco is so powerful that it figured into creation itself is widespread throughout North and South America. Even the tribes

NATIVE AMERICANS

that lack this belief have similar concepts; for example, that the spirits are addicted to tobacco. American Indians view tobacco, almost without exception, as an essential, core element of their religions and rituals. Taken together, these widespread beliefs and practices strongly suggest that tobacco use is a very ancient activity in the Americas, so old and elemental that it probably began very early on, in prehistoric humankind's existence in the Americas.

Evidence for the Early Use of Tobacco

Of the seven species of *Nicotiana* that have been and still are being used by Native Americans, two were domesticated by prehistoric Indians to the extent that the plant species could not survive, beyond a few generations, without the help of people who planted them, weeded them, and otherwise tended to their basic needs. These domesticated species and their regions of use by Native North Americans (exclusive of commercial tobacco and recent introductions) are as follows:

Species	Regions of Use
Nicotiana rustica L.	Eastern U.S. and Canada; MesoAmerica; Southwestern U.S.; probably Caribbean
Nicotiana tabacum L.	MesoAmerica; parts of U.S. Southwest; probably Caribbean

The five other tobacco species, in contrast, are wild plants that can and do thrive from generation to generation without the help of humans, though they do prefer disturbed environments, such as arroyo beds (stream sides), road cuts, and burned over areas, which humans readily provide. The species *Nicotiana quadrivalvis* is somewhere in between domesticated and wild: Two of its varieties (*wallacei* and *bigelovii*) are wild, though they are often cultivated, whereas the other two (*quadrivalvis* and *multivalvis*) are known only in cultivation. The wild species and their regions of use are as follows:

Species	Regions of Use
Nicotiana attenuata Torr.	U.S. Southwest; Great Basin; California; Pacific Northwest; extreme northern Mexico; southwest Canada
Nicotiana quadrivalvis Pursh.	southern California to Washington; Missouri River Valley; Canadian Plains; extreme southern Alaska; upper Columbia and Snake River Valleys
Nicotiana clevelandii Gray	northwest Mexico; possibly southern California
Nicotiana glauca Grah.	Mexico; southern California; western Arizona
Nicotiana trigonophylla Dun.	southwestern U.S.; southern California; Mexico

Archeological evidence from North America indicates the use of several tobacco species for thousands of years. The earliest known tobacco in South America is only a few hundred years old. Earlier evidence is undoubtedly there, since the ancestors of all of these tobacco species originated in South America millions of years ago, then slowly expanded their ranges north through Central America and on into North America or later were carried there. ✦

The archaeological evidence of tobacco comes primarily in the form of carbonized seeds and preserved pollen, which are very difficult to recover and identify. Even the largest tobacco seed is smaller than the period at the end of this sentence, which means that it takes a very fine mesh screen with holes no larger than one-quarter of 1 millimeter across to recover a seed. And while tobacco pollen is fairly distinctive down to the generic level (*Nicotiana*), it is not possible to distinguish among the various species (*rustica* or *tabacum*) based on pollen. Also, the pollen of one of tobacco's close relatives (*lycium*, or Wolfberry) is similar to *Nicotiana*, so the use of pollen can be problematic.

Despite these drawbacks, archeologists in North America have been successful in finding prehistoric tobacco, and there is good evidence for its initial use as early as 1400 B.C.E. in the Southwestern deserts, and by about 180 C.E. in the Eastern woodlands.

The sequence of development, as shown in the map, is summarized as follows. The roman numerals correspond to the map categories.

✦ *See the map in "Origin and Diffusion."*

Species	Description
I. **Desert tobaccos** (*N. attenuata*, *N. trigonophylla*, *N. quadrivalvis*)	Ancestral South American species slowly expanded their ranges naturally, reaching Mexico after the end of the Pleistocene, when conditions warmed enough to allow them to spread north. Helped northward to present extent by human activity, beginning no later than 1000 B.C.E.
II. *Nicotiana rustica*	Domesticated 7,000 to 10,000 years ago in Andes Mountains, then taken north by early farmers, reaching American Southwest by 1000 to 1400 B.C.E. and Eastern Woodlands by 180 C.E.
III. *Nicotiana tabacum*	Domesticated several thousand years ago in the Andes Mountains, then taken east and north through the lowlands. May have reached Southwestern U.S. in late prehistoric times.
IV. *Nicotiana glauca*	Introduced accidentally into Mexico, California, Arizona, and Florida in historic times (for example, in the ballast of ships). Since then, the western Navajo, Barona Digueno, and a few other tribes have adopted it and now consider it traditional tobacco.

ANCIENT FARMING. In both the Southwest and the Eastern woodlands, domesticated tobacco first appeared with other cultivated plants as part of a larger horticultural complex that also included wild plants.

NATIVE AMERICANS

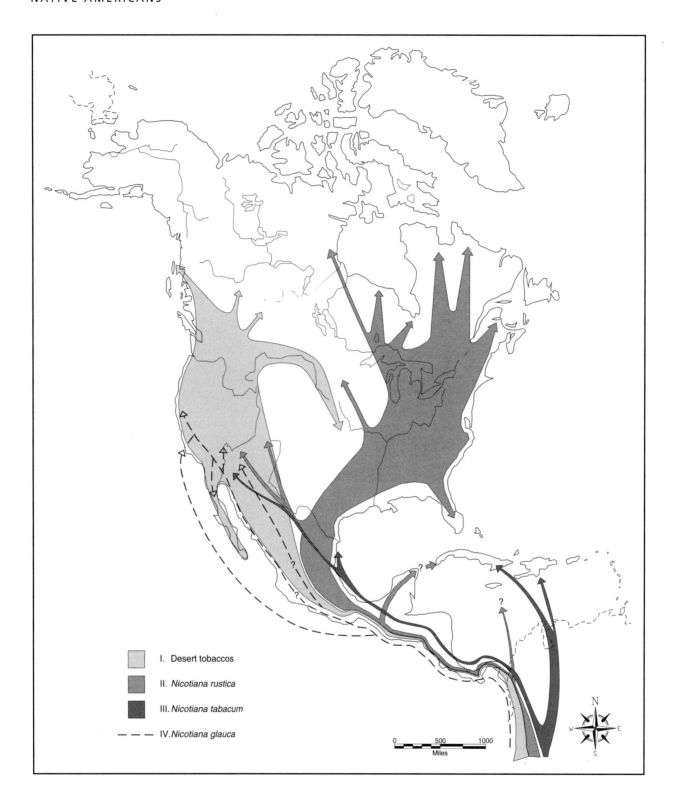

Sequence of Development of Native American Tobacco Use. RONALD STAUBER

In the Southwest, this gardening tradition consisted of cultivated tobacco, and two species of wild tobacco, along with maize, squash, beans, wild and cultivated amaranths, goosefoot and other weedy annuals that were encouraged or at least tolerated in the farm fields. In the East, early gardening focused on cultivated sunflowers, goosefoot,

Native American pipes and cigarettes come in many forms. Upper left: Cochiti Pueblo pipe; upper center and right: Kayenta Navajo "cloud blower" pipes; lower right: Cocopa reed cane cigarettes; lower center: Hopi clay pipe; lower left: Navajo clay pipe. COURTESY OF THE MAXWELL MUSEUM OF ANTHROPOLOGY, UNIVERSITY OF NEW MEXICO. PHOTOGRAPHER DAMIAN ANDRUS.

and marsh elder, with corn and cultivated tobacco added 1,000 years later.

Wild plants were clearly involved in the adoption of corn, tobacco, squash, and beans by the prehistoric Native Americans. In both the Southwest and later the East, maize and tobacco did not arrive out of a vacuum, nor did they drop into one. They were already being grown to the south, in central Mexico, where maize-based agriculture began around 5000 B.C.E., then moved slowly north, as local hunters and foragers added it to their plant husbandry tradition. Or perhaps small agricultural groups expanded their ranges or maybe even migrated from one region to another. However it spread, farming was added to an already existing husbandry complex that involved the encouragement and even planting of a number of wild plants. Two species of wild tobacco, as well as amaranth, goosefoot, purslane, globe mallow, and other plants that preferred disturbed soils, were included in the complex in the Southwest. The early gardening culture in the east grew goosefoot, marshelder and sunflowers, and may have grown wild squash and gourds, maygrass, knotweed and a few other plants.

After the addition of cultivated tobacco, corn, squash, and beans, agricultural societies rapidly evolved throughout North and South America. By the eve of European contact, cultivated tobacco was traded far to the north of its range, into northern Canada, and even the wild *Nicotiana quadrivalvis* variety *multivalvis* was encouraged, if not cultivated, in southern Alaska. Similar processes were at work in South America, and by the time the Europeans arrived, the use and veneration of tobacco was a key, core element of all Native American cultures, with the exception of the Inuit (Eskimo) and Aleut, who were too far away to participate in the tobacco trade system.

From the southern tip of South America to southern Alaska, tobacco was ingested in many forms, including pipes, reed cane and corn husk cigarettes, even in maple and other wild plant leaves. It was also chewed, licked, snuffed, taken as eye drops, and even administered in enemas. Some tribes preferred to smoke tobacco in carved stone, calumet-style pipes, such as those used by the Plains Indians, while

NATIVE AMERICANS

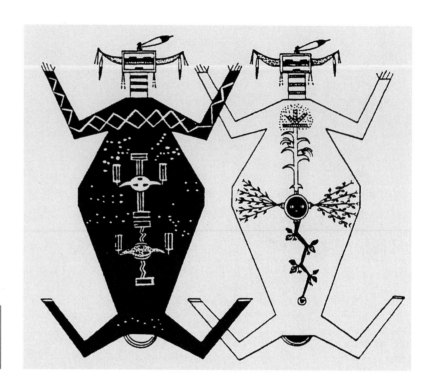

The four sacred plants of the Navajo—corn, beans, squash, and tobacco—growing on Mother Earth. Father Sky and Mother Earth are sometimes shown in sandpaintings and on rugs. COURTESY OF JOSEPH WINTER

others used smaller stone and clay pipes, reed and leaf wrapped cigarettes, or, most simply of all, a wad of tobacco leaves packed between a person's cheeks and teeth or between the lips and teeth.

Tobacco in Native American Religion

Tobacco is the heart of Native American religion and the core of American Indian culture. Tobacco has remained a constant unifying force, linking all tribes together, linking all generations together for thousands of years. Even as Native American religions changed and became organizationally more complex, tobacco use also became more complex, as did the activities of the deities who created it and who were created by it.

snuff a form of powdered tobacco, usually flavored, either sniffed into the nose or "dipped," packed between cheek and gum. Snuff was popular in the eighteenth century but had faded to obscurity by the twentieth century.

Many Native Americans continue to use tobacco in a sacred manner, while others smoke, chew, and **snuff** it in the same manner as non-Indians, as a recreational drug. For the traditionalists, there is nothing recreational about tobacco, for it is considered a sacred plant, a life-affirming force, a food of the spirits, at times a god itself. From southern Chile to Alaska, Native Americans have used and continue to use *Nicotiana rustica*, *N. tabacum*, *N. attenuata*, and several other species of tobacco as a ritual narcostimulant—a psychotrophic, mind-altering substance that serves as a medium between the ordinary world of humans and the super-ordinary world of spirits. Tobacco leaves were and are smoked in pipes, cigars, and cigarettes. Leaves are chewed (often with lime from shells) and sometimes eaten. Resin and concentrates are licked. An infusion is drunk, occasionally with Datura and/or other hallucinogenic plants. Tobacco powder is snuffed. Tobacco smoke is blown on the body and leaves are used medically as a poultice. Tobacco incense is burned. Tobacco offerings are buried, cast on the ground, into the air, onto the water.

This detail from a larger illustration shows four Aztec deities "drinking" tobacco. The full illustration, showing nine panes, was redrawn from a figure in the *Codex Vaticanus.* © BRIGITTE FELIX

TOBACCO AND MEDICINE SOCIETIES. Beginning with individual medicine men and women who ministered to the religious and medical needs of their bands and other groups, American Indian religion became more organized as populations increased, beliefs changed, and outside political and economic relations evolved. After the individual medicine people came the medicine societies, composed of most if not all of the members of the group, with different societies providing different medicines and religious ceremonies. And eventually the societies evolved into priesthoods, whose memberships were restricted and often hereditary, and whose leaders became so powerful that theocracies often emerged, such as the Aztecs and Incas, whose leaders were the highest priests in the land.

But whatever the level and scope of religious power, tobacco was and is still used, with even the medicine people, medicine societies, and priesthoods taking on tobacco-oriented themes and identities. Thus there are Tobacco shamans in South and Central America who ingest the plant almost constantly, not only to heal and bless but also to commune directly with the tobacco spirits.

There are also tobacco medicine societies, such as among the Crow on the upper Missouri, whose sole function is to grow two kinds of sacred tobacco, *Nicotiana quadrivalvis* varieties *quadrivalvis* and *multivalvis*, which are essential for the survival of the tribe. And there are or were even tobacco priesthoods, such as the Cult of Cihuacoatl among the Aztecs, the mother of the other gods, the Snake Woman whose physical manifestation on earth was the tobacco plant and whose chief priest—also called Snake Woman—was second in power only to the great Montezuma himself.

Tobacco shamans, tobacco medicine societies, and tobacco priesthoods were part of an array of Native American religious groups that ranged from the individual medicine-people of tiny bands of Caribou hunters in northern Canada to the deified leaders of huge city-states in Peru that

Tobacco Use During the *Inipi* Sweat Lodge Ceremony

Most contemporary Native American ceremonies involve the use of tobacco. One of the most popular rituals is the *Inipi* purification ceremony of the Lakota, which has been adapted by many individuals and pan-tribal groups throughout the United States. Most tribes have their own sweat lodge purification ceremonies, and the amalgam of Lakota *Inipi* and another tribe's purification rite, such as the Navajo's, is a ceremony that is filled with the smoke of sacred tobacco. In most sweat lodges a Plains Indian–style carved stone and wooden pipe is used; in others, especially in the Southwest, Navajo and Pueblo-style corn husk cigarettes are smoked. All of the participants in the ceremony are purified in two ways: by the steam from the hot rocks and by the smoke from the tobacco. Each participant is given the opportunity to smoke one or more times, and to blow out the smoke and rub it on his (and in some case her) legs, head, and other body parts. It is also puffed in the four directions, and a prayer is often said for one of the participants. There are many variations to this theme, but the overall thrust is that tobacco smoke is a sacred, purifying element that not only cleanses the body and soul but also pleases the Great Spirit and other deities as it wafts its way into the heavens. *The Sacred Pipe: Black Elk's Account of the Seven Rites of the Oglala Sioux* provides detailed descriptions of tobacco use during the *Inipi* and other Lakota ceremonies.

controlled vast empires. All used tobacco as a universal means of communicating with each other as well as with the spirit world.

> *That done they blew the tobacco in all four directions where it appeared as a fog in which they moved away. Those were the sun's inner form, the moon's inner form, and the inner forms of the mountains that had been made. For these the (smoke) ceremony had been performed (to show respect for the inner forms to be). For these, what was to be dark cloud and dark mist, male rain and female rain, sunray, pollens of dawn and evening twilight, rainbow, all of these were laid down before them, in these they clothed themselves (from the Blessingway Songs of Earth's Inner Form, in Wyman, pp. 124–127).*

See Also Caribbean; Mayas; Origin and Diffusion; Shamanism; South and Central America.

▌JOSEPH WINTER

BIBLIOGRAPHY

Bohrer, Vorsila. L. "Flotation Analysis from High Rolls Cave (LA 114103) Otero County, New Mexico." In *Southwest Ethnobotanical Enterprises Report 39*. Portales, N. Mex., 2003. This unpublished work contains the earliest archaeological dates of tobacco. Eventually it will be published by the state of New Mexico.

Brown, Joseph E. *The Sacred Pipe: Black Elk's Account of the Seven Rites of the Oglala Sioux*. Norman and London: University of Oklahoma Press, 1953.

Gerstel, D. U. "Tobacco *Nicotiana tabacum* (Solanaceae)." In *Evolution of Crop Plants*. Edited by N. W. Simmonds. London: Longman, 1976.

Russell, Frank. *The Pima Indians*. Tucson, Ariz.: University of Arizona Press, 1975.

Wilbert, Johannes. *Tobacco and Shamanism in South America*. New Haven, Conn: Yale University Press, 1987.

Winter, Joseph C. "Feeding the Ancestors: The Role of Tobacco in the Evolution of Southwestern Agriculture and Religion." In *La Frontera, Papers in Honor of Patrick H. Beckett*. Paper 24. Las Cruces, N.Mex.: Archaeological Society of New Mexico, 1999.

———, ed. *Tobacco Use By Native North Americans Sacred Smoke and Silent Killer*. Norman, Okla.: University of Oklahoma Press, 2000.

Wyman, Leland C. *Blessingway*. Tucson: University of Arizona Press, 1987.

Nazi Germany

One of the most morally repugnant regimes of the twentieth century was a pioneer in cancer prevention. Following their rise to power in 1933, the Nazis launched what historian Robert Proctor argues was the most dynamic antismoking crusade in the world, at the time. It included a comprehensive range of prohibitions on smoking that would not be rivaled for fifty years or more, and the promotion of pioneering studies on the relationship between smoking and lung cancer, more than a decade before **epidemiological** research in the United Kingdom and the United States identified smoking as a cause of the disease.

epidemiological pertaining to epidemiology, that is, to seeking the causes of disease.

Among the measures the Nazis introduced were bans on smoking in certain public, military, and work spaces including post offices, hospitals, government offices, and the German Air Force, the Luftwaffe. They forbade all uniformed police and special police (SS) officers from smoking on duty; imposed a similar prohibition on midwives; and, in 1944, banned smoking on all civic transport in Germany. Tobacco advertising was also restricted. It was not to appear on billboards, in sports facilities, or on public transportation. It was not to be sent by mail or accepted for publication in the text sections of newspapers or magazines. Advertising was not to associate smoking with sports, driving, or women, or to portray it as healthful or harmless. By contrast, the Nazis stigmatized smoking as the habit of Jews, decadent women, and degenerate intellectuals.

National Socialist opposition to tobacco was consistent with the regime's larger emphasis on racial hygiene. In its view, tobacco was a genetic poison that caused cancer, infertility, heart attacks, and other problems. The particular concern was that tobacco might harm the reproductive performance of the race. The Nazis believed women were particularly vulnerable to the effects of smoking, endangering not only themselves, but also their children, and consequently the German race. In their efforts to protect the race, they created an antitobacco campaign that drew on the broader policy of a doctor-directed health leadership (*Gesundheitsführung*), which emphasized health prevention and elevated the public good over individual liberties, the so-called "duty to be healthy." The Nazis also initiated wide-ranging programs of clinical, experimental, and epidemiological research into tobacco and health. In 1941 they created an Institute for Tobacco Hazards Research at the University of Jena. Its director, the physician and SS member Karl Astel, advocated opposition to tobacco as a "National Social duty" (Proctor).

Such opposition was underpinned, in part, by pioneering research that identified smoking as a major cause of cancer and other ailments. For example, in a 1939 survey of 8,000 publications, the Chemnitz physician Fritz Lickint concluded that tobacco was the cause of cancers in what he called the "smoke alley," or *Rauchstrasse*, the lips, tongue, lining of the mouth, jaw, esophagus, windpipe, and lungs. He argued that tobacco caused arteriosclerosis, ulcers, halitosis, and many other ills. It was a cause of infant mortality; an addiction, akin to morphine. Dr. Lickint suggested that passive smoking (*Passivrauchen*, as he called it) was a danger to nonsmokers.

Other physicians added to the evidence against tobacco. In 1939 the Cologne-based Franz H. Müller published a study "Tabakmissbrauch und Lungencarcinom" comparing the behavior of 96 lung cancer patients with a healthy control group. According to Müller, nonsmokers were more common in the healthy group than in the lung-cancer group; those with lung cancer smoked more than twice as much per day as the members of the healthy group; and 16 percent of the healthy group were nonsmokers, compared with 3.5 percent of the lung-cancer group. In 1943 Erich Schöniger and Eberhard Schairer of the Jena Institute added to this research. They found only 3 nonsmokers

NAZI GERMANY

Despite leading what has been called the most dynamic antismoking crusade in the world at that time, Nazi Germany was not successful in reducing tobacco consumption during its first seven years of rule. Here a Nazi soldier is shown smoking a cigarette.
© BETTMANN/CORBIS

among a group of 109 lung cancer cases, a much lower number than would be expected in the general population. Furthermore, when they sent 555 questionnaires to the families of patients who had died of other kinds of cancer, they also found that smokers were no more likely to develop other forms of cancer than were nonsmokers, suggesting that these victims of lung cancer were not constitutionally predisposed to the disease.

Despite these findings, tobacco consumption rose dramatically for the first six or seven years of Nazi rule. This rise may have symbolized opposition to the regime. However, it was a consequence of poor enforcement, fears of alienating soldiers, economic recovery in the first six or seven years of the regime, and the efforts of the tobacco industry. The latter vigorously opposed the antitobacco crusade, creating a scientific commission to discredit Nazi-inspired antismoking efforts. Tobacco consumption began to fall in 1942, as World War II turned against Germany. Germany's low postwar mortality rate from cancer, therefore, may have had less to do with Nazi policies than the hardships of the war and the postwar years. The one exception may be female lung cancer. The historian Robert Proctor estimates that pressure to stop smoking may have prevented 20,000 lung cancer deaths in German women.

See Also Disease and Mortality; Lung Cancer; Medical Evidence (Cause and Effect).

▌DAVID CANTOR

BIBLIOGRAPHY

Cantor, David. "Cancer and the Nazis." *Science as Culture* 10 (2001): 121–133.

Proctor, Robert N. "The Nazi War on Tobacco: Ideology, Evidence, and Possible Cancer Consequences." *Bulletin of the History of Medicine* 71 (1997): 435–488.

———. *The Nazi War on Cancer.* Princeton, N.J.: Princeton University Press, 1999.

New Deal

As the Great Depression deepened in the months following the 1929 stock market crash, foreign and domestic consumption of all types of American tobacco declined, causing prices to plummet. The average price of **flue-cured** tobacco (the most important variety because of its use by American cigarette manufacturers), which had not dropped below 20 cents per pound from 1920 to 1927, plunged to 8.4 cents by 1931. Revenues for tobacco growers in 1932 were down to only 40 percent of the average received during the 1920s. After 1932, any further slump in prices would have spelled economic ruin for most producers in the tobacco-growing states.

During the 1932 presidential campaign, Franklin Roosevelt endorsed a remedy called the domestic allotment plan. Heavily promoted by the agricultural economist Milburn L. Wilson, the proposal called for the government to pay growers of certain crops to reduce production voluntarily. Farmers would not only benefit from checks financed by a tax on processors, but also from price gains on the crops they did produce. Wilson also desired active grower participation in the programs. Farmers would vote in **referenda** for acceptance of the programs and help oversee implementation by choosing producer committees to ensure compliance and resolve disputes.

After winning the election, Roosevelt worked with Congress to establish the Agricultural Adjustment Administration (AAA)—the New Deal agency established by the Agricultural Adjustment Act of 1933 to administer the production control programs largely based on Wilson's allotment plan. New Dealers modified the programs over the years in reaction to farmer discontent, judicial decisions, and political influences. The result was four distinct tobacco programs for the 1933, 1934–1935, 1936–1937, and 1938–1939 periods, respectively.

flue-cured tobacco also called Bright Leaf, a variety of leaf tobacco dried (or cured) in air-tight barns using artificial heat. Heat is distributed through a network of pipes, or flues, near the barn floor.

referendum an election where voters choose between policies or actions rather than between candidates; for the electorate to vote directly on a law or tax rather than indirectly through representatives.

NEW DEAL

New Deal programs in the 1930s helped tobacco growers weather the hard economic times of the Great Depression. This North Carolina farmer, holding leaf tobacco, represents the American farmers who benefited from government subsidies. © CORBIS

soil conservation the husbandry of the land; making careful use of soils to prevent erosion and maintain fertility and value.

In 1933 the AAA negotiated a marketing agreement with buyers to boost prices. After a requisite number of farmers agreed to reduce their 1934 planted acreage up to 30 percent in exchange for government payments, buyers consented to pay at least 17 cents per pound for flue-cured tobacco. The first tobacco program was a success, as evidenced by the fact that growers in North Carolina (the largest flue-cured tobacco state) received over $85 million for their crop compared to only $35 million in 1932.

The programs changed over the next two years as a majority of farmers rallied behind an effort to compel participation. The fact that a minority of growers did not participate but still benefited greatly from the increased prices on fully planted acreage angered many cooperating producers. Their pressure on the president and Congress resulted in the Kerr-Smith Tobacco Act of 1934. Under this law, producers were assigned a quota based on past production. Tobacco sold over a farmer's limit would have a prohibitive tax placed on it, thus creating a disincentive to overproduce. Meanwhile, the AAA continued payments to growers in exchange for reduced planting.

After the shock wore off from the Supreme Court's January 1936 ruling in the *Hoosac Mills* case (declaring the unconstitutionality of the AAA's production control contracts and processing taxes), Roosevelt supported an important expedient—the **Soil Conservation** and Domestic Allotment Act (SCDAA). Growers would now be paid from the Treasury for planting less "soil-depleting" crops, including tobacco, and planting more "soil-building" crops such as grasses and legumes. The goal was to achieve production control indirectly under the auspices of soil conservation. Because Congress repealed the Kerr-Smith Act after the high court's decision, the SCDAA was a completely voluntary program. It was only marginally effective in reducing surpluses; payments were less than under the first AAA Act, and many growers increased their planted acreage.

With a friendlier pro-New Deal Court in place, Congress passed the Agricultural Adjustment Act of 1938. Building on the existing soil conservation legislation, **marketing quotas** were added during high-surplus periods, subject to a two-thirds approval vote by the growers. This law governed the tobacco programs until World War II.

marketing quota the amount of cured leaf tobacco (usually expressed in pounds) that a tobacco grower may produce and sell in a given year. The United States Department of Agriculture began setting marketing quotas for tobacco in 1933.

By the end of the 1930s, the AAA successfully helped American tobacco farmers weather the storm of the Depression. It enabled growers to better adjust supply to demand while maintaining prices at the 1920s average level. In North Carolina, growers' incomes tripled while land values doubled for farms holding a tobacco allotment. Tenants and sharecroppers in tobacco-growing areas were not as adversely affected by AAA policies as those in the cotton regions, due mainly to the relatively larger labor requirements for raising tobacco. While many cotton regions suffered a tremendous drop in nontenured laborers, North Carolina, for example, experienced only a 10-percent decline in tenants.

In creating the first farm subsidies programs in American history, the AAA helped most tobacco farmers avert economic calamity. The New Deal for tobacco was generally a successful holding operation until the prosperous World War II years. It also had a lasting impact on the nation's tobacco growers. The government's relationship with tobacco producers has never been the same, as farm subsidy programs have become part of America's political economy.

See Also Politics; Sharecroppers; United States Agriculture.

■ KEITH VOLANTO

BIBLIOGRAPHY

Badger, Anthony J. *Prosperity Road: The New Deal, Tobacco, and North Carolina*. Chapel Hill: University of North Carolina Press, 1980.

Rowe, Harold B. *Tobacco under the AAA*. Washington, D.C.: The Brookings Institution, 1935.

Rowley, William D. *M. L. Wilson and the Campaign for the Domestic Allotment*. Lincoln: University of Nebraska Press, 1970.

Nicotine

At the beginning of the twentieth century, lung cancer was a rare medical disease. The "cigarette," a new product, was becoming popular among the wealthy and trendsetters, while in England, Professor John N. Langley of Cambridge University was exploring the effects of nicotine, a powerful chemical and effective pesticide extracted from tobacco. It was known that nicotine could be absorbed through the skin, causing

NICOTINE

sickness in humans. Understanding of how the brain and nervous system work (called "neuroscience" today) was just emerging at the beginning of the century. Little was known about the functioning of the brain or how it sent messages through the body's network of nerve fibers to move muscles or stimulate the heart, or how these nerves transmitted information to the brain. Nicotine would become one of the chemicals used to help unravel these mysteries and jumpstart the field of neurophysiology.

By the end of the twentieth century, a groundswell of scientific research had transformed our understanding of nicotine from being an obscure poison to being an addicting drug responsible for taking millions of tobacco smokers to premature death. Many secrets of the brain and nervous system were also unraveled through the help of nicotine, because nicotine has profound effects on parts of the nervous system (now termed "nicotinic"). Thus, nicotine emerged as a vital laboratory tool in understanding the functioning of the nervous system. From the standpoint of public health, one of the most striking features in the history of nicotine science was the recognition that nicotine was an addicting drug, and that tobacco addiction was among the deadliest addictions in the world. Nearly one-half of daily smokers would die prematurely of tobacco-attributed diseases—primarily cancer, lung, and cardiovascular diseases.

What led to the development and understanding of nicotine as a deadly drug? How does nicotine affect the nervous system, and what role does it play in tobacco use? These are some of the vexing questions that scientists around the world grappled with as they learned about nicotine and its effects on the body.

History of Nicotine

Nicotine derives its name from Jean Nicot, a French ambassador to the Portuguese court from 1559 to 1561. The story is that the thirty-year-old diplomat paid a visit to a famous Portuguese horticulturalist, Damiao de Goes, who gave him leaves from a strange plant reputed to have marvelous effects. Nicot dried the leaves, crushed them, and sent the powder back to the queen mother Catherine de Medici, who suffered from severe headaches. Reportedly, the remedy worked, and the tobacco plant quickly gained popularity in France, making Nicot something of a celebrity. The plant came to be called the Herb of Nicot.

Jean Nicot (c. 1530–1600), French ambassador to Portugal, 1559–1561. Brought use of tobacco to France from Portugal. "Nicotine" and "herba" nicotina, literally, Nicot's herb, are derived from his name. HULTON ARCHIVE/GETTY IMAGES

But it was not until the nineteenth century that the chemical nicotine was identified as a distinct ingredient in tobacco. In 1809, Louis Nicolas Vauquelin (a French chemist) extracted a "potent, volatile, and colorless substance" from tobacco which he named essence de tabac, though it was not pure nicotine that was derived. In 1828 two chemistry students at the University of Heidelberg, Ludwig Reimann and Wilhelm Heinrich Posselt, first isolated nicotine, which they named after Nicot, as the active ingredient in tobacco.

In 1905, John Newport Langley, a British physiologist, discovered that a miniscule drop of nicotine stimulated muscle fibers while a similar amount of another poison, curare, paralyzed them when administered simultaneously to anesthetized birds. Langley correctly concluded that muscles and nerves must contain what he termed "receptive substances" (now called "receptors"). In response to different chemicals, these receptors were either activated or deactivated. Drugs that activate receptors are called "agonists." For example, the deadly poison, curare, exerted its lethal paralysis of muscles, including those working the lungs, by blocking nicotinic receptors. But the right dose of nicotine could reactivate muscles depressed by curare. Nicotine was one of a particularly interesting type of chemicals in which a small amount (called the "dose") could produce activation while a larger dose could produce deactivation. In other words, the strength of nicotine's effects was closely related to the dose administered and repeated dosing led to weaker effects (or tolerance). These discoveries helped to explain how muscles could be stimulated or relaxed by the same nerve. By the end of the twentieth century, thousands more of the body's receptor types and subtypes had been identified, helping to explain many aspects of physical, behavioral, and cognitive functioning. This led to the discovery of medicines for treating hundreds of diseases.

Nicotine's Effect on the Body

Nicotine is the cerebrally acting drug in tobacco that defines its addicting effects, similar to the way cocaine in the coca leaf and morphine in the **opium** poppy define the addictive effects of those substances. Nicotine affects the brain by binding to specific receptors (called nicotine cholinergic receptors) on the surface of brain cells. This stimulates the cells to release neurotransmitters such as **epinephrine** and **dopamine.** Epinephrine provides the fast "kick" to the smoker, causing a release of glucose and an increase in heart rate, blood pressure, and breathing. Dopamine is fundamental to reward and pleasure pathways in the brain and is boosted by other addictive drugs, such as cocaine and heroin, as well as by nicotine.

Nicotine produces an entire range of physical and behavioral effects characteristic of addicting drugs. These effects include activation of brain reward systems (creating behavioral effects and **physiological** cravings that lead to chronic drug use), tolerance and physical dependence, and withdrawal with drug abstinence. Nicotine alters a person's mood, feelings, and behavior, and its effects can be complicated. At very high doses, the effects of nicotine on heart rate and blood pressure can be dangerous, even fatal, but there is no conclusive evidence that modest doses of nicotine—like those received from a nicotine patch—are detrimental to health.

The fast action of inhaled nicotine makes cigarette smoking the most addictive route for administering nicotine, which reaches the smoker's brain less than 10 seconds after inhalation. Because inhaled nicotine reaches the bloodstream so quickly, it produces an intense but short-lived spike in its levels. In contrast, nicotine from a skin patch works its way into the bloodstream slowly, over about three hours, and never reaches the peak levels that inhaled nicotine does, even when the overall dose is the same (nicotine nasal spray and nicotine chewing gum fall somewhere in the middle). Not surprisingly, smokers report that their habit is highly reinforcing (they want to keep repeating the experience), but they do not show the same enthusiasm for the nicotine patch.

Nicotine dependence is far more common than cocaine, heroin, or alcohol dependence following initial use of these drugs. Approximately

opium an addictive narcotic drug produced from poppies. Derivatives include heroin, morphine, and codeine.

epinephrine also called adrenaline, a chemical secretion of the adrenal gland. Epinephrine speeds the heart rate and respiration.

dopamine a chemical in the brain associated with pleasure and well-being. Nicotine raises dopamine levels and intensifies addiction to cigarette smoking.

physiology the study of the functions and processes of the body.

one-third to one-half of those who try smoking increase to more regular or daily use, and most daily smokers become addicted. In contrast, less than one in four persons who try cocaine or heroin develop addiction, and less than 15 percent of alcohol users develop addiction. Nicotine, alone and in combination with other substances, appears to help regular smokers control their mood and body weight and maintain attention when working. Daily smokers will claim that they function best on nicotine. Even a brief period of tobacco abstinence can leave some addicted individuals unable to complete their office- or schoolwork, or to perform adequately.

Tobacco Product Design

Nicotine accounts for approximately 1–4 percent of the weight of a typical tobacco leaf, which is transferred into the bloodstream by chewing products made for oral use or by inhaling the smoke of burning tobacco. Tobacco products can be viewed as nicotine storage and delivery systems. The tobacco industry has used a variety of techniques to enhance the delivery of nicotine to the user by controlling the nicotine dosing characteristics of cigarettes and other products. The modern cigarette is intricately designed, involving numerous patents for cigarette wrappers, filter systems, and processes for making "tobacco filler" from tobacco materials and other substances. William Dunn, a senior Philip Morris scientist, has eloquently described the cigarette's function:

> *The cigarette should be conceived not as a product but as a package. The product is nicotine. Think of the cigarette as a dispenser for a dose unit of nicotine. . . . Think of a puff of smoke as the vehicle of nicotine. Smoke is beyond question the most optimized vehicle of nicotine and the cigarette the most optimized dispenser of smoke*
>
> (CAMPAIGN FOR TOBACCO-FREE KIDS 1998, CITED IN HURT AND ROBERTSON).

menthol a form of alcohol imparting a mint flavor to some cigarettes.

Tobacco is a complex "cocktail" of more than 4,000 distinct chemical substances, some of which can interact to increase the addicting effects of tobacco-delivered nicotine, far above those produced by nicotine alone. For example, buffering compounds in smokeless tobacco products can alter the speed and amount of nicotine delivered in those products. The addition of **menthol** apparently allows smokers to inhale larger quantities of smoke, and nicotine, by making them feel less harsh. Techniques are also employed to control the size of smoke particles allowing the efficient inhalation of nicotine deep into the lungs where absorption is rapid and virtually complete. Among the many chemicals in tobacco smoke, scientists are only now beginning to unravel the many individual chemicals and their combinations that bolster the addictive effects of tobacco.

Nicotine Addiction "Drives" Smoking Behavior

While early antitobacco campaigns warned that cigarette smoking could be habit forming, drawing parallels with narcotics, it was not until the 1980s that leading scientists and health organizations recognized cigarettes to be addicting. The 1988 United States Surgeon General's report focused on the role of nicotine in smoking and concluded that "Cigarettes and other forms of tobacco are addicting," "Nicotine is the drug in tobacco that causes addiction," and "The pharmacologic and behavioral processes that determine tobacco addiction are similar to those that determine addiction to drugs such as heroin and cocaine."

Smokers become very adept at getting the dose that provides the desired effects. This is associated with a phenomenon known as "tolerance," which refers to increasing the amount of drug to experience the same effects once received at lower doses. When tolerance develops and tobacco intake increases, a person typically becomes physiologically dependent. Quitting is accompanied by withdrawal symptoms, including impaired concentration, irritability, weight gain, depressed mood, anxiety, difficulty sleeping, and persistent craving for a cigarette. During withdrawal, resumption of smoking provides rapid relief of withdrawal effects, leading the smoker to believe that smoking is a mood and performance-enhancing substance. However, resumption of smoking prevents withdrawal that occurs because physical dependence results from daily use of tobacco. Although there is individual variation, withdrawal usually peaks within a few days and subsides within a month.

Nicotine and Public Health

The World Health Organization, the United States Public Health Service, and most major health organizations worldwide endorsed efforts to make tobacco abstinence a major health priority by the end of the twentieth century. The overwhelming weight of scientific study has shown that quitting smoking at virtually any age results in a reduced disease risk and better health outcomes if tobacco-attributed disease has already developed. The results of smoking cessation are quite dramatic. For example, the risk of heart disease—the leading cause of death among smokers—is reduced nearly to that of nonsmokers within one to two years of cessation.

Preventing the development of tobacco addiction is vital to the long-term health of generations to come. But the road to longer and healthier lives is in cessation for today's 50 million cigarette smokers in the United States and more than 1.2 billion smokers worldwide. Therefore, major governments and health organizations have launched important initiatives to motivate people to quit smoking. In recognition of the power of addiction and the need for people to quit, these organizations have also made smoking cessation treatments more accessible. Many people can now receive medical assistance to achieve freedom from tobacco by contacting the public health service of their nation, cancer institutes, the World Health Organization, and various voluntary organizations such as local cancer societies and lung health organizations.

See Also Addiction; Genetic Modification; "Light" and Filtered Cigarettes; Toxins.

▮ PATRICIA B. SANTORA
▮ JACK E. HENNINGFIELD

BIBLIOGRAPHY

Benowitz, N. L. "Cigarette Smoking and Cardiovascular Disease: Pathophysiology and Implications for Treatment." *Progress in Cardiovascular Diseases* 46 (July-August 2003): 91–111.

Campaign for Tobacco-Free Kids. Available: <http://www.tobaccofreekids.org/research/factsheets/pdf/0009.pdf>.

Centers for Disease Control and Prevention. Available: <http://www.cdc.gov>.

Fiore, Michael C., Bailey, W. C., Cohen, S. J. et al. *Treating Tobacco Use and Dependence. Clinical Practice Guidelines.* Rockville, Md.: Department of Health and Human Services, United States Public Health Service, 2000.

Hurt, R. D., and C. R. Robertson. "Prying Open the Door to the Tobacco Industry's Secrets about Nicotine: The Minnesota Tobacco Trial" *Journal of the American Medical Association* 280 (1998): 1173–1181.

Kessler, David. *A Question of Intent: A Great American Battle with a Deadly Industry.* New York: Public Affairs, 2001.

National Cancer Institute. Available: <http://www.cancer.gov>.

Royal College of Physicians of London. *Nicotine Addiction in Britain. A Report of the Tobacco Advisory Group of the Royal College of Physicians.* London, England: Royal College of Physicians, 2000.

United States Department of Health and Human Services. *The Health Consequences of Smoking: Nicotine Addiction. A Report of the Surgeon General.* Washington, D.C.: United States Government Printing Office, 1988.

———. *Reducing Tobacco Use: A Report of the Surgeon General.* Rockville, Md.: Department of Health and Human Services, United States Public Health Service, 2000.

United States Department of Health, Education, and Welfare. *Smoking and Health: Report of the Advisory Committee to the Surgeon General of the Public Health Service.* Public Health Publication No. 1103. Washington, D.C.: United States Government Printing Office, 1964.

World Health Organization, Tobacco Free Initiative. Available: <http://www.who.int/tobacco/en>.

Oceania

Covering half the earth's surface and containing thousands of islands, the Pacific Ocean, locus of the world region called Oceania, is so vast and inhabited by such diverse peoples with widely varying histories that almost any generalizations are problematic, and this certainly is the case regarding tobacco use.

Early History

The earliest historical record of tobacco use in Oceania dates from 1616 on islands off the northwest coast of New Guinea. Tobacco cultivation may have been introduced to the Philippines by the Spanish as early as 1575, but it was after large-scale cultivation began to flourish in Europe in the 1590s that the use of tobacco, if not always its cultivation, rapidly spread, with introductions by the Dutch in Java in 1601 and almost immediate diffusion throughout what is now Indonesia, with Halmahera becoming a center of cultivation and export (as was Java) by 1616.

So far as the Western Pacific is concerned, while there are severe limitations in the historical record, especially for the seventeenth and eighteenth centuries, it appears that the adoption of smoking and cultivation of tobacco spread generally eastward, becoming established in most of New Guinea by the time of sustained European colonial presence in the mid- to late-1800s. However, early European sources indicate that tobacco was then still unknown in many parts of eastern New Guinea and on numerous islands of Melanesia, as when German entrepreneurs found it necessary to create smoking schools in the Bismarck Archipelago in 1875. The purpose of the schools was to instruct the people regarding how to stuff and light a pipe, inhale, and then—importantly—blow out the smoke amidst much coughing and choking.

Where tobacco use was established prior to the arrival of Europeans, people in rural areas cultivated it for their own individual use or obtained it through trade from neighbors, as is still true today in most of New Guinea and Melanesia. Moreover, smoking was often highly restricted, usually to adult males and often to ritual contexts. While the

OCEANIA

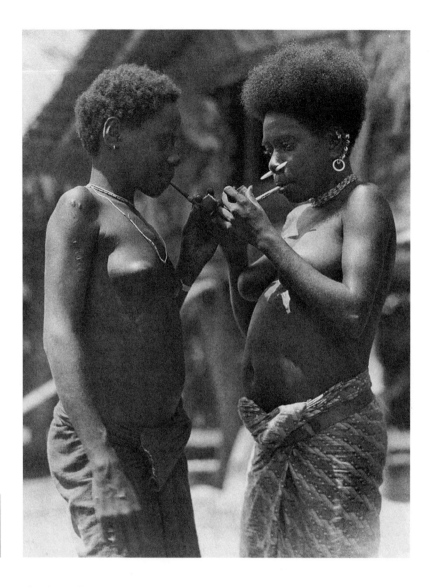

Two young women from the Solomon Islands lighting up pipes, 1950. Pipe smoking is particularly popular among the women on the islands. © HULTON-DEUTSCH COLLECTION/CORBIS

plantation historically, a large agricultural estate dedicated to producing a cash crop worked by laborers living on the property. Before 1865, plantations in the American South were usually worked by slaves.

sharing of a pipe or cigarette was a widespread gesture of sociality, casual recreational smoking appears to have been a product of more modern times and forces.

Diffusion and Trade

Throughout the nineteenth century, traders, whalers, labor recruiters, colonial officials, and missionaries created, or simply amplified, a passion for smoking that soon made commercially produced tobacco (usually in the form of twists or plugs) a nearly universal trade commodity in Oceania. Spaniards had planted tobacco in Tahiti in 1774 and 1775, but by the mid-1800s the smoking of trade tobacco was rapidly becoming promoted and established throughout Polynesia, and by 1850 the island of Guam had become a major supply station for the islands of Micronesia, with large consignments also being sent out of Sydney, Australia, to serve the Western Pacific market by 1848.

Beginning in 1886, the Neu Guinea Compagnie began to establish tobacco **plantations** in colonial Kaiser Wilhelmsland on the northeastern coast of what is now Papua New Guinea. By 1888, tons of tobacco leaf were being shipped to Germany for consumption in Europe, but periodic droughts and health problems among laborers (mostly imported Asians) added to other difficulties, and production for export ceased after 1903.

Throughout Oceania, all manufactured cigarettes were imported until recent decades, with cigarette factories being founded in Fiji first in 1955, then Papua New Guinea, Tonga, and Western Samoa in the 1960s and 1970s. Both domestic and foreign brands are now commonly smoked in those countries as well as in Cook Islands, Kiribati, Solomon Islands, and Vanuatu. In none of these countries, however, is tobacco or cigarette production a major source of export income. In fact, cigarette imports substantially exceed exports in all of the Oceanic nations for which recent information is available. Despite health-related antismoking campaigns by virtually all Pacific governments, in most areas imports of cigarettes have shown steady rates of increase over the past three decades.

Demographics of Consumption

While tobacco consumption (overwhelmingly through smoking) is **ubiquitous** in Oceania today, rates of adult smokers vary considerably. According to the World Health Organization, in 2002 nine of the 100 countries with the highest percentages of adult smokers were Pacific island nations, with Nauru (54%) at the top of the list ranging down to Samoa, in ninety-sixth place with 23 percent adult smokers. In some cases, such as French Polynesia, recent decades have seen a decrease in adult smoking, but more often rates of consumption show steady increases, especially among the younger population.

ubiquitous being everywhere; commonplace; widespread.

World Bank reports indicate that in almost all Pacific nations male smokers outnumber females among young adults, with the highest rates for males appearing in urban Kiribati (95%) and Tonga (60%), corresponding to 63 and 10 percent respectively, for young adult females. However, for much of Oceania, widespread smoking—indeed, smoking itself—is a relatively recent phenomenon.

See Also Philippines.

▮ TERENCE E. HAYS

BIBLIOGRAPHY

Haddon, Alfred C. *Smoking and Tobacco Pipes in New Guinea*. Philosophical Transactions, Series B, no. 232. London: Royal Society of London, 1946.

Marshall, Mac. "An Overview of Drugs in Oceania." In *Drugs in Western Pacific Societies: Relations of Substance*. Edited by Lamont Lindstrom. Lanham, Md.: University Press of America, 1987.

———. "The Second Fatal Impact: Cigarette Smoking, Chronic Disease, and the Epidemiological Transition in Oceania." *Social Science and Medicine* 33, no. 12 (15 December 1991): 1327–1342.

Riesenfeld, Alphonse. "Tobacco in New Guinea and the Other Areas of Melanesia." *Journal of the Royal Anthropological Institute* 81, nos. 1–2 (1951): 69–102.

World Bank. "Economics of Tobacco Control: Country & Regional Profiles and Economics of Tobacco Briefs." Available: <http://www1.worldbank.org/tobacco/countrybrief.asp>.

 Opium

Before concerted international efforts appeared in the first decades of the twentieth century to crack down on opium, the substance was widely available around the world.

In Europe, opium could be dissolved in beer, wine, or vinegar or simply taken as an infusion prepared out of boiled poppy heads, while powdered opium was used in suppositories and raw opium was rolled into pills and preparations such as laudanum. In the Middle East and South Asia, it was more often than not ingested orally. In China and parts of South East Asia, however, opium was frequently mixed with tobacco. Javanese opium, for instance, was blended with roots of local plants and hemp, minced, boiled with water in copper pans, and finally mixed with tobacco. This blend was called *madak*.

Madak was first introduced to Taiwan (Formosa) between 1624 and 1660. The mixture was prepared by the owners of smoking houses and brought prices significantly higher than for pure tobacco. Opium house owners also provided the smoking implement: a bamboo tube with a filter made of coir fibers, produced from local coconut palms. The habit of smoking *madak* spread throughout the coastal provinces of south China, even though it never exceeded the popularity of tobacco.

A precise chronology of *madak* is not possible in the absence of reliable source material, although the first references to the blend date from the early eighteenth century and come from Fujian and Guangdong, the same ports of entry as for tobacco. "The opium is heated in a small copper pan until it turns into a very thick paste, which is then mixed with tobacco. When the mixture is dried, it can be used for smoking by means of a bamboo pipe, while palm fibers are added for easier inhalation." The earliest description of pure opium smoking dates from 1765.

The reasons for a shift away from *madak* toward pure opium after the 1760s are complex. One hypothesis is that pure opium was used to enhance sexual performance. Another explanation is that an early edict against the smoking of *madak* by the Yongzheng emperor in 1729 prompted local users to resort to pure opium instead, the use of which could be justified for medical reasons. It is also possible that the smoking of pure opium served as a marker of social status, as large amounts of money could be spent in one evening on pure opium. The quality of Patna opium—produced in India under British control—improved after poppy cultivation in Bengal was monopolized by the East India Company in 1793, a factor which may also have prompted some *madak* smokers to smoke opium on its own.

Throughout the eighteenth century, however, *madak* remained widespread, as pure opium would only become the norm in the nineteenth century with the lowering of the cost of opium and the spread of local poppy cultivation in China. Tobacco thus allowed opium to become part of a thriving smoking culture well before the "First Opium

Two Indonesian men smoking opium. Before worldwide efforts to reduce opium use began in the first decades of the twentieth century, it was widely available around the world. In China and parts of Southeast Asia, it was frequently mixed with tobacco. © SEAN SEXTON COLLECTION/CORBIS

War" between Britain and China (1839–1842), which revolved partly around the issue of the opium trade.

See Also Alcohol, Tobacco, and Other Drugs.

∎ FRANK DIKÖTTER

BIBLIOGRAPHY

Dikötter, Frank, Lars Laamann, and Xun Zhou. *Narcotic Culture: A History of Drugs in China.* Chicago: Chicago University Press, 2004.

Newman, Richard K. "Opium Smoking in Late Imperial China: A Reconsideration." *Modern Asian Studies* 29 (October 1995): 765–794.

Spence, Jonathan D. "Opium Smoking in Ch'ing China." In *Conflict and Control in Late Imperial China.* Edited by Frederic Wakeman and Carolyn Grant. Berkeley: University of California Press, 1975.

Origin and Diffusion

The tobacco of worldwide commerce belongs to the species *Nicotiana tabacum*. It belongs to the family Solanaceae, which includes the potato, tomato, eggplant, petunia, and many other cultivated and ornamental plants. The genus *Nicotiana* is one of about ninety genera in the family and consists of about sixty-five species in the world, three-fourths of them native to North and South America, one-fourth native to Australia,

ORIGIN AND DIFFUSION

psychoactive a drug having an effect on the mind of the user.

alkaloid an alkaloid is an organic compound made out of carbon, hydrogen, nitrogen, and sometimes oxygen. Alkaloids have potent effects on the human body. The primary alkaloid in tobacco is nicotine.

and a single one, *N. africana*, discovered in the 1970s on a few mountain tops in the Namibian Desert of Namibia. In Africa, the continent where *Homo sapiens* evolved, the human interaction with this genus was nonexistent until the sixteenth century. Humankind became aware of these plants and their **psychoactive** properties about 50,000 years ago when Australia was populated, and approximately 10,000 to 15,000 years ago when the Americas south of the Arctic were being populated.

Perhaps a dozen species of *Nicotiana* have been actively used by humankind, but the remainder have evidently never been seriously used for smoking or chewing, because initial experimentation likely revealed the low nicotine content or the presence of other bitter or more immediately poisonous **alkaloids.**

The various species of *Nicotiana* range along a spectrum from strictly wild species to a completely domesticated one. *Nicotiana tomentosiformis*, which grows in the Andes, is a wild species that grows and propagates entirely on its own, without any deliberate intervention on the part of humans (although human modification of habitats in the last 10,000 years may affect its distribution). Conversely, *Nicotiana tabacum*, which is by far the most widespread and important of the tobaccos in an economic sense, depends entirely on humankind for its continuing existence, and cannot persist for more than a generation or two without being deliberately planted and protected from weeds.

Australia

In Australia, at least the following native species have been used for chewing tobaccos before the arrival of Europeans (and the New World tobaccos): *N. gossei*, *N. ingulba*, *N. simulans*, *N. benthamiana*, *N. cavicola*, *N. excelsior*, *N. velutina*, and *N. megalosiphon*. Agriculture never developed in Australia; but whether or not any of these were sometimes deliberately planted is not known, and none were truly domesticated. In addition, since the mid-1800s *N. glauca* has become widely naturalized and has been used for chewing. In the 2000s, *N. tabacum* is widely cultivated.

North and South America

In North America and Mexico, Amerindians used certain of the native *Nicotiana* species for their psychoactive effects. The range where these wild tobaccos can be found may involve some spread beyond their original native ranges due to human influence. In North America, one species, *N. quadrivalvis* (previously called *N. bigelovii*), was "semi-domesticated," which means it evolved because of human selection for particular traits, but with only a few modifications, so that it could probably exist in the wild; after generations of cultivation by Amerindians, selection had taken place to produce plants with larger flower parts, the parts richest in nicotine. All wild species of *Nicotiana* have the fruit divided into two chambers, but in *N. quadrivalvis*, the number of cells had been increased to three or four. This intensive use also expanded its original range from California eastward to the Great Plains from Texas to the upper Missouri River, and it was the tobacco that the explorers Meriwether Lewis and William Clark encountered being cultivated by the Mandan Indians.

Only two species, *N. rustica* and *N. tabacum*, spread out from a single continent of origin in prehistoric times, and these are the only ones to spread around the world in general cultivation. The introduction of these two into Europe after Christopher Columbus' explorations can be dated quite precisely by both printed descriptions and illustration because they were grown and noted by botanists of the day. *N. tabacum*, for example, was first illustrated in 1571 by Pierre Pena and Mathias de l'Obel (1571), and this provides incontrovertible evidence of the species that had reached Europe by this time.

Investigating the spread of *N. rustica* and *N. tabacum* in America up to the time of Columbus presents serious difficulties because of the lack of a written record, and because these soft-bodied plants are mostly absent from the archeological record except under the most favorable circumstances for preservation. The presence of pipes or representations of smoking on pottery or murals can demonstrate the existence of tobacco (or other plants used for the same purposes) at a certain place or time, but usually not the species being used. Researchers are fairly certain that, in general, by the time of Columbus *N. tabacum* was present in eastern South America, Central America, Mexico, and the West Indies; while *N. rustica* was being cultivated in Mexico and the eastern United States as well as the Andes of South America.

Researchers have proposed a number of theories for the origin of these two species, placing their origin in various areas. The origin of both *N. rustica* and *N. tabacum* in Andean South America had been more or less firmly established by 1954 with the publication of the botanist Thomas Harper Goodspeed's careful treatment of the entire genus, "The Genus *Nicotiana*," based on his thorough knowledge of the morphology of the species, their genetic behavior, and the areas where they were found. An important consideration is that both *N. rustica* and *N. tabacum* have twenty-four pairs of chromosomes, and are termed tetraploids, because they have twice the number of chromosomes as their nearest relatives, termed diploids, which have twelve pairs. Scientists soon realized that these two species must have resulted from the **hybridization** of two other species with the subsequent doubling of the chromosome number in the hybrid.

hybridization the practice of cross-breeding different varieties of plants or animals to produce offspring with desired characteristics.

Nicotiana rustica

With traditional breeding experiments and modern DNA sequence analysis, the origins of *N. rustica* can be more carefully elucidated. Before Columbus, *N. rustica* occurred in the United States and northern Mexico, as well as in the Andes from Ecuador to Bolivia. However, it has never been found as a truly wild-growing plant in Mexico or the United States as it is in the Andes around human habitations. Since it is a hybrid, with subsequent chromosome doubling between *N. paniculata* and *N. undulata*, its area of origin must be within the natural range of these two wild species in Peru, Bolivia, and Argentina (see Map).

N. paniculata is an annual herbaceous or slightly woody species up to 2 meters or more tall, found in a wide altitudinal range, from 300 to 3,000 meters, in western Peru. *N. undulata* is a fleshy, sticky annual herb up to 2 meters tall, from very dry, barren areas, from 2,700 to 4,200 meters altitude in Peru, Bolivia, and Argentina, and is especially common in weedy areas around settlements. The original

ORIGIN AND DIFFUSION

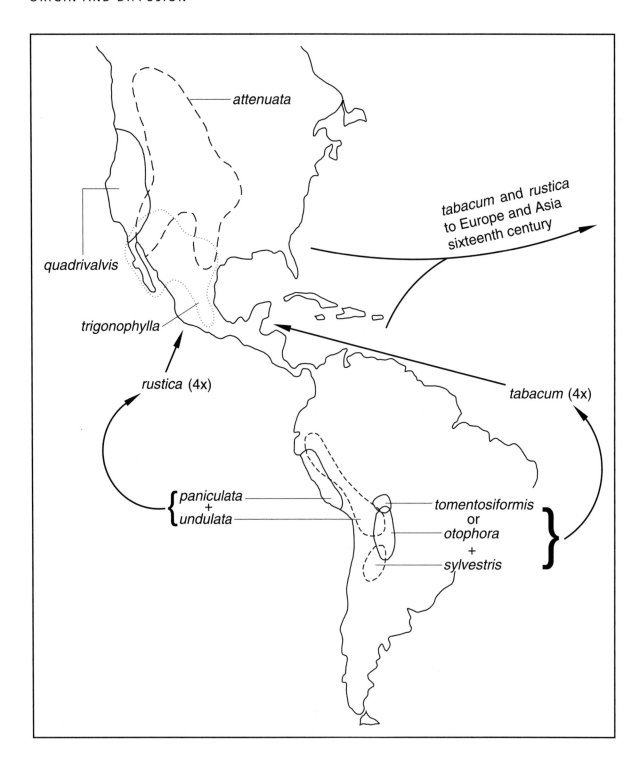

Evolutionary geography of North American tobacco species. RONALD STAUBER

hybridization almost certainly must have taken place in Peru, with subsequent spread due to humans. It is impossible to know whether the hybrid *N. rustica* developed before or after human arrival in the Andes. The fact of a human presence may have inadvertently led to the hybridization of *N. paniculata* with *N. undulata* to create *N. rustica*. Humans may have modified the range of the wild species (for example, by gathering certain specimens) and created exactly the sort of habitats in which *N. rustica* prospers: disturbed soil rich in nutrients (especially nitrogen). Such habitats occur around human habitations and the pens of their animals, and in the 2000s N. rustica grows in such sites without deliberate planting.

The time and route of dispersal of N. rustica north to Mexico and the eastern United States is difficult, if not impossible, to reconstruct, but it was well established in many cultures by the time of Columbus. None of the higher altitude Andean domesticated tubers like the potato (*Solanum tuberosum*) or oca (*Ullucus tuberosus*), for example, ever made it to suitable habitats in the highlands of Central America or Mexico because of the impossibility of passing through the lowland tropical jungles of Panama if the plants were to be travel by the slow route of being traded from village to village. Some domesticated plants that could be grown in the humid tropical lowlands easily passed through this area, the prime example being corn (or maize, *Zea mays*), which spread easily from its area of origin in Mexico throughout the tropical and temperate parts of the Americas, including into South America. More direct dispersion via pre-Columbian Pacific trade routes by boats is the likely means that N. rustica arrived in Mexico from the northern Andes.

Although N. rustica was in use in eastern North America by Amerindians at the time of Europeans' arrival and was even the first commercial tobacco to form the economic basis of the Virginia colony, it would become overshadowed by N. tabacum, which is the overwhelmingly predominant cultivated tobacco throughout the world in the twenty-first century. The prehistory of N. tabacum was somewhat similar to that of N. undulata in that it is a hybrid with chromosome doubling, and had spread widely enough to be the tobacco which Columbus probably encountered in the West Indies on his first trip to the region.

Nicotiana tabacum

One of the parental species of N. tabacum is N. sylvestris, native to northwestern Argentina and southern Bolivia. It is an annual herb with long, narrowly tubular white flowers that open at night and are pollinated by hawk moths. The leaves are large, somewhat similar to those of tobacco, but its only use is as an ornamental in flower beds. The other parent must belong in the *Tomentosae*, a group of six Andean species from Peru to northwestern Argentina which are short-lived shrubs or small trees with short pinkish flowers open in the day and pollinated by bees and hummingbirds (the majority) or open mainly at night and pollinated by bats (N. otophora). The prime suspect for the second parent has been N. otophora, a shrub from central Bolivia to northwestern Argentina, which grows in seasonally dry forests or along washes in more arid areas of desert thorn-scrub. It is the only species of the group that also grows in the range of N. sylvestris. Nicotiana otophora has leaves that look very much like those of tobacco, and in Bolivia they are even used occasionally for smoking when tobacco from N. tabacum is not available. Based on the morphological characters, and on crossing experiments, Goodspeed concluded that N. otophora was the likely second parent of N. tabacum.

However, studies conducted in the 1990s of the DNA sequences have indicated that the second parental species is not N. otophora but rather N. tomentosiformis, a soft-woody shrub or small tree 1.5 to 5 meters tall, from the humid montane forests of northern Bolivia and

southern Peru on the slope facing the Amazon lowlands. *N. tomentosiformis* has large, tobaccolike leaves, and even smells like tobacco both when fresh and when dried. How and when *N. tomentosiformis* and *N. sylvestris* came in contact to hybridize is a mystery, and it is unclear how the hummingbird-pollinated *N. tomentosiformis* would have hybridized naturally with the hawk-moth-pollinated *N. sylvestris*. Even during the climatic changes during the Pleistocene Age, it is highly unlikely that the range of the two species could have allowed them to come into contact naturally.

Scientists have hypothesized that *N. tabacum* originated several million years ago, but this seems almost impossible because of the biology of the species: *N. tabacum* is not known to exist as a wild plant anywhere in the world, despite the fact that it is cultivated on a vast scale worldwide and has the opportunity to escape and become naturalized in innumerable possible habitats. A species formed millions of years ago and capable of persisting until humans could begin cultivating it would certainly continue to exist to this day as a wild plant in some area.

A much more likely possibility is that the second parent of *N. tabacum* is a currently unknown species of *Nicotiana* similar to *N. tomentosiformis*, but growing in southern Bolivia. Vast areas in southern Bolivia have not been explored botanically; there are series of parallel mountain ranges where the more humid forested ridges could easily harbor a species of *Tomentosae*, which would be in position to occasionally hybridize with *N. sylvestris*, known to grow in the intervening dry valleys.

In the 1990s, researchers discovered that two other cultivated plants have their origin in this region. The tree tomato, *Solanum betaceum* (also known as *Cyphomandra betacea*), was discovered in the 1990s to grow wild in this region, finally solving the question of the origin of a species which had long been cultivated in the Andes of Ecuador and Colombia. The origin of the peanut, *Arachis hypogaea*, presents a parallel situation to that of tobacco since it is a cultivated plant not known in the wild, and botanists have long known that it must be a tetraploid hybrid of two wild diploid species. It was only in the 1990s that researchers definitively established the origin of the peanut. The peanut is descended from *A. duranensis*, a widespread species from northwestern Argentina, southeastern Bolivia, and westermost Paraguay, and from *A. ipaensis*, a species only known from two collections in southern Bolivia made in 1971 and 1977.

See Also Missionaries; Sailors.

■ MICHAEL NEE

BIBLIOGRAPHY

Goodspeed, Thomas H. "The Genus *Nicotiana*." *Chronica Botanica* 16 (1954): 1–536.

Krapovickas, Antonio, and William C. Gregory. "Taxonomía del Género *Arachis* (Leguminosae)" (Taxonomy of the genus *Arachis* [Leguminosae]). *Bonplandia* 8 (1994): 1–186.

Peterson, Nicolas. "Aboriginal uses of Australian Solanaceae." In *The Biology and Taxonomy of the Solanaceae*. Edited by John G. Hawkes, Richard N. Lester, and A. D. Skelding. London: Academic Press, 1979.

Ostracism

Tobacco users have been ostracized at different times and in different contexts since 1492, when the explorer Christopher Columbus and his sailors became the first Europeans to encounter what quickly became known as "the devil's weed."

When Columbus and his crew landed in the New World, the indigenous Arawak Indians offered them gifts of "some dried leaves which are in high value among them" (Columbus 1990). To Columbus's disgust, some of his sailors were soon emulating the Indians and "drinking smoke" themselves. One of them, Rodrigo de Jerez of Ayamonte, Spain, reportedly became the subject of the first legal action against a smoker. De Jerez took a supply of tobacco from present-day Cuba back to his home village. It is said that when he lit up for the first time in public, the townspeople—alarmed by the smoke issuing from his mouth and nose— assumed he had been possessed by the devil and turned him over to the authorities.

A broad fraternity of kings, emperors, popes, and potentates condemned tobacco as a **heathen** import in the sixteenth and seventeenth centuries. Ecclesiastic authorities, both Christian and Islamic, associated the plant with barbarism and idolatry. Smokers faced excommunication, imprisonment, and even death. An imperial edict issued in China in 1638 made the use or distribution of tobacco a crime punishable by decapitation. In Russia, smokers were flogged, the nostrils of repeat offenders were slit, and persistent violators were exiled to Siberia. Sultan Murad IV of Turkey had smokers executed as infidels.

Tobacco had defenders, of course. Among them was Sir Walter Raleigh (1554–1618), who popularized the habit among the upper classes in England. According to legend, when one of Sir Walter's servants saw him smoking for the first time, he assumed he was burning up from within and doused him with a bucket of water. The story illustrates how strange, even alarming, the act of smoking must have seemed to Europeans of the sixteenth century.

By the eighteenth century tobacco was commonplace but it was still far from being universally accepted. Religious leaders denounced the plant as a "dry inebriant"—a substance that could induce drunkenness even through it was smoked rather than swallowed. The link between tobacco, alcohol, and sin became even more pronounced after the emergence of a temperance movement in England and the United States in the nineteenth century. Temperance advocates warned that "Smoking leads to drinking and drinking leads to the devil" (Lawrence 1885).

During the Victorian era (bracketed by the reign of Queen Victoria in England from 1837 to 1901), tobacco users began to provoke censure on the grounds of aesthetics as well as morality. Changing standards of hygiene led to complaints about the smell and detritus generated by pipes and cigars. Chewing tobacco, once the most popular form of tobacco in the United States, rapidly fell out of favor, its exit hastened by anti-spitting ordinances. Cigarettes gained social acceptance partly because they were viewed as less offensive in close quarters than other kinds of tobacco.

heathen any person or group not worshiping the God of the Old Testament, that is, anyone not a Jew, Christian, or Muslim. May also be applied to any profane, crude, or irreligious person regardless of ethnicity.

OSTRACISM

In the late twentieth century, people who did not smoke became increasingly less tolerant of those who did. Smoking came to be seen as antisocial behavior. AP/WIDE WORLD

Cigarettes penetrated into all social classes in southern and eastern Europe in the late nineteenth century, but they were disdained as "beggar's smokes" in western Europe and the United States. Britain's Prince of Wales (later King Edward VII) took up the habit in the 1880s, giving it an aura of glamour in England. In the United States, however, the cigarette was a lowly, disreputable product. Respectable men smoked pipes or cigars; respectable women did not smoke at all. Most Americans would have agreed with Rev. William "Billy" Sunday, the popular evangelist, who once said, "There is nothing manly about smoking cigarettes. For God's sake, if you must smoke, get a pipe" (Sunday 1915).

After World War I, cigarette smoking expanded socially, across gender and class lines, and spatially, into public spaces. It began to seem as if nearly everyone smoked. In fact, cigarettes were still a habit of the minority in most countries. In the United States, for example, only 42 percent of adult Americans smoked cigarettes in 1965, at the height of the Cigarette Age (roughly 1930 to 1970). Although a sizable proportion, this was still a minority. Nonetheless, cigarettes were embedded in the cultural landscape, accepted as emblems of modernity and sophistication even by nonsmokers.

In the late twentieth century, people who did not smoke became increasingly less tolerant of those who did. A new generation of antitobacco activists used popular media to convey the message that smokers damaged not only their own health but also that of others. The act of smoking—once an expression of sociability—was redefined as antisocial behavior. Perhaps more tellingly it was also identified with yellow teeth and foul-smelling breath. "You can't talk to a 15-year-old about getting lung cancer in his or her 50s, but they get it when you say kissing a smoker is like kissing an ashtray," commented Joseph Califano, president of the national Center on Addiction and Substance Abuse at Columbia University (Bowman).

Smokers' rights groups have attempted to counter these trends by associating the freedom to smoke with basic human liberties. They use epithets such as "nanny staters" and "health Nazis" to depict antismoking activists as scolds and busybodies. In this view, tobacco is a marker that separates the tolerant from the puritanical.

See Also Antismoking Movement Before 1950; Antismoking Movement From 1950; Psychology and Smoking Behavior; Smoking Clubs and Rooms; Social and Cultural Uses.

■ CASSANDRA TATE

BIBLIOGRAPHY

Bowman, Lee. "40 Years Ago, Government Linked Smoking to Cancer." *Seattle Post-Intelligencer* (10 January 2004): A2.

Columbus, Christopher. *Journal of the First Voyage of Christopher Columbus*. B. W. Ife, edited and translated by Westminster, England: Aris and Phillips, Ltd., 1990.

Dickson, Sarah A. *Panacea or Precious Bane: Tobacco in Sixteenth Century Literature*. New York: The New York Public Library, 1954.

Lander, Meta [Margaret Woods Lawrence]. *The Tobacco Problem*. Boston: Lee and Shepard, 1886.

Sunday, William. *Omaha Sermons of Billy Sunday, September–October, 1915*. Omaha, Nebr.: Omaha Daily News, 1915.

Tate, Cassandra. *Cigarette Wars: The Triumph of "the Little White Slaver."* New York: Oxford University Press, 1999.

Peace Pipe *See* Calumets; Native Americans; Pipes.

 # Philip Morris

Philip Morris is the largest cigarette maker in the world. The company has historical roots dating to 1847, when a London tobacconist and entrepreneur, Philip Morris, Esq. recruited expert hand rollers of cigarettes from Russia, Turkey, and Egypt. Present cigarette manufacturing methods have little resemblance to those early days when an experienced roller would turn out 2,500 a day. Today, a single cigarette machine in the primary Philip Morris manufacturing plant in Richmond, Virginia, can produce more than 4,000 cigarettes a minute, or two billion a year. That plant alone, one of more than fifty Philip Morris cigarette manufacturing facilities in the world, produces more than 245 billion cigarettes a year.

In 1919, the company was taken over by its American shareholders, and its headquarters moved to the United States. In the 1930s company president Reuben Ellis hired the Milton Biow advertising agency, which created a popular radio advertising campaign featuring a bellhop, Johnny Roventini ("Little Johnny"), and the slogan "Call for Philip Morris." The success of Philip Morris during the twentieth century was brought about by the marketing of a single brand, Marlboro, which was introduced in the late 1920s as a woman's cigarette, with advertising slogans such as "Mild as May" and "Red tips for your pretty lips." In the 1950s, under the direction of marketing expert George Weissman, Marlboro achieved enormous appeal when the company emboldened the package design with a medallion-like chevron and revamped the advertising image to one of rugged cowboys and the wide open spaces, television commercials for which were accompanied by the theme song from the film, *The Magnificent Seven*.

PHILIP MORRIS

Philip Morris' "Little Johnny" bellhop advertising campaign made "Little Johnny" a popular fixture of radio with the "Call for Philip Morris" ads. This advertisement, from 1939, touts the comparative health advantages of Philip Morris over other brands of cigarettes as cited by "a distinguished group of doctors."
© BETTMANN/CORBIS

At the same time, doubtless in response to the growing scientific evidence that cigarette smoking caused lung cancer and other serious health problems, Philip Morris added a filter to Marlboro (and its other brands). Today more than 98 percent of those who smoke buy filtered cigarettes, which confer no health protection over other brands. Joseph Cullman was president of Philip Morris from 1957 to 1978 when Marlboro's popularity skyrocketed. During his tenure the Marlboro box, filter, and leaf mix were developed, and by 1976 Marlboro was the largest-selling cigarette in the United States. Today, Marlboro is the top-selling cigarette in the world.

In the 1960s Philip Morris sponsored many of the most successful television programs including *Perry Mason*, *The Dobie Gillis Show*, *Rawhide*, *CBS News with Walter Cronkite*, and National Football League telecasts. When cigarette ads were banned from television in 1971, Philip Morris created and sponsored sporting events such as the Marlboro Grand Prix auto race, the Marlboro Cup thoroughbred race, and the Virginia Slims Women's Tennis Circuit, the televising of which successfully circumvented the broadcast ban on cigarette advertising. Weissman, who ascended to the chairmanship in 1978, stepped up Philip Morris' sponsorship of fine arts, and the company's logo began appearing in association with operas, ballets, and art exhibitions.

Phililp Morris was the first cigarette manufacturer to recognize the need to shape its identity through diversification. In 1957 it purchased a producer of flexible packaging. It acquired the Miller Brewing Company in 1969 and the 7-Up Bottling Company in 1976 (both since sold). Moving aggressively into consumer packaged goods, Philip Morris acquired General Foods in 1985 and Kraft Foods in 1988 by means of hostile takeovers. Such diversification enabled Philip Morris to regain clout with television networks, which were covetous of the enormous outlay of advertising dollars for the company's many food products. By 1990, in an effort to further downplay its identity as primarily a cigarette manufacturer, Philip Morris had dropped the word *tobacco* from its name. In 2002 the company renamed itself Altria, diminishing the profile, on paper at least, of the domestic and international Philip Morris cigarette manufacturing divisions. Despite diversification, the company continues to earn half its profits from cigarette sales. During the 1990s, profit from the Marlboro brand alone exceeded the combined profit of the 3,000 Kraft and General Foods products.

Today, Marlboro accounts for nearly 40 percent of all cigarettes sold in the United States, and Philip Morris' market share of total U.S. cigarette sales is nearly 50 percent. Marlboro is the largest-selling cigarette in the world. Perhaps the biggest threat the company faces is what it describes in its 2001 annual report as "management of our litigation challenges," namely, lawsuits brought by state attorneys general, the U.S. Department of Justice, and numerous personal injury attorneys representing persons claiming to have been made ill by smoking. Although the Master Settlement Agreement negotiated with the major tobacco companies by the state attorneys general in 1998, as well as other cash settlements, resulted in a major financial outlay, the good news for the company was that it resulted in a legitimizing financial relationship with the states and provided a measure of stability for shareholders well past the year 2020. Altria has also admitted to the harmfulness of smoking on its corporate website, while also increasing its contributions to charity. Lone among the cigarette companies, it has campaigned for regulation by the Food and Drug Administration, a strategy that could result in greater security for the company by inhibiting the marketing campaigns of competing cigarette manufacturers.

▌ALAN BLUM
▌LORI JACOBI

BIBLIOGRAPHY

Altria Group, Inc. 2002 Annual Report. New York: Altria Group Inc. 2003.

Blum, Alan. "The Marlboro Grand Prix: Circumvention of the Television Ban on Tobacco Advertising." *New England Journal of Medicine* 324 (1991): 913-917.

Kluger, R. *Ashes to Ashes: America's Hundred-Year Cigarette War, the Public Health, and the Unabashed Triumph of Philip Morris.* New York: Vintage Books, 1997.

Philip Morris Companies Inc. Annual Report, 1985. New York: Philip Morris Companies, Inc., 1986.

"PM's Art Programs Tour the World." *Philip Morris News* 17, no. 7 (November 1976).

White, Larry. *Merchants of Death: The American Tobacco Industry.* New York: Beech Tree Books, 1988.

# Philippines

Named after King Philip II of Spain, the Philippines is a sprawling, irregular shaped archipelago, located some 500 miles off the southeast coast of Asia, that consists of approximately 7,100 islands, with a total land area of 114,830 square miles. Roughly two-thirds of the land mass is found on two large islands, Luzon (40,420 square miles, in the north) and Mindanao (36,537 square miles, in the south). Bounded on the west and north by the South China Sea, on the east by the Pacific Ocean, and on the south by the Celebes Seas and the coastal waters of Borneo, the Philippines is a mountainous country with narrow and interrupted coastal plains. Its physical geography was an important factor in shaping tobacco's role in the islands. In the lowlands, where tobacco is cultivated, there is an abundance of water and differences in temperature are slight, favoring the plant's growth. The soil is fertile and the tropical climate is suitable for the cultivation of tobacco and other agricultural products. While many local rainfall patterns exist, in general, there are two seasons—dry in winter and spring and rainy in summer and autumn—typical to Monsoon Asia.

In the twenty-first century, tobacco is cultivated, but it is a minor export item that pales in comparison with its past. The trends in the Philippine tobacco industry contrast starkly with historical evidence; for example, employment in the sector from 1975 to 1997 has fallen from 4 percent to 1 percent of the total employed population.

Tobacco's Introduction

The Spanish introduced a number of species of the genus *Nicotiana* in the Philippines from America in the last quarter of the sixteenth century. The islands were first encountered by a European expedition that recorded the first successful circumnavigation of the globe in the early sixteenth century. Led by Ferdinand Magellan, a Portuguese captain, this Spanish expedition inaugurated a long period of contact, exchange, and colonial relationship between Spain and the Philippines that would last from the early sixteenth century to the end of the Spanish-American War in 1898. This early contact was driven by Spain's political, economic, and religious objectives to compete with Portugal and claim territory, wealth, and Christians around the globe. These two European and Christian powers sought to monopolize and control the access to spices—the cloves, nutmeg, and mace—found to the south of the Philippines in the Indonesian Archipelago. In this context of European contest and, particularly for the Spanish in the southern islands of the Philippine Archipelago, the Christian Europeans encountered some indigenous societies that had fervently embraced Islam and would contest and resist the Christian presence.

The introduction of tobacco in the Philippines occurred, in all probability, after the Spanish established the city of Manila on the island of Luzon in 1571. Two primary species of tobacco were introduced at this time: *N. rustica* and *N. tabacum*. Based on the subsequent diffusion of tobacco from Luzon to China, it was *tabacum*, the typical species from America, that was more widely diffused and accepted on Luzon. Spanish imperial governmental officials and missionaries are the two agents

A barefoot worker sorts tobacco leaves to be sent to the rolling department in the la Flor de la Isabel factory in suburban Paranaque, south of Manila, July 1997. La Flor is one of the biggest of the Philippines cigar companies. AP/WIDE WORLD PHOTOS

that have been named as responsible for tobacco's introduction into the Philippines. While both hypotheses are plausible, the missionaries probably were more responsible because of their New World experience in organizing its cultivation on landed estates and incorporating its commercialization to support their activities.

Habits, Diffusion, and Consumption

The Philippine indigenous population practiced a number of social rituals and intergroup exchange relationships, which may have aided in the rapid acceptance and diffusion of the smoking habit and the use of tobacco in the islands. They possessed the habit of chewing betel and quickly incorporated the habit of smoking and the exchange of tobacco in receiving visitors, the passing of time, and the provision of pleasure. The name betel applies to two different plants—the fruit (or nut) of the areca palm and the leaf of the betel pepper or pan—that are combined with lime (chunam) and perhaps an aromatic spice such as cardamom. People throughout the

Philippines and many parts of Asia habitually chewed (and chew) betel. When masticated, it produces a flow of brick-red saliva that usually temporarily dyes the user's mouth, lips, and gums an orange-brown hue. It is likely that betel chewing paved the way for tobacco consumption.

Tobacco became highly esteemed. It was included as partial payment for indigenous auxiliaries, the Christian or non-Muslim troops and manpower levies from neighboring islands that provided military support to Spanish efforts to control Muslim areas of the southern Philippines. And, with the scarcity and rampant debasement of copper coin, cigars circulated in the hinterland as money as late as the early nineteenth century. Travelers' accounts and Spanish colonial administrators' reports suggest that the habit of smoking was embraced by both genders and by all age groups. Prior to the introduction of a monopoly on tobacco in the eighteenth century, one Spanish official estimated that tobacco consumption included as many as 1 million persons. Forty years later, the number of tobacco consumers was estimated at 3 million persons. Both figures are an extremely high percentage of the islands' population.

Production, Commercialization, and Trade

Early-sixteenth- and eighteenth-century Spanish accounts laud the excellent soil and climatic conditions for the cultivation of high quality leaf tobacco for the production of cigars, cigarettes, and **snuff** throughout the Philippines. There were, however, limits to Spain's control over the islands and the method of tobacco production. On Luzon, for example, where the Spanish were present in greater numbers, tobacco was produced on landed estates in the flat plains. Entrepreneurs and different religious orders owned these estates. In the mountainous regions, which were difficult for the Spanish to control, there were exchanges between smaller lowland indigenous producers and highland consumers. On some of the southern islands, Mindanao in particular, which had already embraced Islam prior to the arrival of the Spanish and resisted the Europeans, indigenous growers produced tobacco that compared favorably in quality with the Spanish Manila or Luzon product. Mindanao encountered limitations in the commercialization of tobacco because it did not operate a large maritime trading fleet and could not attract the same level of external interest as Manila with its availability of silver. In addition to supplying domestic markets, tobacco from the Philippines competed with production from China and Java in inter-Asian exchanges and markets, primarily in the South China Sea. In general, these differing regional tensions and methods of production and commercialization continued until the mid- to late eighteenth century, when the Spanish began to exert greater control.

Revenue and Colonial Governance

From its inception as a Spanish colony, the Philippines depended on financial support from Spain, delivered annually via the remission of New World silver on the Acapulco to Manila galleon. Accordingly, Spain's colonial administrators sought to control, diminish or eliminate this financial drain on the Crown's Treasury. They attempted to monopolize tobacco on Luzon in the early seventeenth century, but the scheme caused such discontent that it was revoked in less than seven months. By the mid-eighteenth century, numerous American colonies implemented

snuff a form of powdered tobacco, usually flavored, either sniffed into the nose or "dipped," packed between cheek and gum. Snuff was popular in the eighteenth century but had faded to obscurity by the twentieth century.

"It [the fire] originated from some tobacco; cursed be it, and the harm, that that infernal plant has brought, which must have come from hell. The wind was brisk, and blowing toward the convent. In short, everything was burned, though we saved the silver and whatever was possible."—Father Juan de Medina, O.S.A., on smoking tobacco and the fire that destroyed the Convent of San Nicolas in March 1628

■ BLAIR AND ROBERTSON, VOL. 24, PP. 145–146

tobacco monopolies, and Spanish promoters of fiscal reform identified tobacco as the only commodity in sufficient demand to justify a monopoly. They hoped it would become a source of revenue that would offset the chronic drain on the Spanish Crown's Treasury caused by subsidizing the administration of the Philippines.

The monopoly on tobacco was established in 1781. It prohibited and reserved for the Spanish government the sale, traffic, and manufacture of tobacco. Proprietors or growers sold their entire crop at contract prices to the monopoly. Subsequent governors occupied themselves with defusing resistance to the tobacco monopoly, increasing the area planted and improving the tobacco **plantations,** and implementing administrative and accounting controls to diminish fraud. It became the most important source of revenue and, in general, it temporarily resolved the Spanish Crown's imperial financial problems. Its implementation impinged upon the local societies' freedom (which they had enjoyed until then) to cultivate without restriction a plant they had been accustomed to using since childhood. The monopoly also forced consumers to pay a higher price for a commodity, which until then had been inexpensive. The monopoly was in place, approximately, a century. An unpopular measure, in the short term, it produced the desired increase in fiscal income but its profitability decreased over time as the expenses of administration grew, as did contraband, corruption, and evasion. With other sources of fiscal income growing from a monopoly on **opium** and the expansion of the exportation of sugar, hemp and other commodities, the opponents of the tobacco monopoly succeeded in obtaining its repeal.

In the nineteenth and early twentieth centuries, cultivation grew and markets for Philippine cigars and leaf tobacco were expanded or developed in China, Japan, the East Indies, the United Kingdom, Spain, and Australia. After sugar and abaca (Manila hemp), tobacco was the third largest export earner during this period. Tobacco manufacturing was an early leader in providing industrial employment. Most of the workers were women, and in the mid-nineteenth century, approximately 30,000 people were employed making cigars and cigarettes in the province of Manila.

See Also China; Origin and Diffusion.

■ GEORGE BRYAN SOUZA

plantation historically, a large agricultural estate dedicated to producing a cash crop worked by laborers living on the property. Before 1865, plantations in the American South were usually worked by slaves.

opium an addictive narcotic drug produced from poppies. Derivatives include heroin, morphine, and codeine

BIBLIOGRAPHY

Blair, Emma Helen, and James Alexander Robertson, eds. and trans. *The Philippine Islands 1493–1803.* 55 vols. Cleveland, Ohio: Arthur H. Clark, 1903–1909. Reprint, Taipei: International Association of Historians of Asia, 1962.

Corpuz, Onofre D. *An Economic History of the Philippines.* Quezon City: University of Philippines Press, 1997.

Jesus, Edilberto. C. de. *The Tobacco Monopoly in the Philippines: Bureaucratic Enterprise and Social Change, 1766–1880.* Quezon City, Philippines: Ateneo de Manila University Press, 1980.

Larkin, John A. *Sugar and the Origins of Modern Philippine Society.* Berkeley: University of California Press, 1993.

Legarda, Benito Justo, Jr. *After the Galleons: Foreign Trade, Economic Change, and Entrepreneurship in the Nineteenth-Century Philippines.* Madison: Center for Southeast Asian Studies, University of Wisconsin–Madison, 1999.

# Pipes

Long before the coming of Europeans to North and South America, long before they discovered what would later be called tobacco, it was custom in primitive communities to breathe the smoke of burning roots, palm leaves, aromatic plants, and herbs for their narcotic or intoxicating effects. Later, humans chose to inhale this smoke in a device fashioned out of crude materials that would eventually bear the name "pipe."

In the fifteenth century Native Americans introduced tobacco to Europeans, and soon the social custom of smoking tobacco gained acceptance and became a fashionable pastime. As tobacco's popularity increased, and its use spread around the globe, smoking spawned an industry of artisans who created an assortment of utensils, accessories, and accoutrements for tobacco's use, storage, preservation, and display. One of the most elegant, intriguing, and artful utensils for smoking tobacco is the pipe, a utensil whose use waxed and waned in the eighteenth and nineteenth centuries because **snuff** became a celebrated habit in the eighteenth century, and the cigar and cigarette were introduced in the nineteenth century. Today, cigars, cigarettes, and pipes peacefully coexist, whereas snuff-taking, at least in the United States, is largely a thing of the past.

snuff a form of powdered tobacco, usually flavored, either sniffed into the nose or "dipped," packed between cheek and gum. Snuff was popular in the eighteenth century but had faded to obscurity by the twentieth century.

What Is a Smoking Pipe?

The pipe is a smoking device that consists of a tube with a mouthpiece on one end and a bowl on the other. Anyone who has studied the history of pipes in depth, however, would say that this definition is shallow and bland because this complicated smoker's utensil, expressed in a variety of formats around the world, defies a simple generic definition. In 1965 Carl Weber, an American pipe maker, opined:

> *The pipe has survived its threatened eclipse by cigar and cigarette for a number of reasons, but the primary one is simple. It is the most attractive, most effective means yet devised by which the smoker can obtain full pleasure from tobacco.*

For historians of tobacco culture, educators, archaeologists, craftsmen of smoking pipes, and, particularly, pipe collectors, the following quotation from E. R. Billings's *Tobacco: Its History, Varieties, Culture, Manufacture and Commerce* (1875) describes the pipe's historical importance and its positive cultural impact on our world:

> *Of all the various branches of the subject of tobacco, that of the history of pipes is one of the most interesting and one that deserves every attention that can possibly be given. Whether considered ethnographically, historically, geographically or archaeologically, pipes present food for speculation and research of at least equal importance to any other set of objects that can be brought forward.*

Tobacco pipes have been made from just about every natural and man-made material. During the expansion of tobacco culture around the world, the pipe evolved as a national expression, appealing to each country's culture, employing available indigenous materials, and taking

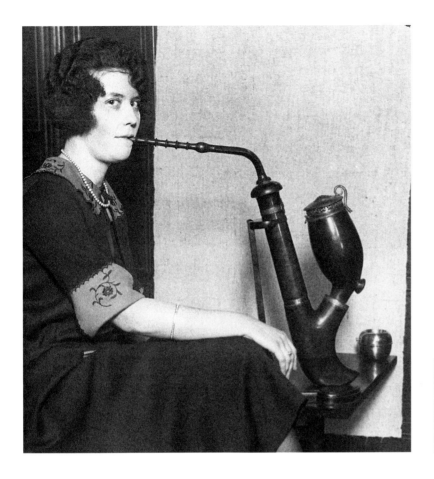

A nineteenth-century German wood *gesteckpfeife* (a pipe in parts) is shown here on display at a New York import house. This outsized pipe was manufactured as a trade sign or for shop display. © UNDERWOOD & UNDERWOOD/CORBIS

shape and form within the region or locality in which it was made. To describe them all would require a lengthy itinerary traveling nearly the entire globe, because as the habit of smoking encircled the earth, nearly every race has adopted the pipe in some form.

Native Americans produced pipes using such materials as steatite, **argillite,** limestone, pipestone, and catlinite (a soft red siltstone named after George Catlin), and most of these pipes were used principally for spiritual ceremonies, given as gifts, or for barter with European explorers. However impractical (because ivory cannot withstand dramatic changes in temperature), the walrus tusk, native to many parts of Alaska, Greenland, and Siberia, has been used as the principal medium for the pipe. Another early example was the trade pipe, a simple utensil made of tin or iron. European voyagers to the New World offered these trade pipes to the Native Americans in exchange for local goods. Although far-fetched, at an early time in Europe after tobacco was introduced, some even assembled a do-it-yourself smoking pipe made of half of a walnut shell and a chicken bone.

For about 400 years, skilled craftsmen in the Western Hemisphere produced pipes that were both beautiful in design and exhibited excellent smoking qualities. Artisans experimented with a wide range of materials, such as pottery, stoneware, amber, antler horn, bone, **gutta-percha,** gold, and silver, but these mediums were not ideally suited as smoking pipes because either they did not withstand heat or they produced an offensive odor or taste when smoked. Many of these pipes were regarded as eccentric or offbeat folk art, but those that have survived are often

argillite A smooth, black sedimentary rock. American Indians sometimes carved tobacco pipes from argillite.

gutta-percha a form of hard rubber made from the sap of a Malaysian tree. Widely used in the nineteenth century, gutta-percha was largely replaced by plastics in the twentieth century.

remarkable examples of inspired imagination, individual innovation, and creativity.

The most eye-appealing and pleasant-smoking pipes were made from four mediums: clay, wood, porcelain, and meerschaum. Between 1600 and 1925, millions of pipes were manufactured of these four materials, a considerable percentage of which depicted classical and dramatic subject matter, as well as whimsical, fanciful, bizarre, and, to the delight of some collectors, erotic and scatological motifs. Thus, the earliest tobacco pipes, once utilitarian and commonplace utensils, mere conveyances for holding tobacco, eventually evolved over several hundred years into an art form. These four mediums predominated in the Western World. Of these four, two—wood and meerschaum—have survived the test of time, and are still being produced today in large quantities for smokers and collectors alike. Each of these is explicated in the following sections. (The corncob pipe, an unattractive, extremely inexpensive, yet distinctively American form, invented after the American Civil War, and still being manufactured today, has never had much of a following.)

Pipes of the Western World

CLAY. The clay pipe was the first practical smoking pipe, introduced at the end of the sixteenth century in England, and its usage soon spread to the European continent, where factories were later established in Belgium, France, Germany, and Holland. Fragile, yet cheap to produce, the cost of a clay pipe was markedly less expensive than the price of the earliest commercially sold tobacco, so, accordingly, the makers produced small pipe bowls. As tobacco became more readily available, the size of the pipe's bowl was commensurately enlarged. The very earliest clay pipes were plain and utilitarian in appearance, but by 1750, when clay pipe manufacture for domestic use became a thriving industry in Europe, a status it enjoyed for the next two centuries, some factories began producing pipes embossed with various decorative designs on the bowl and stem to distinguish one maker from another and, of course, as marketing one-upmanship.

In the nineteenth century, as some pipe craftsmen experimented with other materials, clay pipes began to feature ornate designs of people, animals, plants, and symbolic motifs in a variety of styles, shapes, sizes, and finishes. The French were undoubtedly that century's nonpareil clay pipe artisans; three French clay pipe manufacturers—Duméril-Leurs, Fiolet, and Gambier—collectively designed and manufactured more than 5,000 different clay pipe motifs while their factories were in operation. A majority of these pipes exhibited fanciful raised decor and were fire-glazed in brilliant colors, each pipe bearing the raised letters of the company name and an identifiable model number on the shank.

Because Colonial America did not have a noteworthy clay pipe industry of its own, it imported almost all of its clay pipes from Europe. After the American Revolution, potters in the United States began molding pipe bowls in both earthenware and stoneware, producing typically less ornate, more functional clay pipes, but occasionally making some featuring faces, animals, or simple decorative designs. In the latter part of the nineteenth century, clay pipes imported from Europe stimulated American makers to copy foreign styles and to create original designs.

PIPES

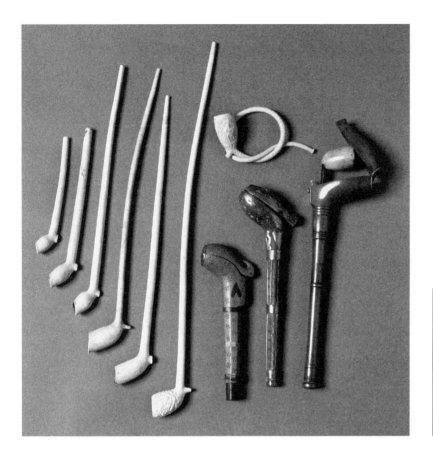

Shown left to right is an assortment of six English and Dutch clay pipes with graduated-size bowls and stems dating from about 1600 to about 1890. In the center is an early nineteenth-century Dutch "knotted stem" clay pipe. On the right are three decorated wood clay pipe cases from Holland from the early nineteenth century. PRIVATE COLLECTION. PHOTOGRAPHY COURTESY OF THE BRANDYWINE RIVER MUSEUM

One memorable clay maker in the late 1800s was A. Peyrau, a French immigrant living in New York City, who made a series of terracotta pipe bowls featuring bizarre, yet comedic, caricature heads of contemporary celebrities, among them P. T. Barnum, Joseph Pulitzer, and William March "Boss" Tweed. Today, Peyrau's pipes are considered some of the finest clay pipes ever produced in the United States.

Although a few clay pipe producers continue in operation today in North America and Europe, their collective output is not significant because the clay pipe is no longer popular. It has neither the cachet nor the smoking appeal that the meerschaum pipe and the **briar** pipe have today.

briar a hardwood tree native to southern Europe. The bowls of fine pipes are carved from the burl, or roots, of briar trees.

PORCELAIN. Porcelain pipes, unpleasant to smoke because of the nonporous material's inability to breathe, are nonetheless remarkable examples of sculptors' and molders' dexterity. In the mid-1700s a few European factories such as Meissen, Mennecy, Nymphenburg, and Sèvres created polychrome pipe bowls in baroque, neoclassical, and Romanesque styles.

Later examples, illustrating mythological, entomological, and floral subjects, were produced at many European porcelain factories. Between 1850 and 1870, of the approximately 18.7 million pipes produced in one pipe-making center in Ruhla, Germany, 9.6 million were porcelain. The bowls frequently exhibited hand-painted portraits, landscapes, hunting scenes, or commemorative events, and were fitted with three- and four-foot stems of hardwood, ivory, or horn. After the Franco–Prussian War and until World War I, a pipe format known as

PIPES

the Regimental was produced in large quantities, particularly in Germany, Austria–Hungary, and Denmark, and presented to soldiers to honor their military service. The Regimental was a unique and vibrantly colorful style of pipe, with its porcelain bowl depicting martial scenes and accompanied by an exceptionally long cherry wood stem. Today, a few German potteries produce porcelain pipes for domestic use, but many are bought by tourists as mementos of their visit.

POTTERY. Another variety of pipe closely aligned with the evolution of clay and porcelain pipes was the pottery pipe, the most notable of which were produced at potteries in Staffordshire and Whieldon, England. These elaborate showpieces, known as puzzle pipes, were amusing whimsies, not functional pipes, a product of excess clay and spare time. They are distinguishable by their unusual design: long, polychrome-painted, soft-paste coils fashioned into twisted and looped designs.

European-designed porcelain and pottery pipes were exported to America, but the annals of the U.S. tobacco industry indicate that no American company produced either porcelain or pottery pipes, probably because Americans, in general, never were able to acquire a taste for smoking tobacco in such pipes.

WOOD. In their search for a durable, non-breakable, and pleasant-tasting material, wood turners during the eighteenth and nineteenth centuries experimented with more than twenty-five different domestic and exotic woods as possible substances for producing pipes, as shown in the following list.

This unusual porcelain pipe bowl is a full figural of a pasha or emir dressed in the *haute couture* of a caftan. The bowl is accented with appliquéd gilt buttons. It was made in the late nineteenth century, and probably is French. FROM THE COLLECTION OF DR. SARUNAS PECKUS

The Early Woods

Acacia	Alder	Ash	Beech	Birch
Blackberry	Boxwood	Buckthorn	Cedar	Cherry
Chestnut	Dogwood	Elder	Elm	Hazel
Heather	Hornbeam	Lava	Linden	Maple
Morello	Mulberry	Oak	Olive	Peat
Poplar	Rosewood	Sycamore	Walnut	

By the mid-nineteenth century, one variety of the heath shrub, *erica arborea*, native to the Mediterranean coast and commonly known as briar, a porous and lightweight wood, proved to have exceptional qualities for smoking tobacco, and its superior grain inspired hand-crafted pipes executed in ornate and delicate shapes. History recounts that the briar pipe industry began in the French village of St. Claude where, by 1892, more than sixty different briar pipe factories thrived.

Today, briar pipes are made in almost every European country, Japan, and the United States, from mass-produced pipes at the low end of the price spectrum to exquisite, limited-production, one-of-a-kind, handcrafted specimens costing thousands of dollars. The briar rivals the meerschaum as the better of the two readily available and popular smoking pipes.

MEERSCHAUM. The aristocrat of smoking pipes, known by such appellations as "Venus of the Sea," "Queen of Pipes," and "White Goddess," is

PIPES

Cherry was one of more than twenty-five woods popular with pipemakers during the eighteenth and nineteenth centuries. This unusually shaped cherry wood pipe bowl, carved in bas-relief and high-relief rococo style, is from France (c. 1860). FROM THE COLLECTION OF DR. SARUNAS PECKUS

made of meerschaum, the German term for "sea foam." Known to geologists as sepiolite, this claylike mineral's composition is hydrated silicate of magnesium. In addition to the ease with which this substance can be intricately carved, pipe enthusiasts prize meerschaum's ability to mellow, mutate, and metamorphose over time through a range of colors from its original white to hues of brown as it absorbs the byproducts of tobacco.

The discovery of meerschaum's qualities as an excellent pipe material is shrouded in mystery and myth. But since the mid-eighteenth century, tons of this substance—mostly from mines in Anatolia, Turkey—have been converted into exquisite smoking implements. Early meerschaum pipe manufacturing centered principally in Berlin, London, Paris, Prague, Venice, and Vienna. These cities contained warrens of **ateliers** bustling with skilled artisans working alongside craftsmen of related guilds—such as jewelers, metal smiths, and wood turners—who made the pipe stems, mouthpieces, wind covers, and other pipe fittings.

atelier a small workshop or studio.

Soft and pliant, meerschaum became the medium of choice for the more dexterous craftsmen who executed precise facsimiles of works by contemporary sculptors, muralists, illustrators, etchers, and engravers. Some carvers, however, used their own imagination for the images they sculpted. In its golden age, from 1850 to 1925, meerschaum was used not only for pipes but also for cigar and cigarette holders.

Information about the evolution and growth of the American meerschaum pipe industry is sparse. As one early-twentieth-century writer reported, the American meerschaum trade began approximately in 1855 when a New Yorker, Frederick W. Kaldenberg, met an Armenian named Bedrossian, who brought two cases of raw meerschaum from Asia Minor to the United States (Morris 1908).

> *It was not long before these two cases of meerschaum were turned into pipes of special shape and design, which brought the literati, the artistic and the mercantile **nabobs** of the great City of New York, to the workshops of the artisan who had wrought the first meerschaum into pipes in the United States.*

nabob a very wealthy person, often having political and social influence.

Tobacco in History and Culture
AN ENCYCLOPEDIA

419

PIPES

Meerschaum pipe, standing nude, bas-relief—carved ram's head on front of bowl, inset coloring bowl, amber mouthpiece, 8.5-inch length, 5-inch height, probably American, c. 1875. PHOTO BY GARY KIEFFER

Smoking Pipes in Other Regions

THE NEAR EAST. As mentioned previously, the pipe in many quarters of the world is a national expression, and this is especially true in the Near or Middle East, where two customary pipe formats are found—the chibouque (or chibouk) and the water pipe. The chibouque is peculiar to countries bordering the Mediterranean Sea, such as Turkey, Egypt, and Syria, and is best described as a long pipe comprised of a baked terracotta clay bowl shaped like a cup or bowl and, most often, a long pipe stem made of a jasmine branch or other fragrant wood, some as long as five feet.

Known in the west as the hubble-bubble, the hookah (called nargileh, shisha, or kalian in different countries) is a class of pipes from the Islamic world that originated in India as a tradition, fashioning a water pipe from an empty coconut husk. The style evolved into a device found in two configurations, one for personal use at home and one for travel. The typical hookah consists of a base, a "chillum" that holds the leaf tobacco, a stem, and a flexible tube. The base is the component on which the craftsman expends his artistic energy, and the bases of the better hookahs fabricated of glass, ceramic, or silver can be exceptionally ornate and elaborate.

The hookah uses a small charcoal tablet to gently heat tobacco that rarely burns, but is filtered as it is drawn through the water-filled base and inhaled through the tube. The tobacco might be mixed with a special blend of fruit shavings or flavored molasses to produce a deliciously fruity aroma, or it might be cultivated tobacco that yields a strong aroma.

Smoking a hookah is a ceremonial experience shared in the company of friends. Both the chibouque and the hookah have transcended national boundaries and are now found in the West, where they are for rent at many bars, coffeehouses, and hotels, and where anyone can partake in this social endeavor.

THE FAR EAST. For several centuries, the Orient has had at least two distinctive styles of tobacco pipe: the Japanese *kiseru* and its lengthier counterparts, known by different names in Korea and China; and a different type of water pipe, used in China, Cochin China, and Annam (now Vietnam).

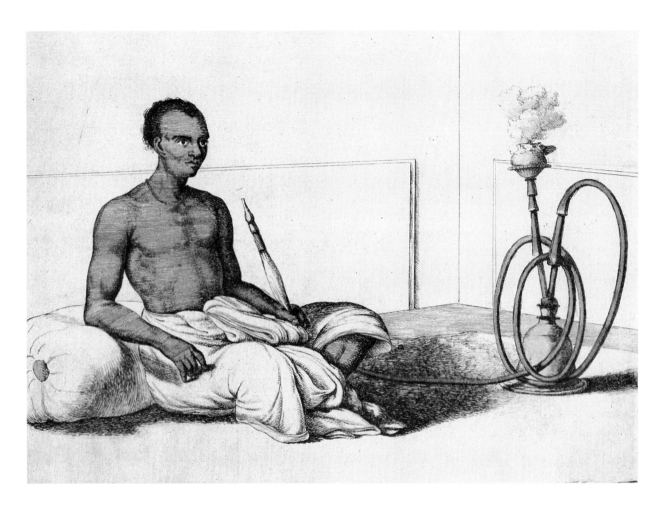

The hookah, also known as the hubble-bubble and by other names, remains popular today in most countries of the Near East, Middle East, and South Asia. The long, flexible tube of the hookah allows freedom of movement for the smoker. Modern-day hookahs, like their older counterparts, are often made to be more portable and convenient, that is, with smaller bodies and shorter hoses.
© STAPLETON COLLECTION/CORBIS

The common *kiseru* is a three-part pipe, consisting of a metal bowl, a metal mouthpiece, and a bamboo or wood stem that connects the two metal components. Some, made especially for the Imperial family, shoguns, and local lords, were ornate masterpieces worthy of being characterized as art. The classic Chinese water pipe is a boxy metal contraption that is functionally similar to the hookah in that the tobacco is filtered through water.

Nowadays, the cigarette has all but replaced traditional pipes except for the occasional tea ceremony and private use in the home. Countries such as Borneo, Indonesia, and Brunei have their own national pipe expressions but, collectively, theirs have never had an impact on or influence in pipe design beyond their own borders.

AFRICA. Because Africa is a continent of many countries, diverse peoples, and myriad tribes, each with its own customs and culture, one must ascribe to Africa myriad assorted pipes made of different materials, each attributable to a different place in this land mass. It is difficult to generalize about the form or functions of African pipes other than to state that the calabash gourd (botanically *Lagenaria vulgaris* or *Lagenaria siceraria*), assorted woods, terracotta and other earthenware, bronze, brass, tin, iron, bone, ivory, and assorted other materials have been fabricated into pipes for not only smoking tobacco, but also kief, hemp, dagga, and various herbs and roots. So few serious studies have been

Who Smokes a Pipe?

Pipe smoking is a common practice among both genders, the young, the middle-aged, the old, people from every walk of life. In the past, surveys have attempted to determine the mean age, gender, economic stratum, and epicenters of pipe smokers. Depending on when the survey was conducted, the results and conclusions always vary, because the number of pipe smokers ebbs and flows with each generation. The price range of pipes, attitudes about smoking, and health issues all play a part in this ebb and flow.

Most agree that there are millions of pipe smokers around the world, and the burgeoning number of local, regional, and national pipe clubs in the last twenty-five years evidence this. Pipe smokers are a brotherhood, bound by a common love of this ubiquitous utensil, and as affiliates or associates in this elite club, nationality notwithstanding, they come together frequently at various trade shows, exhibits, and pipe-smoking competitions around the world to share their experiences with their pipe-smoking brethren, ogle new products for the smoker, taste new tobacco blends, and trade anecdotes about this or that pipe. Despite differences in language, nationality, income bracket, and education, they are bound by a simple device made of wood, or meerschaum, or clay, a universal symbol of camaraderie.

conducted about the pipes of Africa that, even on careful inspection of its construct or composition, it is hard to determine whether a particular pipe was specifically designed for tobacco, or another intended use. What is certain is that the peoples of Africa continue to produce an endless assortment and variety of pipes in a broad array of mediums, each with the character and personality of its maker and its locale.

The Gentle Art

What can be said of all this? It is a fact that smoking is a worldwide cultural phenomenon, and pipe smokers are a rather unique group who attribute a special aura to the pipe, claiming that it denotes the "gentle art." Art is an apt description because not only can the pipe be an art form, it is also represented in works of art, stories, and songs that document, celebrate, and, occasionally, condemn it.

The tobacco pipe occupies center stage in the engravings of the seventeenth-century Dutch artists Jan Steen, David Teniers, and Adriaen van Ostade; in the eighteenth-century illustrations of British painter and printmaker William Hogarth; and in the caricatures of George Cruikshank, James Gillray, and Thomas Rowlandson, also of England.♦ The pipe plays a significant role in the nineteenth century's *trompe l'oeil* works of America's William Harnett and John Frederick Peto, and in the twentieth century's canvasses of Russia's Marc Chagall and the Spanish cubist Pablo Picasso.

♦ *See "Visual Arts" for a Jan Steen illustration that portrays tobacco in an unfavorable light.*

Although criticized in some art and literature, the pipe has been praised in hundreds of published poems, couplets, rhymes, and paeans penned by the well known and the anonymous and in many languages during the last three centuries. Nonfiction literature abounds on the history and manufacture of the pipe, but one of the most famous fictional works about man's love for the pipe is *My Lady Nicotine: A Study in Smoke*, written in 1890 by Sir James M. Barrie, the author of *Peter Pan*, *The Admirable Crichton*, and *Margaret Ogilvy*. Even the occasional musical score has been written as a tribute to the pipe, such as "Put On Your Slippers and Fill up Your Pipe" (c. 1916).

To conclude, pipe smokers around the world uniformly agree that the pipe, whatever its shape, style, format, or medium—for no single pipe is the perfect pipe for all—is the most perfect way to smoke tobacco. And pipe collectors, a complementary group who may or may not be pipe smokers, derive equivalent pleasure for yet a different reason: owning antique, vintage, or new pipes, elegant miniature masterpieces in wood, meerschaum, clay or porcelain, each spawned from imagination, each crafted with skill and dedication, each executed by some master artisan in his time.

See Also Africa; Archaeology; Calumets; China; Connoisseurship; Consumption (Demographics); Islam; Japan; Literature; Middle East; Music, Classical; Music, Popular; Native Americans; South Asia; South East Asia.

■ BEN RAPAPORT

BIBLIOGRAPHY

Armero, Carlos. *Antique Pipes: A Journey Around the World.* Madrid: Tabapress, 1989.

Billings, E. R. *Tobacco: Its History, Varieties, Culture, Manufacture and Commerce.* Hartford, Conn.: American Publishing Company, 1875.

Crole, Robin. *The Pipe: The Art and Lore of a Great Tradition.* Rocklin, Calif.: Prima Publishing, 1999.

Dunhill, Alfred H. *The Pipe Book.* London, England: A. & C. Black, 1924. Reprint, London, England: Arthur Barker Limited, 1969.

Ehwa, Carl, Jr. *The Book of Pipes and Tobacco.* New York: Random House, 1974.

Fresco-Corbu, Roger. *European Pipes.* Guildford, England: Lutterworth Press, 1982.

Goes, Benedict. *The Intriguing Design of Tobacco Pipes.* Leiden, The Netherlands: Pijpenkabinet, 1993.

Liebaert, Alexis, and Alain Maya. *The Illustrated History of the Pipe.* (Translated and adapted from the French, *La Grande Histoire de la Pipe* (1993) by Jacques P. Cole.) Suffolk, England: Harold Starke Publishers, Ltd., 1994.

Morris, Fritz. "The Making of Meerschaums." *Technical World*, April 1908, pp. 191–196.

Rapaport, Benjamin. *A Complete Guide to Collecting Antique Pipes.* Exton, Pa.: Schiffer Publishing, Ltd., 1979; Reprint, Atglen, Pa.: Schiffer Publishing, Ltd., 1998.

———. *Collecting Meerschaums: Miniature to Majestic Sculpture, 1850–1925.* Atglen, Pa.: Schiffer Publishing, Ltd., 1999.

Weber, Carl. *The Pleasures of Pipe Smoking.* New York: Bantam Books, 1965.

Wright, David. *The Pipe Companion: A Connoisseur's Guide.* Philadelphia, Pa.: Running Press, 2000.

Plantations

Tobacco has been one of the major plantation crops of the Americas. It has been especially important in North America and at different times in other parts of the Americas, including Cuba and Venezuela. In contrast to other plantation crops, especially sugar, tobacco plantations

PLANTATIONS

tended to be relatively small; tobacco was also grown on smaller holdings as well. While enslaved Africans and Amerindians were used as workers on tobacco plantations and farms, they often worked alongside their masters, indentured workers, and others. The crop required careful tending of the tobacco leaves and thus close supervision.

Europeans had acquired their knowledge of tobacco cultivation from Amerindians. This apprenticeship was rapid and short, and in places like Trinidad and Venezuela it seems that both groups worked side by side. Tobacco was often raised in fields that were separate from other crops. In some fishing and hunting communities, tobacco was the only crop actually cultivated. Very quickly, however, tobacco was transformed into a European commodity. In the region of the Chesapeake, for example, Amerindian ties with tobacco were undermined, and even the memory of an association was erased from the traditions. At the same time, European settlers transformed production as they began to supply European tobacco and smoking devices, such as clay pipes, to the native population.

Expansion of output essentially required knowledge of the crop, suitable land, and sufficient labor, which did not necessarily mean that the crop would be grown on plantations. Unlike sugar and some other crops, the differences in productivity on relatively large holdings were not dramatically different from production on small holdings, but there were nonetheless some economies of scale. In fact, during the seventeenth and the eighteenth centuries, tobacco was grown in Europe, with production almost exclusively confined to peasants and small farmers. Similarly, in the seventeenth century, the Dutch tried unsuccessfully to introduce this style of peasant production in their South American colonies.

Despite these isolated examples, the plantation model for production, as adopted in the Tidewater region of the Chesapeake, became the model for the Americas, with little variation among European colonies. There, as in many other places where the combination of open land and the lack of free labor defined the possibilities of expansion, bonded labor, whether under indenture or as slaves, became the basis of labor supply. In North America, for example, the Virginia Company sponsored the migration of indentured workers, both men and women, who were expected to work off the price of their passage across the Atlantic and other debts that had been accumulated through a system of quasi-coerced labor. Normally, such indentures lasted from four to seven years, after which the workers were free to establish themselves as independent farmers or otherwise work on their own account. Hence indentured labor eased the early stages of production during the clearing of land and the expansion in production, but such labor was inadequate as a long-term source of labor. Once the indenture was finished, there was little reason for individuals to continue to work for their masters; with the wide availability of land, it was more usual for people to become independent producers themselves or find other employment. The use of slave labor, particularly enslaved Africans, was a response to this labor shortage.

Culture and Methods of Cultivation

The cultivation and care of tobacco involved a well-defined sequence of steps from seed to market. Successful production depended on the management of the interaction between human activity and the natural

PLANTATIONS

After 1680, African slaves replaced European indentured servants as the primary form of labor on tobacco farms. The black population rose dramatically and tobacco farms were transformed into plantations such as the one shown in this photo. Wealth became measured by the number of slaves on the plantation. © MEDFORD HISTORICAL SOCIETY COLLECTION/CORBIS

conditions. Tobacco cultivation required a constant input of different skills through each stage, each one requiring a great deal of care.

In the first stage, the seed was planted in a seedbed. This method was common practice, the persistence of which to the present shows its superiority over planting in the open field. The seedbed increased the survival opportunities for the young plants, allowing for more careful use of fertilizers. Careful scientific study has determined that the act of transplanting has the beneficial side effect of enhancing growth. Apparently, Chesapeake planters had come to understand this effect, although the extent of experimentation needed to reach these results is unknown. However, the use of seedbeds marked an important difference between the way tobacco had been cultivated by Amerindians and on plantations.

The seedbed stage was not onerous in work, but the transplanting to the open fields was labor-intensive. Furthermore, determining when transplanting should take place was a crucial decision of management. The shift had to be done in favorable conditions, usually after a heavy rain. The complete operation of transplanting usually took many weeks. This stage demanded a lot of people at the precise time, because the field needed to be prepared and the transplanting performed quickly in order to prevent the plants from drying out; each plant had to be transplanted individually. It is likely that some of the biggest Virginian planters cultivated more than 100,000 plants, with several thousand per acre.

In the next stage, weeding, the fields were cleaned of any growth that menaced the development the tobacco plants. Weeding required a

PLANTATIONS

great deal of time and energy. As the plants grew, they had to be topped, which prevented the plant from producing flowers and thus permitted the growth of the leaves. Secondary branches were trimmed in order to allow the plant to concentrate all growth in the principal leaves. This selective action was called suckering and was done plant by plant. Suckering and topping were crucial for the final quality of the leaves and hence to the yield. However, when the plants became mature, the most difficult decision of the planters was to choose the exact time to cut. In the end, the attention shifted from the plant itself to the individual leaves, which could be easily damaged.

The cutting of the leaves was the point at which the tobacco was transformed from a botanical item into a salable commodity. After cutting, the leaves had to be cured, which entailed allowing the tobacco stalks with their leaves to dry out naturally in barns specially built for this purpose. After the tobacco was cured, the leaves were stripped from the stalks, and the main stem of each leaf was removed. Depending on the number of plants, stalking and stemming could be done in twenty-four hours. Once finished, the leaves were packed into barrels in a process called prizing. By the time prizing was completed, the seedbeds for the next calendar year needed to be underway and the production cycle begun anew.

However, time was not the only aspect of the Chesapeake culture that was influenced by the rhythms of the tobacco plant. The human and material geography of the region were also shaped by the demands of the plantations. In North America, the combination of open land and short labor force necessary to supply the necessities of the farming developed a decentralized spatial organization. From the settlements on the James River, the English colonies extended to the north, where rivers like the York, the Rappahannock, and the Potomac created rich wetlands could be brought into production at low cost. The spread of tobacco cultivation was also favored by the fact that the Chesapeake Bay region is a myriad of rivers, inlets, and tributaries penetrating the maritime plain. And this water system was deep enough to allow the entrance of the largest vessels of the colonial trade.

In fact, the river system was the usual means of transportation within this region. The river system enabled planters to move tobacco to market and ultimately to England without major expense or risk. Moreover, because of the geography of the region, people lived in rural areas, scattered along the rivers. Until the first quarter of the eighteenth century, despite legislation passed by the Virginia and Maryland Assemblies to encourage settlement in towns and villages, there were fewer than ten small villages containing between fifty and a hundred residents in the region.

In the seventeenth century, the Chesapeake colonies remained essentially European, and the number of Africans remained small. Despite changes in the organization of labor, tobacco defined the particular expressions of Chesapeake culture. Until the second half of the seventeenth century, most of the tobacco farms were small properties and often had no more than one bound worker. On the larger estates, five workers were able to take care of ninety or a hundred acres planted in tobacco. Nevertheless, this age of the small planter lasted only until the end of the seventeenth century.

After 1680, Chesapeake society was completely transformed. The labor regime changed from indentured servitude to slavery and from a European to an African base. The social hierarchy, which had been characterized by relatively minor class differences, became an elitist system based on wealth, race, and power.

The population of the Chesapeake colonies increased dramatically in the eighteenth century. Although the black population at the end of the seventeenth century was small in comparison to the white population, the number of blacks increased rapidly, both absolutely and relatively. Whereas blacks accounted for only 13 per cent of the population in 1700, by the end of the century the figure reached 40 per cent in a total population of almost 800,000 people. Tobacco farms were transformed into tobacco plantations. Planters turned to slave labor instead of servant labor, and wealth became measured by the number of slaves on the plantation. Among several alterations introduced in the organization of tobacco cultivation, the most important were the increasing size of the unit and the growing complexity of administration on the plantations. This process led to the formation of a gentry, a new class of planters. For them, the combination of slavery and tobacco cemented a social system of dominance that came to shape their material welfare and the symbols of power.

At the same time that the plantations grew in size, the slave labor force expanded, and the demand for supervision increased, the gentry became the repository of the almost mystical understanding of the tobacco business. If the possession of slaves became the measure of wealth, the quality of the tobacco was a measure of self-esteem. The techniques of harvesting became a well-kept secret passed from father to son. Some of these rich men were obsessed by tobacco. It was the basis of their culture of debt, which linked them with the English merchants in a consignment system that allowed them to accumulate manufactured goods and European foodstuffs.

Plantations in the Caribbean and Brazil

With the exception of Jamaica, tobacco was the first or one of the first crops grown in the British and French Caribbean. As was the case in the Chesapeake, the islands initially relied on indentured European labor rather than enslaved Africans. During the seventeenth century approximately 60 per cent of the British emigration was bound for the Caribbean.

The mortality rate was huge among the European population in all of the British-controlled islands. The initial rise in the white population and the subsequent, equally dramatic decline occurred in tandem with the growth and collapse of tobacco cultivation. In many aspects, what happened on the English Caribbean islands also happened in the French Antilles. The decline in the white population was in both cases was related to the abandonment of tobacco as a major crop, by the shift to cheaper African slave labor, and by the consolidation of landholdings.

In Brazil, tobacco cultivation was concentrated in Bahia, in the northeast of the country. The production cycle in Brazil was shorter than it was in the Chesapeake. In addition, the Bahia fields did not entail a long-fallow system, since fallowing there was combined with a routine application of animal manure. In both regions, however, the economies of scale in tobacco cultivation were limited. The expansion of production depended primarily upon a proportionate increase in inputs. In Brazil, by the seventeenth century, most of the labor force was made up African slaves, in part because Brazil was not settled with indentured servants and in part because the Bahian region developed an important sugar industry supported by a slave system in the late sixteenth century. Furthermore, much of the Bahian tobacco was used to buy slaves on the West African shores rather than to satisfy a European market.

See Also Labor.

■ CARLOS FRANCO LIBERATO

BIBLIOGRAPHY

Fairholt, F. W. *Tobacco: Its History and Associations.* London: Chapman and Hall, 1859. Reprint, Detroit: Singing Tree Press, 1968.

Goodman, Jordan. *Tobacco in History: The Cultures of Dependence.* London: Routledge, 1994.

Narid, Jean Baptiste. *O fumo brasileiro no periodo colonial.* São Paulo: Brasiliense, 1996.

Ortiz, Fernando. *Cuban Counterpoint: Tobacco and Sugar.* New York: Alfred A. Knopf, 1947.

Robert, Joseph C. *The Story of Tobacco in America.* Chapel Hill: University of North Carolina Press, 1967.

Politics

Because the tobacco industry makes a product that is deadly and generates substantial controversy and opposition, its survival depends on being an active and effective political player at all levels of government. The industry fares better when there are low taxes on its products: little or no regulation on tobacco products and advertising; and few restrictions on its legal liability for the death and disease it causes, or where smoking is permitted. Conversely, enactment of public health policies to reduce the burden of disease and death that the tobacco industry causes involves political action on the part of public health advocates, which the industry works to block.

Since the tobacco industry is held in low esteem nearly everywhere in the world, it often exercises its political influence in the background, working through third parties (from the liberal American Civil Liberties Union to the conservative Cato Institute) and through front groups that it creates and secretly funds (such as "hospitality associations" that oppose clean indoor air laws) to press its agenda. It also exerts more direct influence on individual politicians and political parties through large and strategically placed campaign contributions.

Public health advocates have been most effective in countering the tobacco industry's influence when they can move the field of play from the national or state to the local level. There, the resources that public health advocates can muster are adequate to the task and the tobacco industry's superior resources and national political connections are less effective. The tobacco industry works to neutralize local political action with "preemption," whereby national or state government restricts the right of subordinate political bodies, which are closer to the public and more willing to implement the popular will for tobacco control, than units higher in the political system.

Early Battles on Smoking and Health

The tobacco industry's heavy involvement in politics began in the mid-1950s after the *Reader's Digest* published an article titled "Cancer by the Carton." As a result, the public began to embrace scientific research linking smoking to lung cancer. A wave of public concern led to debates by political units at all levels on restrictions on the sale of cigarettes and on cigarette advertising and promotion. The tobacco industry responded by creating the Tobacco Industry Research Committee (TICR, later renamed the Council for Tobacco Research), a nominally independent scientific body, to get to the bottom of the "smoking and health controversy," and the Tobacco Institute, a lobbying organization, based in Washington, D.C., that was created to allow the cigarette manufacturers to present a unified front to Congress and other political decision makers at all levels. Both organizations were tightly controlled by industry executives and lawyers.

Public awareness of the evidence that smoking was dangerous increased in 1964, when the U.S. Surgeon General, acting on behalf of the United States government, released a report concluding that tobacco use was linked to lung cancer, chronic bronchitis and emphysema, cardiovascular diseases, and other forms of cancer. Public interest in the report—and concerns by the industry that it would adversely affect them—was so strong that release was delayed until after business for that week when the New York Stock Exchange was closed. The industry also feared that public health groups, most notably the American Cancer Society, would use the publication of the report to severely restrict the industry.

The resulting wave of public concern led to several legislative proposals and in 1965 Congress passed the Cigarette Labeling and Advertising Act. While health forces were pleased that the act added warning labels to cigarettes, the combined political and economic power of the tobacco industry and the strength of the constituency of tobacco farmers kept the warning label small and weak. More important, the act prevented states (and localities) from taking any further action on cigarette labeling or advertising. While the labeling law marked a small step forward at the time, preemption prevented strong local and regional action against cigarettes permanently. Indeed, half a century later, in 2001 in the case of *Lorillard Tobacco et al. v. Reilly, Attorney General of Massachusetts et. al.*, the Supreme Court cited the 1965 act in striking down strong advertising and labeling legislation enacted by Massachusetts that would have prohibited tobacco advertising within 1,000 feet of schools, including both outdoor advertising and advertising

POLITICS

FIGURE 1

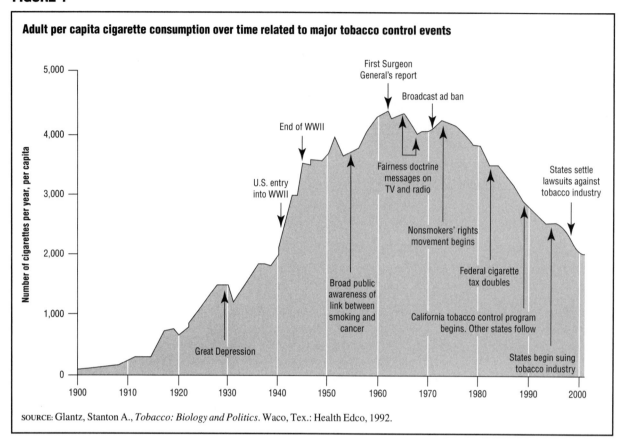

Adult per capita cigarette consumption over time related to major tobacco control events

SOURCE: Glantz, Stanton A., *Tobacco: Biology and Politics*. Waco, Tex.: Health Edco, 1992.

within retail stores that could be viewed from outside the store. The industry would use the strategy of giving a little, such as agreeing to a modest warning label, to get preemption and arrest future progress on tobacco control at both the national and state levels in the coming decades.

Most political battles during the late 1960s and early 1970s over tobacco control continued at the national level. Tobacco control forces won a substantial victory in 1967 when, in response to a lawsuit brought by a law professor at Georgetown University, the Federal Communications Commission (FCC) ruled that cigarette advertising was "controversial," requiring radio and television stations to give free time to broadcast antismoking advertisements. These advertisements, produced by health groups, ran at the rate of approximately one antismoking advertisement for every three protobacco advertisements. As shown in Figure 1, this counter-advertising campaign led to a 5 percent decline in cigarette consumption per person in the United States, in 1975, the first sustained drop in cigarette consumption.

The tobacco industry responded by going back to Congress and supporting legislation to eliminate cigarette advertising on radio and television. While ending broadcast cigarette advertising was viewed as a public health positive, the fact that the cigarette advertisements were off the air meant that the antismoking advertisements also disappeared as of 2 January 1970. Unlike the health groups, however, the tobacco industry had the resources to continue to expand its advertising efforts in magazines, billboards, and other media.

The Rise of Nonsmokers' Rights

The focus of tobacco politics shifted away from the national level to the states in the mid-1970s. The tobacco industry had effectively contained legislation on cigarette advertising at the national level, but there was growing awareness that secondhand smoke was dangerous to nonsmokers. In 1975, after some limited legislation in Arizona restricting smoking in most public places such as government buildings and health facilities, freshman Minnesota Representative Phyllis Kahn introduced the first comprehensive state clean indoor air law. This legislation, which passed with relatively little opposition, prohibited smoking in public places except in smoking-designated areas, and required barriers and ventilation for smoking areas. While modest by twenty-first-century standards, the Minnesota Clean Indoor Air Act represented a real step forward and stimulated efforts to enact similar legislation elsewhere, particularly in California and Florida.

In California, a small group of local activists worked to pass local legislation modeled after the Minnesota Clean Indoor Air Act, and passed the first such law in April 1977 in Berkeley. Following this success, they worked for several years to enact a state law through both the legislature and the initiative process, through a law is enacted by a direct vote of the people. Between 1977 and 1980 the tobacco industry spent more than $10 million opposing these efforts. Recognizing that they could not win in the state legislature or in an expensive state initiative campaign, which, in a large state like California, is essentially an advertising contest, Californians shifted their efforts to enacting local ordinances. The organization they created, which later became known as Americans for Nonsmokers' Rights (ANR), took the lead in grass roots organizing against the tobacco industry.

Across the country in Florida, local advocates were also passing local clean indoor air ordinances. In 1985 in Florida, in contrast to California, the tobacco industry, with the naive support of some health advocates, was able to pass a weak statewide law. This law appeared to address the problem of secondhand smoke but included preemption, which overturned the then-existing local tobacco control laws, effectively stopping local restrictions in Florida until 2002, when health advocates overturned the 1985 law with a voter-enacted initiative after a major state political campaign.

In the mid-1980s, as the local clean indoor air movement was gaining momentum, the tobacco industry responded with a national effort to pass weak state laws that preempted local tobacco control activity. The industry was successful in passing some form of preemption in 22 states. At the same time, however, local tobacco control advocates prevented preemption in the remaining 28 states. Despite increasingly sophisticated and aggressive use of third parties and front groups, initially in the hospitality industry, then expanding to gambling interests to fight local tobacco control laws, the tobacco industry often lost efforts to enact these laws. Between the early 1980s and 2004, the nonsmokers' rights movement has helped to pass clean indoor air ordinances in 1,675 municipalities across the United States (see Figure 2).

However, the tobacco industry did not limit its pursuit of preemption to clean indoor air ordinances. The tobacco industry co-opted a federal effort designed to make it more difficult for children to purchase tobacco, so-called youth access, as another vehicle to preempt local

POLITICS

FIGURE 2

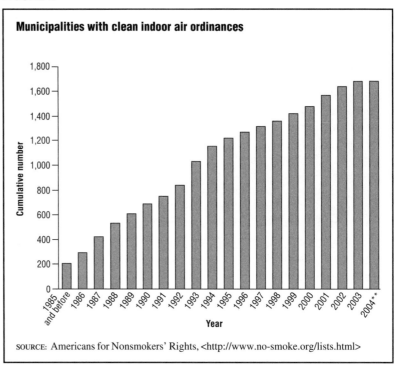

Municipalities with clean indoor air ordinances

SOURCE: Americans for Nonsmokers' Rights, <http://www.no-smoke.org/lists.html>

tobacco control efforts. In 1992, Congress passed the Synar Amendment requiring states to reduce the illegal sale rate of tobacco to minors to less than 20 percent of attempted sales or risk losing federal substance abuse block grants. In 1996, when the implementing regulations were issued, the tobacco industry pushed "compliance bills" that included preemption of more aggressive local youth access laws in many states. In some cases, the industry managed to use the debate over youth access to pass broad preemption that also preempted local clean indoor air ordinances.

National Politics

While most of the successes in tobacco control during the 1980s and 1990s were at the local or state level, there were still several important debates at the national level, where the tobacco industry continued to dominate the process through a combination of campaign contributions and well-connected lobbyists and allies. As of 2004, tobacco remained the only substance ingested by humans that was exempt from any federal regulation as a food, drug, or consumer product.

Contributions from tobacco interests (including contributions to federal candidates, political parties and noncandidate committees) increased from $7.8 million in 1997–1998 to $8.7 million in 1999–2000 to $9.4 million in 2001–2002, the most recent complete election cycle for which data are available. Tobacco companies spent an additional $91.1 million on lobbying between 1999 and 2003.

These expenditures have been effective investments for the tobacco industry because the industry has continued to prevent any meaningful action at the federal level. Bills were introduced in 2001 and debated during the 107th Congress to give the U.S. Food and Drug Administration the authority to regulate tobacco products. A weak proposal supported by Philip Morris was sponsored by Senate Majority Leader Bill Frist (R-Tenn.) and Representative Tom Davis (R-Va.). Frist accepted more than $2.2 million from the tobacco industry between 1999 and 2002 in his capacity as chair of the National Republican Senatorial Committee, and Davis accepted more than $2.5 million between 1999 and 2002 in his capacity as chair of the National Republican Congressional Committee, in addition to $14,000 in contributions for his personal reelection campaign. At the same time, Representatives Greg Ganske (R-Iowa), John Dingell (D-Mich.), and Henry Waxman (D-Calif.) supported a stronger bill that granted substantial authority to the FDA to regulate tobacco. The 127 supporters of this bill accepted an average of $613 in tobacco industry campaign contributions compared with an average of $12,707 in tobacco industry contributions among the 17 supporters of the Frist/Davis bill. Neither bill passed.

In June 2000, tobacco industry dollars once again secured political allies during debate on funding for a U.S. Department of Justice lawsuit against the tobacco companies. The intent of the lawsuit was to recover tobacco-related health care costs paid for by the federal government, similar to the successful state lawsuits that recovered tobacco-related healthcare costs incurred by state **Medicaid** programs. On 19 June 2000 during the first vote to provide funding for the lawsuit, 207 members of the House of Representatives who voted against the funding had accepted an average of $9,712 in tobacco industry contributions. In contrast, the 197 representatives who voted to approve funding for the lawsuit accepted an average of $1,750 in contributions since January 1997. This proposal also did not move forward, but the lawsuit continued.

Medicaid a public health program in the United States through which certain medical expenses of low-income persons are paid from state and federal funds.

State Politics

The states, sandwiched between the federal level where the tobacco industry dominated the political process and the local level where health advocates often prevailed, were battlegrounds in three areas: preemption, taxation, and, beginning in the mid-1980s, large scale tobacco control programs.

In 2002, Delaware became the first state to overturn preemption of local tobacco control laws after a long campaign by public health advocates. Voters in Florida enacted a state clean indoor air law through direct voter initiative that made all workplaces and public places (except bars) smoke-free, and other states began enacting state clean indoor air laws. As is true at the national level, the tobacco industry fought back using well-connected lobbyists and campaign contributions. Occasionally the tobacco industry's largesse backfired on them. For example, in 2003 in Connecticut, public health advocates successfully brought public attention to financial ties between the tobacco industry and the Speaker of the House who was blocking a bill to repeal preemption in that state. The result of the controversy was that the Speaker of the House introduced and championed a statewide clean indoor air law, thereby skirting the issue of preemption, which passed and went into effect in the fall of 2003 for restaurants and April 2004 for bars and cafes.

Similar to preemption, tobacco taxation is another area where the tobacco industry works at multiple levels. Federal tax increases on cigarettes, although beneficial in financial and health terms, rarely occur;

POLITICS

the last increase in the federal cigarette tax was in 2001 when it only increased from 34 cents per pack to 39 cents per pack. However, after many years of relatively small tobacco tax increases, the fiscal difficulties that engulfed much of the United States in the early twenty-first century forced states to substantially increase the cigarette tax. Between 2001 and 2003, 33 states increased their cigarette tax; five states made two increases during that time period. One such state, New Jersey, had the highest tax rate at $2.05 per pack. The tobacco industry was unable to stop these taxes, but, despite the proven effectiveness of large-scale tobacco control programs, little of the money the taxes raised was devoted to helping smokers to quit or to prevent young people from starting. While the industry fights these tax increases, it also often uses them to mask price increases that exceed the tax, ensuring continued revenue growth for the companies.

Large-Scale State Tobacco Control Programs

Perhaps the most important innovation in tobacco control at the state level has been the emergence of large-scale tobacco control programs. Such programs represent the first real challenge to the tobacco industry's monopoly of the advertising medium since 1970 when the industry effectively removed antismoking advertising from television and radio when it had Congress enact the broadcast advertising ban that went in to effect. Minnesota developed the first state-funded antsmoking campaign in the United States in 1983 and implemented the program in 1985. The tobacco industry worked from the early stages of conception of the program to defeat it through campaign contributions and lobbying efforts and portraying the program as ineffective. It even developed its own "youth smoking prevention program," "Helping Youth Decide," that carefully avoided talking about the health dangers of smoking or the fact that nicotine was an addictive drug. In Minnesota, as elsewhere in the world, the industry presented its own ineffective programs as an alternative to meaningful tobacco control measures run by public health professionals. In addition, third party allies, such as the Teamsters Union, were recruited to defeat the tobacco control campaign. The tobacco industry also created a lobbying team made up of former state legislators and state employees with access to the legislative decision-making process, which allowed the industry to stay a step ahead of all plans for implementation of the program. While the industry did not prevent the program from beginning in 1985, its efforts to chip away at the program began to succeed in 1990 when the state legislature cut the program's budget from $1.5 million to $1 million. The election of Republican Governor Arne Carlson, whose ties with the tobacco industry included campaign contributions from industry lobbyists, and a 1996 outing in Australia financed by Philip Morris during Minnesota's case against the tobacco industry, led to the fall of the program in 1993. Carlson used inflated claims of a fiscal crisis, saying that the state was running out of money even as he was cutting taxes on the grounds that the state had more money than it needed; then he used the money "saved" by eliminating the tobacco control program for tax rebates.

The largest and longest surviving state program was created in California in 1988. After a hard-fought election campaign between health advocates and the tobacco industry, voters enacted an initiative known as Proposition 99, that increased the state tobacco tax by 25 cents per pack and allocated some of the revenues to fund tobacco education and prevention programs. The state Department of Health mounted an aggressive campaign that combined tough antismoking ads, many of which confronted the tobacco industry's decades of manipulation of the public and built a statewide infrastructure to support local tobacco control activities, particularly clean indoor air and encouraging organizations to refuse tobacco industry money. The program reduced cigarette consumption so rapidly that it produced a corresponding drop in smoking-induced heart attacks.

Encouraged by California's success, public health activists used the initiative process in several other states—Massachusetts, Arizona, and Oregon—to enact programs modeled on California.

The tobacco industry, however, did not accept these developments. In addition to fighting these initiatives at the polls, the industry used its considerable political muscle in state legislatures to hobble these programs by forbidding them from attacking the industry or working on policy change. In California the industry increased its campaign contributions dramatically and, working through allies in the California Medical Association (the political deal was that the tobacco industry would help the Medical Association enact favorable legislation on malpractice in exchange for its help in shutting down the tobacco control program), nearly destroyed the program. Only an aggressive attack on the California Medical Association and the governor led by the American Heart Association and Americans for Nonsmokers' Rights, combined with several lawsuits defending the program, saved it. Other states did not fare so well; despite the fact that the tobacco control programs were enacted by the voters and had demonstrated effectiveness, the tobacco industry lobbying efforts led to state budget cuts that virtually shut down the Massachusetts program in 2002 and the Oregon and Florida programs in 2003.

The federal government recognized that public policy interventions at the state and local level were the most promising way to lower the burden of disease and death cased by the tobacco industry. Between 1991 and 1999, the National Cancer Institute carried out a large-scale trial of this proposition through the American Stop Smoking Intervention Study (ASSIST) program. ASSIST provided funding to 17 states, awarded after a national competition, to build a local infrastructure to enact policy changes, including increased tobacco taxes and local clean indoor air laws. ASSIST represented the first large scale (although at a state level not as large as the large-scale tobacco control programs the states mounted themselves later), and represented a serious threat to the tobacco industry. Secret tobacco industry documents acknowledged that "ASSIST will hit us in our most vulnerable areas—in the localities and in the private workplace. It has the potential to peel away from the industry many of its historic allies" (<http://www.gaspforair.org/ 2000>) and "the antitobacco forces have developed a more sophisticated and well-funded structure to address local government affairs. . . . [ASSIST] guarantees that local matters will take increasing portions of our time and effort. . . . Thus our local plan is crucial" (<http://www.tobaccodocuments.org/ 1998>).

The industry mobilized aggressively against ASSIST by organizing tobacco vendors, company sales people, restaurateurs, grocers, convenience

store owners, and other business organizations. These organizations were used to create accusations and divert attention from reducing tobacco use in the population to claims of "illegal lobbying" and used massive and targeted requests made under the Freedom of Information Act (FOIA) to burden tobacco control advocates with the task of copying documents rather than pursuing tobacco control. The results were also used in attempts to smear the work of ASSIST and other tobacco control organizations by claiming that these parties were using funds for "illegal lobbying." The industry also used its allies in Congress to put restrictions into law that restricted ASSIST's activities designed to promote public policy change to promote the public health.

Despite these attacks—and the corresponding reduction in effectiveness of ASSIST because it reduced its policy-related activities—ASSIST was successful in reducing tobacco use. ASSIST was associated with a decline in prevalence that could have resulted in 278,700 fewer smokers between 1991 and 1999 across the United States if ASSIST had been implemented nationwide.

Lawsuits Against the Tobacco Industry

A new front opened in the political battles between public officials and the tobacco industry in 1994 when the Attorneys General of Mississippi, Minnesota, and other states sued the tobacco industry to recover the costs of smoking paid by taxpayers and to stop other industry practices, particular predatory marketing against children. (These lawsuits were separate from private suits that had been in litigation with little success for years.) The industry opposed these suits not only in court, but also through the political process. In many states, the industry succeeded in preventing the attorneys general from spending state funds on the litigation (that led to the cases being pursued in cooperation with and financed by private lawyers). In Mississippi, the pro-tobacco governor even sued the attorney general in an unsuccessful effort to stop the suit, claiming that the suit was illegal since he never consulted with the governor.

None of these cases went all the way to a court verdict. Instead, all the cases were settled out of court, the first four states (Mississippi, Florida, Texas, and Minnesota) individually, and the remaining 46 states in the jointly negotiated Master Settlement Agreement (MSA) in 1998. These settlements imposed some restrictions on cigarette advertising (most notably, ending large billboards), established state antismoking programs in the states that settled individually, required the release of about 40 million pages of previously secret internal tobacco industry documents, and, most important, provided hundreds of billions of dollars to states into the indefinite future to partially reimburse the states for the costs of smoking through a complex formula based on cigarette sales.

The MSA and other settlements created the opportunity for every state to build a successful large-scale tobacco control program based on the successes of California, Massachusetts, ASSIST, and others. It remained up to the state legislatures—where the tobacco industry still wielded substantial political clout—to allocate some of the MSA money for tobacco control programs. In the early years after the settlements, some states did use the settlement dollars for tobacco control, but as of

2002, states were using less than 25 percent of the MSA payments for tobacco prevention and an even smaller portion of the state's total tobacco-related revenues. Most of the money went to anything but tobacco control, including capital projects, public works, and health services. By 2004, however, the threat of even these modest programs to the tobacco industry had greatly diminished. Programs in Florida and Minnesota established by their individual state settlements were eliminated (in large part because of the failure of health advocates in those states to mount the kind of aggressive defense that had rescued the California program a decade before), and funding in many other states had been cut.

Even worse from a public health perspective, the MSA created an unexpected alliance between the tobacco industry and some of the states that were more interested in protecting the cash flow of the MSA than reducing tobacco use. In the spring of 2003, 37 attorneys general (many of whom had been directly involved in negotiating the MSA) filed a brief of amici curiae in support of Philip Morris, which was attempting to avoid posting a $12 billion appeal bond after losing a private class action lawsuit in Illinois, claiming that its marketing of "light" and "mild" cigarettes defrauded the public. The attorneys general accepted the claim that posting a bond of this magnitude would jeopardize Philip Morris' ability to make its annual MSA payments to the states.

Beyond general support from the attorneys general due to financial interests, the tobacco industry built a solid alliance with members of the Republican Attorneys General Association (RAGA). RAGA was conceived by Alabama Attorney General Bill Pryor in 1999 (Pryor was named a federal judge by President George W. Bush in 2004), as a means of defending against the alliances that some attorneys general had formed with private lawyers to sue the tobacco industry, providing an unfair advantage against the tobacco industry and threatening the entire business community. The links between the tobacco industry and RAGA were well hidden because RAGA was not required to report campaign contributions.

In many ways, the politics of the post-MSA era have marked a return to those of the late 1980s and early 1990s, with a resurgence of local tobacco control activities concentrated on clean indoor air. Probably the most important legacy of the litigation against the tobacco industry is the fact that more than 40 million pages of previously secret tobacco industry documents are now available to the public on the Internet. During its lawsuit against the industry, Minnesota Attorney General Hubert Humphrey III doggedly pursued release of the tobacco industry's secret internal correspondence. He forced this material to be made public as part of the Minnesota settlement. Later, the MSA required that this material be placed on the Internet (<http://legacy.library.ucsf.edu>). These documents give an unprecedented view into the inner workings of the tobacco industry and its involvement in politics at the local, state, national, and international levels. While the practices of the tobacco industry have not drastically changed over time, access to the tobacco industry's internal documents has allowed the public to see how the tobacco industry does business and allowed health advocates to do a better job of countering the industry's activities.

POLITICS

International Politics

Three multinational corporations operating worldwide—Altria (Philip Morris), British American, and Japan Tobacco—dominate the tobacco industry. The combined opportunities created for these multinational corporations through globalization and through the reduction of smoking in the United States and other developed countries by tobacco control advocates have led the tobacco industry to increasingly focus its efforts on the developing world. As in the United States, these tobacco companies, and their smaller cousins, aggressively use politics to protect and promote their interests, using the same techniques as in the United States: well-connected lobbyists, political money, and third party front groups and allies.

For example, in the mid-1980s, the tobacco industry faced a significant challenge in the European community, which was proposing legislation to end tobacco advertising and promotions. The tobacco industry recognized the economic consequences of such an action and began to forge alliances with third parties, including the International Chamber of Commerce, the Union of Industrial and Employers' Confederations of Europe, and several members of the communications and business communities, as well as friendly governments, most notably Germany, against the legislation. While these strategies were not successful in preventing the passage of the ban on advertising and promotion, a case brought by Germany led the European Court of Justice to strike the legislation in 2000.

Similar to the tactics of the tobacco industry in the United States, Philip Morris and Brown & Williamson developed an international network of scientists secretly funded by the tobacco industry and managed by Covington and Burling, the law firm that handles much of the tobacco industry's political work in the United States, to conduct research to refute the dangers of secondhand smoke in the mid-1990s. This network successfully delayed the spread of clean indoor legislation to Latin America and other parts of the world.

The industry faced its strongest international challenge in the late 1990s, when the World Health Organization began using its treaty-making powers for the first time to create the first international public health treaty, the Framework Convention on Tobacco Control (FCTC). The goal of the FCTC is to create a framework to be implemented at all levels of government to reduce the prevalence of tobacco use and exposure to secondhand smoke, thereby decreasing the health, social, environmental, and economic consequences. To accomplish this goal the FCTC envisions bans on tobacco advertising, promotion, and sponsorship, including requiring the placement of prominent and strong health warnings on all tobacco packages, banning the use of deceptive terms such as "light" and "mild," protecting the public from exposure to secondhand smoke in all public places, increasing tobacco taxes, and working to prevent cigarette smuggling, which is often organized with the active participation of the multinational cigarette companies as a way to penetrate new markets and bypass national tobacco control laws.

The tobacco industry actively monitored the treaty development and worked through national governments in sympathetic countries, most notably the United States (particularly after the pro-tobacco George W. Bush administration took power), Germany, and Japan. The industry also applied standard tactics such as working through third parties and front groups to lobby for weakening the treaty by eliminating key

European Union Commissioner for Health and Consumer Protection David Byrne signs the World Health Organization (WHO) Framework Convention on Tobacco Control (FCTC) during a signing ceremony at the WHO headquarters in Geneva, Switzerland, Monday 16 June 2003. AP/WIDE WORLD PHOTOS

provisions. In the end, a concerted effort led by countries in the developing world and nongovernmental organizations in the United States, Europe, and elsewhere forced the United States, Germany, and Japan to back down, and the treaty was approved by the World Health Assembly in 2003. However, the treaty must be ratified by the participating countries for it to go into effect, and this process will take substantial time and will face opposition in many countries orchestrated by the tobacco industry (often acting through other third parties), including the United States.

Conclusion

The politics surrounding tobacco are often characterized as a tug-of-war between the tobacco industry and the public health community. The tobacco industry works to pull the issue up the hierarchy of the political system, knowing that its greatest chance for victory is at the federal level since it is the most concentrated area of government, furthest from the people, and most susceptible to tobacco industry lobbyists and campaign contributions. Working down the political system to state and local governments, health advocates increase their chances of making progress in tobacco control as the tobacco industry cannot be in all places at all times and because, in a highly visible public political fight, local politicians are more sensitive to the public's desire to be protected from the tobacco industry than to tobacco industry money. The industry has responded by increasing its efforts to stay in the background and work through other organizations such as "hospitality associations." Local tobacco control advocates tend to be successful when they can expose these connections and frame the issue as local citizens against Big Tobacco.

See Also Doctors; Insurance; Toxins.

▌JENNIFER K. IBRAHIM
▌STANTON A. GLANTZ

BIBLIOGRAPHY

Aguinaga-Bialous, Stella, and Stanton A. Glantz. *Tobacco Control in Arizona, 1973–1997*. Tobacco Control Policy Making: U.S. Paper AZ1997. San Francisco, Calif: Center for Tobacco Control and Research at University of California, 1997.

Americans for Nonsmokers' Rights. Available: <http://www.no-smoke.org>.

Campaign For Tobacco-Free Kids. Available: <http://www.tobaccofreekids.org>.

Committee of Experts on Tobacco Industry Documents, World Health Organization, "Tobacco Company Strategies to Undermine Tobacco Control Activities at the World Health Organization" (July 1, 2000). *Tobacco Control. WHO Tobacco Control Papers.* Paper WHO7. <http://repositories.cdlib.org/tc/whotcp/WHO7>.

Dearlove, J. V., Stella Aguinaga-Bialous, and Stanton A. Glantz. "Tobacco Industry Manipulation of the Hospital Industry to Maintain Smoking in Public Places." *Tobacco Control* 11 (June 2002): 94–104.

Gasp for Air. Available: <http://www.gaspforair.org/gasp/gedc/artcl-new.php?ID=74>.

Givel, M. S., and Stanton A. Glantz. "Tobacco Lobby Political Influence on United States State Legislatures in the 1990s." *Tobacco Control* 10 (June 2001): 124–34.

Glantz, Stanton A., et al. *The Cigarette Papers*. Berkeley, Calif.: University of California Press, 1996.

Glantz, Stanton A., and Edith Balbach. *Tobacco War: Inside the California Battles*. Berkeley, Calif.: University of California Press, 2000.

Minnesota Smoke-Free Coalition. Available: <http://www.smokefreecoalition.org/utils/printArticle.asp?id=358>.

Pringle, Peter. *Cornered: Big Tobacco at the Bar of Justice*. New York: Henry Holt & Company, 1998.

Siegel, M., et al. "Preemption in Tobacco Control. Review of an Emerging Public Health Problem." *Journal of the American Medical Association* 278 (10 September 1997): 858–863.

Tobacco Scam. Available: <http://www.tobaccoscam.ucsf.edu>.

UCSF Legacy Library. Available: <http://legacy.library.ucsf.edu>.

Portuguese Empire

Tobacco from Brazil became an important trade product in the Portuguese Empire during the 1620s, and held a leading role in trade throughout the seventeenth and eighteenth centuries in Portugal, West Africa, and the Portuguese colonies in Asia. In fact, the revenues from state tobacco monopolies became a crucial element in the finances of the *Estado da India* (Portuguese colonies in East Africa and Asia) during the second half of the seventeenth century. Meanwhile, tobacco played a major role in building the Brazilian economy, as it was exchanged for slaves on the western coast of Africa. Thus, tobacco helped establish the connection between the Indian Ocean and the South Atlantic regions in the seventeenth and eighteenth centuries, breaking up the compartmentalized vision of the Portuguese Empire. To some extent, Portugal's finances depended on tobacco revenues until the end of the monarchy in 1910, and the debate over liberalization of the tobacco state monopoly played a decisive role in the military coup d'état of 1926, which established a forty-eight-year dictatorship in Portugal.

Consumption

In 1555, Pedro Fernandes Sardinha, the first bishop of Brazil, launched a campaign against the addiction to smoking in the Portuguese colony on the grounds that it was a **heathen** habit that was improper to Christians. His campaign, which involved the excommunication of several colonizers—including Vasco Fernandes Coutinho, who held the captaincy of the Espírito Santo region—caused a major commotion in the colony, forcing the governor to intervene, appease the outraged captain, and calm the Portuguese community. The outcome was that the bishop was forced to return to Portugal. During the voyage, however, his ship was wrecked on the Brazilian coast, and, although the bishop was initially rescued by the natives, they ultimately ate him. There are no records of any other attempts to punish or restrict the consumption of tobacco, in either Brazil, Portugal, or other parts of the Portuguese Empire. Instead, the Portuguese authorities very quickly discovered the importance of this addictive new stimulant and tried to maximize the fiscal and financial benefit that they gained from it.

heathen any person or group not worshiping the God of the Old Testament, that is, anyone not a Jew, Christian, or Muslim. May also be applied to any profane, crude, or irreligious person regardless of ethnicity.

The story of the intolerant bishop demonstrates the extent of European tobacco consumption in Brazil during the middle of the sixteenth century. Within Europe itself, the plant had been introduced into Portugal around 1542, and was cultivated in the royal gardens, more specifically in the nurseries of a princess, the Infanta D. Maria. Jean Nicot, the French ambassador in Lisbon, later sent tobacco seeds on to Queen Catherine of Medici and to Cardinal Lorrène, specifically in 1569–1570. The first authors to mention Brazilian tobacco were the French Franciscan André Thevet, the Huguenot Jean de Léry, the Jesuit missionaries Manuel da Nóbrega, José de Anchieta, and Fernão Cardim, and historians Damião de Góis and Jerónimo Osório. They all emphasized the medicinal properties of the plant in healing wounds and skin diseases, its capacity as a painkiller (for such ailments as headaches), and its use as a drug to help bear the effects of hunger and thirst. In addition, they also described the social and ritual uses of the plant among the American natives in honoring guests and foreigners, treating the sick, reinforcing collective ties in village or tribal assemblies, and stimulating the **shamanistic** practices in divining and healing rituals. In contrast, the transformation of tobacco for smoking, drinking, chewing, and inhaling was mainly targeted at the pleasure of Europeans.

shamanism an ancient religion based on commune with animal spirits and characterized by magic, healing, and out-of-body experiences.

Portuguese medical and pharmaceutical treatises from the late sixteenth and seventeenth centuries—by Leonel de Sousa, Zacuto Lusitano, António da Cruz, Francisco Soares Feio, Manuel de Azevedo, Gabriel

Grisley, João Curvo Semedo, and Duarte Madeira Arrais—all noted that tobacco possessed healing properties. It was considered under the traditional theory of humors, because the smoke was supposed to prevent excess humidity in the body. There was also a widespread belief that tobacco could heal skin diseases, and that it had properties as an antiseptic, sternutatory (stimulator of sneezing), and emetic. When transformed into a beverage, powder, or plaster, tobacco could also be used to treat insomnia, asthma, worms, stomach pains, flux, bleeding, colic, and coughs. Although this reputation could have helped the use of tobacco to spread, attributing to it some distinctive qualities and avoiding any connection with the native rituals, its widespread consumption was not due to its healing proprieties. At the beginning of the eighteenth century, André João Antonil (a pseudonym of the Jesuit Giovanni Antonio Andreoni) published the book *Cultura e opulência do Brasil* (Culture and Opulence of Brazil), which evaluated the importance of tobacco production in Brazil and its consumption by all social groups. During this period, it is known that the annual quantity of tobacco consumed in Portugal and the Portuguese-controlled Atlantic islands (Madeira and the Azores) was about 294 tons. Within Portugal, by far the heaviest consumption was in Lisbon, followed by the southern regions.

snuff a form of powdered tobacco, usually flavored, either sniffed into the nose or "dipped," packed between cheek and gum. Snuff was popular in the eighteenth century but had faded to obscurity by the twentieth century.

The consumption of tobacco in Portugal began with people smoking pipes or rolls of dried leaves, the predecessors of cigars. During the eighteenth century, the habit of inhaling tobacco powder—which required the development of **snuff**-producing factories—became the preferred means of tobacco consumption, a position it held until the beginning of the nineteenth century. Cigarette production, which started in Portugal in the first decades of the nineteenth century, then gave the tobacco industry a new boost. However, it was the combination of cigarette and match production, introduced into Portugal at the beginning of the nineteenth century, that transformed the market.

Production

The production of tobacco in Brazil increased throughout the seventeenth and eighteenth centuries. The quantity of tobacco shipped to Portugal between 1676 and 1700 can be used as an indicator of this production, even accepting that smuggling reached significant levels. During this period, the quantity of tobacco that arrived from Brazil increased from 720 tons per year to 1,542 tons, while in the following decades these imports more than doubled, reaching 3,520 tons by 1710 and 1720. The tobacco was normally grown on small and medium-sized properties, using little slave labor and providing sustenance for a diverse group of landowners and workers, mainly in the captaincy of Bahia. As the production and trade in Brazilian tobacco was not under a royal monopoly—the *estanco do tabaco* only covered its trade in Portugal and India—production and the Atlantic trade developed enormously. However, the king of Portugal decided to intervene in one delicate matter: the direct trade between Brazil and the West Coast of Africa. This extremely important bilateral trade was responsible for the development of the Brazilian economy, since slaves were exchanged mainly for tobacco, but also for brandy made from sugar cane. The king was worried that the tobacco traded in Africa might be re-exported to Europe, as

Portugal had lost control of the Guinea coast in 1637 when the Dutch conquered the fort of Mina. Consequently, in 1698, King Pedro II established a limit of 60.8 tons of tobacco that could be exported every year from Brazil to the Guinea coast, to be carried in twenty-four ships. Moreover, he only authorized the export of the lowest quality tobacco, mixed with sugar syrup.

The colonial pact explains why, after the royal monopoly of transformation and distribution of tobacco was established on the mainland and in the Atlantic islands, the production was prohibited in Portugal. The prevailing idea until the Pombal administration (1750–1777) was that the main products from the colonies—pepper from India, cinnamon from Sri Lanka, sugar from Madeira and Brazil, tobacco from Brazil—should not be planted in other regions, so as to protect the established monopoly leases and the customs duties. Planting tobacco in Portugal and the Azores was excluded for fiscal reasons: It was easier to control imports from Brazil. Nonetheless, smuggling was estimated to make up half the total volume by the end of the seventeenth century. The logic of the colonial pact also meant that factories had to be concentrated on the Portuguese mainland, while tobacco processing was prohibited in India, for instance. In 1674, the creation of the *Junta da Administração do Tabaco* (Council for the Administration of Tobacco) by the crown was followed by a huge campaign to eradicate tobacco planting in Portugal.

Despite these measures to control the production, transformation, and trade in tobacco, the plant was introduced into the Azores during the eighteenth century, leading to the creation of several small factories in the 1740s. The colonial pact, whose logic had been shaken by the new policies introduced by Pombal, was disrupted by the independence of Brazil in 1822. Nevertheless, tobacco imports from Brazil did not go into immediate decline. The production of tobacco—never completely suppressed in the northwestern Minho region—was immediately stimulated in the Azores by the government and also introduced into the Douro region when a crisis struck the area's vineyards in the middle of the nineteenth century. In 1888, the tobacco crop in Portugal reached 90 tons, while the production in the Azores increased tremendously throughout the second half of the nineteenth century, giving rise to the creation of new factories. The climatic conditions and soil explain why the region continues to be conducive to the production of cigarettes and cigars.

Tobacco was also introduced into Angola, a Portuguese colony until 1975, in the eighteenth century, mainly in the southern Moçamedes region. In the 1830s, the Portuguese government did try to stimulate production, but it only really started to acquire any significant scale in the 1870s. In 1929, the export of tobacco from Angola to Portugal reached 642 tons, and in the following decades, the average exports to Portugal stabilized at around 300 tons. This contrasted with the situation in the Cape Verde islands, where despite government efforts to encourage production, the people continued to grow the small quantities of tobacco they had produced since the eighteenth century. Curiously, the islands also maintained a tradition of women smoking pipes. The situation in Mozambique, another Portuguese colony, was different again. Tobacco growing there became significant in the early twentieth century, and by the 1940s, exports to Portugal varied between 33 and 88 tons.

Finances

In 1639, the Crown established a royal tobacco monopoly that controlled the transformation and distribution of Brazilian tobacco in Portugal and the Atlantic islands. The monopoly was leased in return for an annual rent, payable to the king, of 8 million *réis*. (*Real*, *réis* in the plural, was a coin and account unity: roughly 400 *réis* were equivalent to one cruzado.) The contract did not cover customs duties, which meant that imported Brazilian tobacco still had to pay 15 percent on the fixed value of 100 *réis* per pound, plus an additional 3 percent when re-exported. However, war broke out in 1640 when Portugal sought to regain its political independence from Spain. This led to new leases in 1641, with an annual rent of 12.8 million *réis*, and again in 1644, with an annual rent of 25.6 million *réis*. This system of leases meant that the state fixed a price for the producers in Brazil, the export duties to be paid to Brazilian customs, import and re-export duties paid to Portuguese customs, and sales prices, although these were often left up to the holder of the lease. In addition, they established measures to prohibit tobacco growing in Portugal and other parts of the empire except Brazil, to ban distribution outside the lease, and to allow subcontracts between the central lessee and small companies operating at the level of *comarcas* (judicial districts).

This contracted-out royal monopoly saw a significant increase in annual rents between 1651 (28.4 million *réis*) and 1674 (34.8 millions *réis*). However, this growth was not linear and, as with other leases, companies went bankrupt or renegotiated the value during the period when Portugal was in war with Spain (1641–1668). Once the war was over, the *Junta da Administração do Tabaco* was created, expropriating the existing factories and centralizing the entire industrial capacity in a single large factory located in Lisbon. This new institution directly controlled the purchase, transformation, and distribution of tobacco in Portugal, and ushered in a period of increased revenues, despite ongoing smuggling. By the end of the seventeenth century, the debate regarding tobacco revolved around the best way to rationalize this major source of income: through royal monopoly (either under subcontract or directly managed by a central institution) or through free trade and the payment of duties. The debate would continue on into the early twentieth century.

While the debate continued, in 1700 the king decided to return to the system of leases, which was cheaper in administrative terms. The annual rent agreed on was 614.4 million *réis*, which reveals both the expansion of tobacco consumption and the increase in Brazilian production. The annual rent paid for the tobacco contract rose successively—despite several short-term falls and disruptions due to bankruptcies—to 720 million *réis* in 1722, 764 million in 1741, 884 million in 1759, 960 in 1783 (including Macao), and 1.06 billion in 1800 (again including Macao). This was followed by a period of political turmoil brought by the Napoleonic invasions (1807–1812), the liberal revolution in 1820 and the civil war of 1832–1834. Nonetheless, a new contract was established at 1.44 billion *réis* in 1816, and despite falls in revenue in the 1820s and 1830s, the value of the tobacco contract had reached 1.521 billion *réis* by the beginning of 1860s.

During these years, the customs revenue from tobacco reached 234 million *réis*. In fact, calculations show that customs represented between 10 percent and 30 percent of the state's total revenues from tobacco, only in Portugal.

In 1865, the Portuguese parliament decided to liberalize the system, leading to the creation of new factories and trading companies. The new system certainly proved profitable, as the state increased its revenues from 2.3 billion *réis* in the first years to 3.9 billion *réis* by the late 1880s. The creation of a state monopoly was decided on in 1887, which maintained the profit levels but forced the government to pay huge amounts of money as compensation for expropriating factories. However, in 1890, the state's serious financial crisis forced a return to the system of leases, negotiated with a progressive value that started at 4,250 million *réis*. A similar contract was renegotiated in 1906 for a twenty-year period, this time starting at 6,000 million *réis*.

It was the parliamentary debate in 1926 on renewing the tobacco lease that triggered the military coup d'état of 28 May, which in turn established a fascist-style dictatorship that lasted for forty-eight years in Portugal. The *Companhia do Tabaco*, a joint Portuguese and French venture that had held the tobacco contract since the 1890s, managed to maintain the system. However, there was greater intervention from the state, which forced the company to share the market with another company, *Tabaqueira*, created by Alfredo da Silva's new industrial group (CUF). This new company set up modern factories in the protected market, and by the 1960s had reached a dominant position, surviving the liberalization that followed the establishment of democracy in Portugal after the 1974 revolution.

The importance of tobacco for public finances was not restricted to Portugal. In the *Estado da India*, the royal monopoly had been established in 1623 with a first set of leases that governed the tobacco trade in the territories of Goa, Bardez, and Salcete. Those contracts, arranged almost exclusively with Indian merchants, initially stipulated an annual rent of 2 million *réis*, increasing to 12.9 million *réis* in 1634. The latter year also brought contracts covering the northern territories of Bassein (8.3 million *réis* per year) and Chaul (3.5 million *réis*). In total, all these leases reached 24.7 million *réis* in 1634 , or 8 percent of the total revenue of the *Estado da Índia*. In 1687, over fifty years later, the total income from the tobacco contracts had increased to 46.5 million *réis*, more than the total income from the customs, and 19 percent of the *Estado da India*'s total revenues. This means that tobacco played a significant role in the finances of the Portuguese Empire in India, even after the loss of the main trading posts and territories (such as Hormuz, Malacca, Sri Lanka, and Cochin) between the 1630s and the 1660s. Indeed, tobacco was of such importance that in 1680, the king decided to replicate the *Junta da Administração do Tabaco* in India to control the contracts and prohibit the production of powdered tobacco, thus protecting the mainland's industry. Despite the absence of complete data, the value of the tobacco leases in India evidently rose throughout the eighteenth century. For example, in the Goa region, the contract for 1709–1712 stipulated an annual rent of 32.4 million *réis* rising in 1756 to 57.3 million.

Commerce

The circulation of tobacco established the connection between the two main axes of the Portuguese Empire: the Indian Ocean and the South Atlantic. There are no reliable data for the global trade between these regions, because tobacco had to pass via Lisbon before being re-exported to Goa and direct trade between Bahia and Goa was only officially carried on after 1770s. Nor are there precise data on the importance of local production, which was quite widespread in southern India throughout the seventeenth century. Certainly, when the *Junta da Administração do Tabaco* was established in Lisbon in the late 1600s, the first reports recognized the importance of tobacco exports to India, and proposed isolating this region from the rest of the empire and establishing specific trade contracts. It was against this background that Manuel Lopes de Lavre proposed a contract worth 160 million *réis* per annum to distribute tobacco in Asia, a sum that corresponded to three-quarters of the *Estado da Índia*'s total revenues. Several other reports indicate that the average volume of tobacco exported from Portugal to India was around 240 tons per year.

In addition, there is no series of figures on the quantities of tobacco shipped from Brazil to Portugal, but only data for specific periods, such as from 1676 to 1700, when the total volume increased from 720 to 1,542 tons. Of this, only an average of 294 tons was consumed in Portugal, Madeira, and the Azores, while the rest was re-exported to Spain, which generally consumed 442 tons, Italy (Genoa), the Netherlands, and France.

In social terms, the main holders of the tobacco leases in Portugal and India belonged to the upper elite of bankers and top financiers who were intimately involved in the Crown's main operations in Europe and Asia. However, the two groups did not connect with one another. The contractors in Portugal were almost all Portuguese, with some Castilians, particularly in the early eighteenth century. The increasing need for a solid financial backing to pay for the huge contracts in the late nineteenth century explains why French capital joined the company led by the Portuguese banker Burnay. In contrast, the lease holders in India were almost all Hindu bankers and merchants, who controlled most of the *Estado da Índia*'s financial operations throughout the seventeenth and eighteenth centuries.

While tobacco helped create this financial elite, which was boosted under the Pombal administration, the constant increase in the tobacco trade also supported a large group of ship owners, merchants, and vendors. The tobacco subcontractors in Portuguese districts played an important role at the regional level, and the agents of the small shops (mainly women in Lisbon) formed an important network. In fact, there were hundreds throughout the entire mainland, with around sixty in Lisbon alone during the late seventeenth century. The tobacco factories, which became the most profitable industrial units between 1850 and 1925, employed thousands of workers (between 2,000 and 4,000), who launched the strongest strikes during the Constitutional Monarchy and the First Republic.

See Also Brazil; British Empire; Dutch Empire; French Empire; Smuggling and Contraband; Spanish Empire; Trade.

▌FRANCISCO BETHENCOURT

BIBLIOGRAPHY

Ames, Glenn Joseph. *Renascent Empire? The House of Braganza and the Quest for Stability in Portuguese Monsoon Asia, c. 1640–1683*. Amsterdam: Amsterdam University Press, 2000.

Bethencourt, Francisco. "O Estado da Índia." In *História da Expansão Portuguesa*, Vol. 2. Edited by Francisco Bethencourt and Kirti Chaudhuri. Lisbon: Círculo de Leitores, 1998.

Cheis, Maria da Conceição Franco. *O tabaco do Brasil nos quadros da economia portuguesa do século XVII* [Tobacco from Brazil in the framework of the Portuguese economy in the 17th century]. M.A. thesis, Lisbon, Faculdade de Letras, 1967.

Gonçalves, Paula Alexandra Grazina. *Usos e costumes do tabaco em Portugal nos séculos XVI e XVII* [Uses and habits of tobacco in Portugal, 17th–18th centuries]. M.A. thesis, Lisbon, Universidade Nova de Lisboa, 2003.

Hanson, Carl A. "Monopoly and Contraband in the Portuguese Tobacco Trade, 1624–1702." *Luzo-Brazilian Review* vol. XIX, no. 2 (1982): 149–168.

Lapa, José Roberto do Amaral. "O tabaco brasileiro no século XVIII: anotações aos estudos sobre o tabaco de Joaquim Amorim de Castro." [Brazilian tobacco in the 18th century: Comments on the studies on tobacco by Joaquim Amorim de Castro]. *Studia* 29 (1970): 57–145.

Lugar, Catherine. "The Portuguese Tobacco Trade and Tobacco Growers of Bahia in the Late Colonial Period." In *Essays Concerning the Socio-Economic History of Brazil and Portuguese India*. Edited by Dauril Alden and Warren Dean. Gainesville: University Press of Florida, 1977.

Nardi, Jean Batiste. "Retrato de uma indústria no Antigo Regime: o estanco real do tabaco em Portugal (1570–1830)." [Portrait of an industry in the ancien régime: the monopoly of tobacco in Portugal, 1570–1830]. *Arquivos do Centro Cultural Português* XXVIII (1990): 321–339.

———. *Le tabac brésilien et ses fonctions dans l'ancien système colonial portugais (1570–1830)* [Tobacco and its functions in the old Portuguese colonial system (1570–1830)]. 5 vols. Thèse de doctorat, Aix en Provence, Université de Provence, 1991.

Santos, Raúl Esteves dos. *Os tabacos: sua influência na vida da nação* [Tobacco and its influence in the life of the nation]. 2 vols. Lisbon: Seara Nova, 1974.

Processing

Processing is what farmers do to a crop after harvest. Nurturing the living plants is part of the agrarian cycle. But after harvesting, the agrarian cycle is complete, and further handling is properly called processing. Some crops require little processing. Indeed, many fruits and vegetables can go from field to table with a simple washing. For example, after picking and packing, the apple growers' work is done. Not so with tobacco growers.

Forms of Processing

Raw tobacco contains moisture that must be removed before manufacturing can begin. As the leaves are the only plant part that is sold, tobacco processing focuses on drying or curing the leaves. As much as 90 percent of the weight of raw tobacco is lost in curing. Proper curing is important. An average field crop can be greatly improved by careful curing. Conversely, a fine field crop can be ruined in the curing barn. Thus, tobacco growers must be as skilled in curing as they are in cultivating.

PROCESSING

This eighteenth-century illustration called "Curing, Airing, and Storing Tobacco" depicts tobacco processing in early America. Moisture from the raw tobacco must be removed before manufacturing can begin. © CORBIS

air-curing the process of drying leaf tobacco without artificial heat. Harvested plants are hung in well-ventilated barns allowing the free circulation of air throughout the leaves. Air-curing can take several weeks. Burley tobacco is air-cured.

air-cured tobacco leaf tobacco that has been dried naturally without artificial heat.

There are three curing methods in common use in the United States: **air-curing,** fire-curing, and flue-curing.

AIR-CURING. Air-curing is the oldest form of tobacco processing. Burley, Maryland, and Connecticut Valley tobaccos are **air-cured.** Maturity of the living plant is judged by color, and plants are ripe when the leaves change from dark green to light green to yellow. In air-curing cultures, growers harvest the entire plant. Stalks are cut at the bottom, laid on the ground, and allowed to wilt. The stalks are then spiked—or pierced near the bottom with a metal tool—and sticks are inserted through the stalks. Typically, five or six plants are hung on a stick. Some Burley growers hang the tobacco sticks on outdoor racks to hasten curing, but most Burley is barn-cured. After spiking, growers carry the plants to the curing barn and hang them in tiers, leaving air spaces in between.

Historically, air-curing barns have been designed to provide protection from rain and wind but afford ample air exchange and circulation. The plants cure naturally without artificial heat, but fans are sometimes used to improve air movement. Curing times vary with environmental factors like humidity and temperature, but plants usually remain in the curing barn for four to six weeks. As they cure, the leaves continue to change from light green to light brown, mahogany, or gold.

When the leaves have thoroughly dried and coloring is complete, leaves are stripped from the stalks. Growers then bulk the leaves, forming them in piles and covering them with fabric for protection. Many growers scatter the stripped stalks in the fields and thus return their substance to the earth. When preparing the leaves for market, growers sort or grade the cured leaves by color and tie them in small bundles. More recently, some tobaccos are pressed into bales. The tobacco is stored carefully to remain in order—that is, moist enough to be pliable yet dry enough not to mold or mildew. At marketing time, growers

448 | Tobacco in History and Culture
AN ENCYCLOPEDIA

carry their tobacco to a warehouse for auction or, more recently, directly to a purchaser by prearrangement.

Air-curing produces flavorful tobaccos highly valued by the trade. Most Burley is consumed in cigarettes, but some Burley is blended into pipe tobaccos and other smoking products. Connecticut Valley leaf, the most valuable tobacco in the world, is used in premium cigars. Approximately 40 percent of the total U.S. tobacco crop is air-cured.

FIRE-CURING. Fire-curing is practiced along the Tennessee–Kentucky border and in some parts of Virginia. Fire-curing is a variation of air-curing in which small fires are built on the floors of the curing barns to aid drying. Great care is taken to maintain the correct temperature and humidity and thus affect a proper cure. Sometimes, very little firing is needed, and dry weather can delay firing for several days. In damp weather, however, growers light a series of low fires. Hickory and oak are the fuels of choice, and fires are sometimes fed with sawdust so they smolder rather than flame. Several firings may be needed to completely cure and smoke the leaves. Purchasers value the rich, smoky flavor of dark-fired tobacco, and a high smoke volume is maintained during the final curing stage. When the tobacco is finished, roof ventilators purge the heat and smoke.

As in air-curing cultures, fire-cured tobacco is stripped and bulked. When ready, the leaves are graded and tied. Leaves are assorted by color or stalk position, four or five grades being typical. The rustic flavor of fire-cured leaf is popular in chewing tobacco, **snuff,** pipe tobacco, and in certain European cigars. Less than 10 percent of the American crop is fire-cured.

FLUE-CURING. Also called Bright leaf, **flue-cured** varieties account for about half of American tobacco production. Areas of Virginia, North Carolina, South Carolina, Georgia, and Florida flue cure tobacco. Most flue-cured tobacco is consumed in cigarettes. A major cultural difference from other types is how Bright leaf is harvested. Tobacco leaves do not all ripen at once. The bottom leaves ripen first, then those next to the bottom and so on to the uppermost leaves. Waiting for the upper leaves to ripen ensures that the bottom leaves are overripe and nearly worthless. Therefore, Bright leaf growers harvest the leaves as they ripen, breaking off a few leaves every week until all leaves have been harvested. Unlike Burley and dark-fired tobacco, the stalk is not harvested. This method assures all leaves are harvested at the proper time.

Historically, flue-curing evolved from fire-curing. Fire-cured leaves are strongly flavored and coated by smoke and soot. By the 1880s, however, demand for milder, more aromatic cigarette tobaccos drove the development of flue-curing. Artificial heat flows from an outside furnace through a network of stovepipes, or flues, running parallel to and a few inches above the floor of the curing barn. An exhaust pipe vents smoke and soot outside. Thus, the leaves are cured rapidly by artificial heat free of ashes, soot, and smoke. Moreover, even temperatures are easier to maintain with flues, resulting in a more uniform cure. Temperatures as high as 71 degrees Celsius (or 160 degrees Fahrenheit) are applied, and the tobacco is fully cured in a few days rather than several

snuff a form of powdered tobacco, usually flavored, either sniffed into the nose or "dipped," packed between cheek and gum. Snuff was popular in the eighteenth century but had faded to obscurity by the twentieth century.

flue-cured tobacco also called Bright Leaf, a variety of leaf tobacco dried (or cured) in air-tight barns using artificial heat. Heat is distributed through a network of pipes, or flues, near the barn floor.

weeks. After curing, leaves are bulked—there are no stalks to strip—and stored until marketing time.

See Also Architecture; Chewing Tobacco; Cigarettes; Cigars; Pipes.

■ ELDRED E. PRINCE JR.

BIBLIOGRAPHY

Gately, Iain. *Tobacco: The Story of How Tobacco Seduced the World*. New York: Grove Press, 2001.

Goodman, Jordan. *Tobacco in History: The Cultures of Dependence*. London: Routledge, 1993.

Hirschfelder, Arlene B. *The Encyclopedia of Smoking and Tobacco*. Phoenix, Ariz.: Oryx Press, 1999.

Kluger, Richard. *Ashes to Ashes: America's Hundred-Year Cigarette War, the Public Health, and the Unabashed Triumph of Philip Morris*. New York: Vintage Books, 1997.

Tilley, Nannie M. *The Bright-Tobacco Industry, 1860–1929*. Chapel Hill: University of North Carolina Press, 1948. Reprint, New York: Arno Press, 1972.

Product Design

The cigarette is a uniquely successful drug delivery device. It provides an effective vehicle (inhaled smoke) for nicotine to travel deep into the lungs, resulting in the most rapid and efficient possible route to the brain (approximately eight seconds). It enables the smoker to manipulate smoke delivery—and therefore nicotine dose—easily with each successive puff, constantly adjusting delivery to individual needs and circumstances. And it facilitates a host of secondary behaviors and cues tied psychologically to smoking "satisfaction," including physical and oral manipulation of the cigarette, smoke aroma and taste, and sensory impact (bite) at the back of the throat and mouth preceding the delivery of nicotine to the brain.

The design of the cigarette appears uncomplicated at first glance: Tobacco is rolled in paper, then burned at one end and inhaled at the other. However, research by tobacco manufacturers, particularly since the mid-twentieth century, has resulted in a highly engineered product drawing on an increasingly sophisticated understanding of product design factors (such as filter, paper, ventilation, and additives) and their effects. Manufacturers have developed technologies to alter smoke delivery including the form and availability of nicotine, to adjust smoke sensory cues such as impact, and to facilitate smoker manipulation of delivery. Design factors can also alter delivery of specific smoke constituents, including nicotine analogues and other components affecting addiction, and may increase or reduce smoke toxicity and subsequent health risk.

PRODUCT DESIGN

This photo, taken in 1928, shows a woman displaying a selection of tobacco products the United States was sending to the International Exposition at Seville. Products included fragrant perfectos and rose-scented snuff. © UNDERWOOD & UNDERWOOD/CORBIS

The Early Cigarette

Today's cigarette may be a highly engineered product, but its roots are more humble. Records from the sixteenth century indicate that Mexicans smoked tubes of reed or cane packed with the aromatic balsam of liquidambar (a deciduous tree growing in Central America) incense and tobacco. Spanish colonists introduced the product to Europe when they brought back small cylinders of tobacco wrapped in covers of vegetable matter or leaves. By the seventeenth century, the vegetable wrapping had been replaced by fine paper, creating so-called "papalettes." In Spain and other countries of southern and eastern Europe, a market developed among the affluent classes for these paper cigarettes, hand-rolled by girls or women with expensive tobaccos from Turkey or Egypt, known as Oriental leaf. The habit spread further during the Crimean War (1853–1856) when British soldiers were introduced to cigarettes by their Turkish allies and Russian enemies. The first known British cigarette manufacturer dates from this period, when Robert Gloag manufactured cigarettes in London, using Russian tobacco, yellow tissue paper, and a cane mouthpiece. The handmade cigarettes of the time came in different shapes and sizes; in Austria, so-called double cigarettes were three times as long as modern cigarettes and came with a mouthpiece at each end. They were designed to be cut in two before smoking.

Cigarette manufacturers also began to cater to the emerging female market in the late nineteenth century by producing small, dainty cigarettes for upper-class society women. These were often scented and flavored and some had gold or colored tips. Although there are many social and financial reasons why more women began to smoke from the late nineteenth century onward, the fact that cigarettes were milder, easier to smoke, and smelled less offensive than pipes and cigars undoubtedly contributed to the trend.

However, there was a good deal of prejudice extended to the cigarette among male smokers as the quality of cigarettes was perceived as inferior to that of cigars and pipe tobaccos, while their size and name—the diminutive "ette" on the end—led to charges of effeminacy. In London, cigarettes were commonly associated with the immigrant population and evidence suggests that the workers making cigarettes were predominantly foreign.

In the United States, cigarettes were practically unknown until the mid- to late nineteenth century, and were again predominantly associated with immigrants. Antismoking literature warned of "cigar-butt grubbers" in New York, boys and girls who scoured the streets for stumps of discarded cigars, which they dried and sold to be used for making cigarettes (Lander 1886). However, most cigarettes were produced legitimately by the immigrant population and were taxed by the government from 1864 onward. As in Europe, these were hand-rolled with expensive Oriental tobaccos and sold to affluent city dwellers.

There was also a market for hand-rolling tobacco, which the smoker rolled into cigarettes, a cheaper option. The tobacco used for this in the United States was predominantly the domestically produced Bright tobacco, a Virginian leaf dried by indirect heat from flues run through storage barns. This process resulted in a golden-colored tobacco that produced a mellower smoke and was easier to inhale. The nicotine was therefore absorbed more readily into the body than with traditional pipe or cigar tobacco and was more likely to lead to addiction.

Mechanization

The standard product one associates with cigarette smoking in the 2000s came about with the introduction of mechanized production. While several people developed machines to make cigarettes in the late nineteenth century, the most well known and successful was the Bonsack machine. This was designed and patented by James Bonsack in the early 1880s and exhibited at the Paris exhibition of 1883. This machine could produce cigarettes at the rate of 300 per minute, reducing the costs of production and making it possible to supply an emerging mass market with a standardized product. The rights to the machine were bought by an English firm, W.D. & H.O. Wills, in 1883 and by James Buchanan Duke of the American Tobacco Company in 1884. Despite some initial mechanical problems, it proved a worthwhile investment for both firms as it brought cigarettes within the price range of the lower classes and vastly expanded the potential market for cigarettes.

PRODUCT DESIGN

U.S. firms further changed cigarette production by introducing Virginia tobacco into ready-made cigarettes, creating an affordable and convenient factory-made cigarette. This tobacco could be blended or used alone, and its porous nature meant that it was particularly suitable for additives and flavorings.

As cigarettes gained in popularity, the number of brands proliferated and manufacturers looked for new ways to distinguish their brands from the rest. Key selling points by the interwar period included the mildness and purity of the smoke, achieved through quality blends. Mildness was an important quality because of concerns about "smoker's throat" and the irritation caused by inhaling tobacco smoke. The effects of nicotine were also a consideration; in the 1920s, "denicotinized" tobacco and cigarettes were available in Britain and the United States, but historians do not know how popular they were. Innovations in product design included cork-tips, longer cigarettes, and the addition of **menthol**. Cork-tips, unlike the later introduction of filters, served an aesthetic rather than a health purpose, keeping loose strands of tobacco off the lips. They also maintained cigarette length while avoiding the waste of tobacco leaf in the unsmoked cigarette end. Some manufacturers, for example, De Reske in England, made cigarettes with colored tips for women to conceal lipstick stains. Advertising sometimes made a virtue out of a process common to all manufactured cigarettes—Lucky Strike cigarettes were sold with the slogan "It's toasted." Applying heat during the drying and sterilizing process was common to all leaf tobacco production, but the idea of toasting suggested a warm and appetizing, as well as flavorsome, product.

Economic factors can affect product design. In Denmark these cigarette-sized cigars, made by the Nobel Cigar Company, can be made more cheaply than conventional cigarettes due to Danish tax laws. © LEIF SKOOGFORS/CORBIS

menthol a form of alcohol imparting a minty flavor to some cigarettes.

PRODUCT DESIGN

tar a residue of tobacco smoke, composed of many chemical substances that are collectively known by this term.

By the 1940s and 1950s, manufacturers were concerned to salvage the tobacco stem and dust that went to waste during the production of cigarettes. They developed reconstituted tobacco sheet (RTS) by grinding the tobacco waste to a pulp and then pressing it. The RTS was then shredded and blended with tobacco leaf, allowing financial savings, and additives were used to improve the taste. The blending process and the addition of additives also allowed tobacco companies to control how fast cigarettes burned, how easy smoke was to inhale, and nicotine and **tar** levels. This is controversial as some additives, particularly ammonia, may increase the speed with which the cigarette delivers nicotine to the brain.

From the 1960s onward, following publication of major reports on smoking and health in Britain and the United States, health concerns became a key factor in cigarette production. In the decades that followed, concerns about health led to the increasing popularity of filtered, low-tar, low-nicotine cigarettes and to the development of the highly engineered products called cigarettes in the twenty-first century.

The Modern Cigarette

The modern cigarette can be broken into four major components: the tobacco column, filter, paper, and ventilation. Each of these components may be modified with direct effects on smoke delivery. Likewise, they may be used to control sensory perception, to reduce the degree of effort required by the smoker to obtain a given amount of smoke, or to control other important product factors such as feel, taste, and aroma. The manufacturer utilizes computer-based design models as well as chemical and physical analyses to control all aspects of the finished product.

flue-cured tobacco also called Bright Leaf, a variety of leaf tobacco dried (or cured) in air-tight barns using artificial heat. Heat is distributed through a network of pipes, or flues, near the barn floor.

air-cured tobacco leaf tobacco that has been dried naturally without artificial heat.

The primary component of a cigarette is tobacco. Burning tobacco generates nicotine and other smoke constituents, which are then inhaled by the smoker and absorbed into the body. Different tobaccos have unique physical and chemical characteristics, such as burn rate, tar, and nicotine delivery, flavor, and aroma. Thus, the choice and blending of tobaccos is critical to the final product. Tobacco used to manufacture cigarettes traditionally differs by region. **Flue-cured tobaccos** predominate in the United Kingdom, Finland, Canada, Japan, China, and Australia; **air-cured tobaccos** are preferred in France, parts of Germany and Italy, and South America; and sun-cured (Oriental) tobaccos are used in Turkey and Greece. In the United States and parts of Western Europe, a blend of these different tobaccos in combination with reconstituted and expanded tobaccos is typical, incorporating the different characteristics of each. Since the 1980s the U.S. blended cigarette has become widespread internationally.

Processed tobaccos (that is, reconstituted and expanded tobaccos) may constitute as much as one-third of the total tobacco used in a modern cigarette. This is due in part to their reduced cost, as well as their ability to impart unique qualities to the finished product. For example, reconstituted tobaccos are generally processed with additives, often at high temperatures that induce further chemical changes. This process may increase the amount of nicotine available in freebase form, and alter smoke impact and sensory perception. Expanded tobaccos, developed in the late 1980s to reduce nicotine levels, have a high filling power (less tobacco is needed to fill the cigarette) and increase the speed at which the cigarette burns between puffs. Additives may be used to introduce new

PRODUCT DESIGN

smoke constituents such as nicotine analogs, smoke smoothing agents, or bronchodilators (agents which facilitate inhalation).

The majority of cigarettes today, including 98 percent of cigarettes sold in the United States, use filters that may reduce delivery of some smoke constituents to the smoker. Different filters are common in different cigarette markets. The cellulose acetate filter typical of U.S. style cigarettes are most effective at reducing smoke particles ("tar"), while charcoal filters common elsewhere (Japan) are intended to filter out gases present in cigarette smoke. The effects of filter differences on overall health risk are not easily measurable, but in all cases the "filtered" smoke remains toxic.

Ventilation holes (small holes in the paper cigarette wrapping around the filter) are commonly introduced in filtered cigarettes, diluting smoke with air by as much as 95 percent. Filter ventilation is the most critical design component in the development of lower delivery cigarettes ("lights" and "ultralights"). However, it is commonly accepted that a smoker will simply inhale more deeply in order to compensate for this reduction in delivery. In addition, since ventilation holes are often invisible, they may be unconsciously blocked by a smoker's lips or fingers, reducing dilution and leading to increased smoke delivery. At higher ventilation levels, it becomes extremely difficult for the smoker to draw smoke from the cigarette, leading to consumer unacceptability.

Cigarette paper porosity is likewise an important factor in overall smoke delivery. A more porous paper allows air to be drawn into the tobacco column with each puff, reducing the amount of smoke generated. The porosity may be increased by adding tiny holes to the paper either electrostatically or mechanically. Cigarette paper is also generally coated with additives that are used to control the rate at which the tobacco burns. Most cigarette papers contain between 20 percent and 30 percent chalk, in order to cause the formation of an attractive white ash.

In combination with these major design components, physical parameters such as length, circumference, density, and the coarseness (cut) of the tobacco are used to fine-tune smoke delivery. The manufacturer adjusts the character of the smoke (including smoothness, body, impact, irritation, and flavor), reducing undesirable components and increasing those (such as nicotine) with "desirable" effects. The number of puffs per cigarette, the burn rate, and the delivery per puff are all carefully monitored.

The modern cigarette has reduced irritation to allow deeper inhalation; provides enough sensation in the throat to "cue" the smoker regarding delivery; and facilitates the absorption of nicotine through increased freebasing of nicotine and other chemical changes. Particular care is given by manufacturers to how the product affects puffing behaviors, in order to allow the smoker increased control over the cigarette dose, and maximizing the delivery produced from a minimum of effort.

See Also Additives; Cigarettes; Fire Safety; Genetic Modification; "Light" and Filtered Cigarettes; Marketing; Menthol Cigarettes; "Safer" Cigarettes; United States Agriculture.

■ ROSEMARY ELLIOT
■ GEOFFREY FERRIS WAYNE

BIBLIOGRAPHY

Alford, B.W.E. *W.D. & H.O. Wills and the Development of the UK Tobacco Industry 1786–1965*. London: Methuen, 1973.

Arents, George. *Tobacco: Its History Illustrated by the Books, Manuscripts and Engravings in the Library of George Arents, Jr. Together with an Introductory Essay and Bibliographic Notes by Jerome E. Brooks*. New York: The Rosenbach Company, 1937.

British American Tobacco. "Cigarette Design." Bates: 400132742-400132776. Available: <http://tobaccodocuments.org/product_design/3194.html>.

Browne, C. L. *The Design of Cigarettes*, 3d ed. Charlotte, N.C.: Filter Products Division, Hoechst Celanese Corporation, 1990.

Lander, Meta. *The Tobacco Problem*. Boston: Cupples, Upham & Co., 1886.

National Cancer Institute. *Risks Associated with Smoking Cigarettes with Low Tar Machine-Measured Yields of Tar and Nicotine*. NIH Publication No. 02-5074. Bethesda, Md.: U.S. Department of Health and Human Services, National Institutes of Health, 2001. Available: <http://cancercontrol.cancer.gov/tcrb/monographs/13/index.html>.

Prohibitions

In all societies at all times, prohibitions exist without provoking controversy. In fact, many are essential for humankind to live together: Most would agree that the prohibition "Thou shall not kill" is a worthy one. It is when prohibitions fall on products or behavior desired by but harmful to their users that they become problematical and often contentious. An extreme form of government intervention, prohibition is often alluring because it seems a more straightforward way to deal with social ills than persuasion or education. Governments act all-powerfully when they legislate out of existence activities of which they disapprove.

Prohibition: A Most Peculiar Policy

History shows that this type of government intervention is not effective. In presence of a popular demand, enforcement proves very difficult if not impossible. Prohibition does not eradicate the banned product; it just drives it underground, giving rise to smuggling and illegal black markets. A gap is created between the legislations in the books and the reality of daily life. On top of these effects, the credibility and legitimacy of the state may be undermined as these laws are largely disrespected.

At the heart of the debate on prohibition lies the crucial moral issue of personal liberty. Should the state protect individuals from harming themselves or should individuals be left to decide for themselves?

The liberal (some would say the libertarian) view of prohibition was perhaps best expressed in 1859 by the famous British philosopher and economist John Stuart Mill (1806–1873): "The only purpose for which power can be rightfully exercised over any member of a civilized community, against his will, is to prevent harm to others. His own good, either physical or moral, is not a sufficient warrant. He cannot

---

Sumptuary Regulations

Great philosophers like Baruch Spinoza (1632–1677) recognized the ineffectiveness of government prohibition more than three centuries ago: "All laws which can be broken without injustice to another person are regarded with derision and intensify the desires and lusts of men instead of restraining them; since we always strive for what is forbidden, and desire what is denied.... He who tries to determine everything by law will foment crime rather than lessen it." Many economists of the twenty-first century, most notably the Nobel Laureate Gary Becker, still make the same argument.

Vice President Al Gore gestures toward Jessica Goh, of Jacksonville, Florida, during a news conference on Friday 27 February 1998 in the Old Executive Office Building in Washington, D.C., where he announced the federal government campaign aimed at cutting underage smoking. The campaign was designed to target retailers that sell tobacco and to warn those retailers that selling tobacco to underage teens would now be a federal crime. AP/WIDE WORLD PHOTOS

rightfully be compelled to do or forbear because it will be better for him . . . because it would be wise, or even right. These are good reasons for . . . reasoning with him, or persuading him . . . but not for compelling him. . . . Over himself, over his own body and mind, the individual is sovereign" (Mill 1956).

To this position, supporters of prohibition reply that in the presence of addictive substances, the individual is not truly "sovereign" and that freedom is an illusion. On the contrary, they believe that abstinence would liberate users (smokers or drinkers or drug users) from their addictions, allowing them to have better and freer lives. Reaching this outcome voluntarily is of course preferable but human nature might be too weak and consequently has to be strengthened by the law.

Prohibiting Tobacco Use: Three Big Waves

In the 2000s, especially in North America, smoking is so strongly stigmatized that prohibition seems close. Many believe the phenomenon is relatively new. Without denying the radical shift since the 1970s, the fact is that the users of tobacco have been ostracized almost right from its introduction in Europe in the sixteenth century.

In the four centuries since tobacco's introduction, the world witnessed three big waves of tobacco prohibition, each in symbiosis with its time. The first one almost covered the globe in the seventeenth century following the Great Explorations with its cortege of exciting but

> ### A Philosophical Warning
>
> Ludwig Von Mises (1881–1973) said in 1949:
>
> But once the principle is admitted that it is the duty of the government to protect the individual against his own foolishness, no serious objections can be advanced against further encroachments.... Why limit the government's benevolent providence to the protection of the individual's body only? Is not the harm a man can inflict on his mind and soul even more disastrous than any bodily evils ? Why not prevent him from reading bad books and seeing bad plays, from looking at bad paintings and statues and from hearing bad music....
>
> If one abolishes man's freedom to determine his own consumption, one takes all freedoms away. The naïve advocates of government interference with consumption delude themselves when they neglect what they disdainfully call the philosophical aspect of the problem (Von Mises 1949).

intrusive novelties such as tobacco. The second wave at the turn of the twentieth century was geographically much more limited. Cigarette bans were enacted only in the United States and belonged to the temperance movement of the Progressive era. Humankind is living in the heart of the third wave, which began in the 1970s and continues to the present. Originating in North America, the antismoking movement is largely a phenomenon of the Western world. Clean air and healthiness are preoccupations of wealthy societies. Antismoking proponents argue that eliminating tobacco smoke is one of the easiest steps to improve a heavily polluted environment.

THE FIRST WAVE: THE SEVENTEENTH CENTURY. Tobacco was introduced in Europe following Christopher Columbus's "discovery" of America and rapidly spread around the world through trade. It was immediately controversial; its supporters saw it as a panacea and its enemies as a pure evil. In the first half of the seventeenth century, many states and cities enacted prohibitions, some with quite spectacular penalties if historians are to believe European travelers' accounts, the only sources of information for that period.

There were two different types of prohibition. The first, proscribing tobacco smoking or snuffing in public, was based on moral, religious, and cultural grounds. The other, banning domestic cultivation or manufacturing of tobacco or sometimes importation, was based on financial grounds. Governments always eager for money wished to maximize revenues from the tobacco habit. In order to do so, they needed some control over production, manufacturing, and sale.

What appears to be the earliest interdiction took place in Mexico in 1575 when the Catholic Church issued an order forbidding the use of tobacco in churches throughout Spanish America. This first edict was mostly aimed at the converted Indians who were used to smoking in their ceremonies, but later orders concerned priests as well. The Church also banned smoking or snuffing during or before Mass in Europe. A series of papal bulls under Urban VIII and Innocent X from 1624 to 1650 threatened excommunication to tobacco users in churches.

In the first half of the seventeenth century, governments around the globe multiplied edicts and proclamations banning the use of tobacco, especially in public. Western and Central European measures were much milder than Eastern and Asiatic rules. In the latter, penalties were spectacular and terrifying, reflecting the autocratic nature of the political regimes. In all cases, they were unable to stop the spread of the habit in the various populations across cultures and religions.

Probably the most extreme case of punishment for tobacco use can be found in the Sultan Murad IV (1623–1640) of the Ottoman Turkish Empire who decreed death penalty for smoking tobacco in 1633. There might have been as many as fifteen to twenty daily executions. As Count Corti wrote, "Even on the battlefield the Sultan was fond of surprising men in the act of smoking, when he would punish them by beheading, hanging, quartering. . . ." (Corti 1931). In Persia as well under the reign of the Shah Abbas I (1587–1625) there were eyewitness accounts of torture or death inflicted upon tobacco smokers or sellers. Russia was not less fierce. When the patriarch of the Russian Church placed tobacco use in the category of deadly sins, Michael Feodorovitch, the first of the Romanoff czars, prohibited smoking in 1634 under penalties for first offense like slitting of the nostrils or whipping and the death penalty for persistent offenders.

In India, the Mogul emperor Jahangir outlawed tobacco in 1617 with slightly more restrained penalties as smokers were merely to have their lips slit. The Chinese authorities perceived the use of tobacco as subversive of the national interest. A succession of imperial edicts forbade the planting, importation, and use of tobacco in 1612, 1638, 1641 (this time under threat of decapitation) until as late as 1776. Their frequency and repetitive character show how ineffective they were. In Japan, the repeated attempts by the shogun to prohibit the growing and use of tobacco lasted only two decades and were all lifted in 1625. By 1640, tobacco accompanied the tea ceremony and was part of daily life.

In Europe, the most serious attacks against tobacco were in the Holy Roman Empire after the Thirty Year War (1618–1648): Cologne prohibited tobacco use in 1649, Bavaria in 1652, Saxony in 1653, Zurich in 1667, and Berne in 1675. The same type of bans on smoking in public was imposed in some North American colonies (Massachusetts in 1632, Connecticut in 1647, New Amsterdam in 1639).

Tobacco was a foreign novelty and smoking an outsiders' habit. In Europe, the outsiders were the American Aborigines, seen as "Savages." In the Middle East and in Asia, the outsiders were the Infidels and the Westerners. In both cases, those outsiders were highly suspicious, if not threatening. Unsurprisingly, this foreign intrusion provoked strong reactions.

Looming also very large was the morality issue. Not only did outsiders with very different cultures introduce tobacco, but also tobacco provided gratification and pleasure to its consumers. The habit was quickly labeled a vice and a sin. Both Catholic and Protestant churches, the loudest being the Calvinists and the Puritans, condemned indulgent pleasure as immoral and contrary to a good life and a good society. The same was true for Muslims. Even though the Koran did not expressly mention tobacco, it condemned intoxication, and tobacco was considered an intoxicant.

A Royal Enemy

James I was King of England from 1604 to 1625. In the first year of his reign, he wrote the most famous work in English on the subject of tobacco. His *A Counterblaste to Tobacco* (published anonymously) concluded that smoking was: "a custome lothsome to the eye, hateful to the Nose, harmefull to the braine, dangerous to the Lungs, and the blacke stinking fume thereof, neerest resembling the horrible Stigian smoke of the pit that is bottomeless" (James I, p. 36).

Finally, there were some more practical considerations behind the interdictions. Smoking increased the risk of fires, which, in those days of wooden towns, were highly destructive. Tobacco cultivation used land that could have been used for growing foodstuffs—a high opportunity cost for these societies.

A number of states never prohibited tobacco smoking. Instead, they adopted mercantilist measures to regulate, control, and tax tobacco production and trade. A good example is England. In spite of the ferocity of King James I's antitobacco position, tobacco consumption was always legal. Tobacco domestic cultivation was prohibited for a long period but this was done to protect the government income from the **tariffs** on imports of tobacco, which were set up from the beginning at very high rates.

France used a different and very lucrative strategy for more than three centuries: monopoly control of tobacco at every stage (cultivation, fabrication, and sale). From 1674 until 1791, the king sold the monopoly rights to private authorities; since 1810, this has been a state monopoly. Similar regimes were set up in Portugal, Spain, and Italy.

Once they realized how ineffective were their bans, the prohibitionist states joined the mercantilist club by turning to taxation and regulations. Except in China, all prohibitory legislations were abolished before the end of the seventeenth century, as is shown in the following portrait of Western government regulation of the tobacco industry from the mid-seventeenth through the nineteenth centuries (Rogozinski 1990). In this portrait, *state* means public administration of a monopoly; *farmed* indicates that the government granted its monopoly power to a private concessionaire; *private* means no particular regulations, only general laws regulating trade.

tariff a tax on imported goods imposed by the importing country to protect native industry from foreign competition, protect jobs and profits, and raise revenue. Tariffs typically raise consumer prices of affected products.

Country	Cultivation	Imports	Manufacture	Sale
Britain	prohibited	high duties	private	private
France	controlled	farmed to 1791; state from 1810	farmed to 1791; state from 1810	farmed to 1791; state from 1810
Italy	prohibited	farmed; state since 1882	farmed; state since 1882	farmed; state since 1882
Spain	prohibited	farmed	farmed	farmed
Portugal	farmed	farmed	farmed	
Austria	prohibited	farmed to 1784; then state	farmed to 1784; then state	farmed to 1784; then state
Sweden	encouraged (18th century)	prohibited (18th century)	mixed	private
Alsace	private	private	private	private
Bavaria	private since 1717	private since 1717	private since 1717	private since 1717
Prussia	regulated	state 1765–1787, then private	state 1765–1787, then private	state 1765–1787, then private
Switzerland	private	private	private	private

Country	Cultivation	Imports	Manufacture	Sale
Netherlands	private	private	private	private
United States	private after 1776	private after 1776	private after 1776	private after 1776

THE SECOND WAVE: THE TURN OF THE TWENTIETH CENTURY. By the nineteenth century, the use of tobacco was generalized among men. A few antitobacco voices could be heard occasionally: for instance, Dr. Benjamin Rush, signer of the Declaration of Independence, who wrote a tract in 1798 titled *Observations upon the influence of the Habitual use of tobacco upon health, morals, and property*, or Horace Greely, publisher of the *New York Tribune*, who once described the cigar as "a fire at one end and a fool at the other" (Tate 1999). But their warnings went unheeded until the cigarette made its apparition in the 1880s.

In contrast to the first wave of prohibition, the second wave was largely confined to the United States. There was some organized opposition to cigarette smoking in Britain (where Queen Victoria considered the habit an "abomination" and an offense against good manners) and in Canada (where a national ban was seriously contemplated and regularly debated until World War I) but they did not succeed in passing legislation.

Beginning in the late 1890s, cities and states in the United States passed acts to prohibit the sale, manufacturing, and use of cigarettes (but not pipes or cigars). The statute in Illinois was the shortest lived, being declared unconstitutional by the Illinois Supreme Court the same year it was adopted. The following table (Tate 1999) summarizes the various forms the laws concerning adults took in 15 states (key: S = sale; M = manufacture; G = giving away; P = possession; A = advertising of cigarettes). Such laws were on the political agenda of 22 other states, in some cases several times. More widespread were the cigarette laws prohibiting sales to minors. By 1890, 26 states prohibited sales to minors, and in 1940 all states except Texas had such laws.

State	Adopted	Repealed	Ban Content
Washington	1893	1895	S, M
Washington	1907, 1909	1911	S, M, P
North Dakota	1895	1925	S
Iowa	1896	1921	S, M
Tennessee	1897	1919	S, G
Oklahoma	1901	1915	S, G
Indiana	1905	1909	S, M, P
Wisconsin	1905	1915	S, M, G
Arkansas	1907	1921	S, M
Illinois	1907	1907	S, M
Nebraska	1909	1919	S, M, G
Kansas	1909	1927	S
Kansas	1917	1927	+ A, P
Minnesota	1909	1913	S, M
South Dakota	1909	1917	S, M, G
Idaho	1921	1921	S
Utah	1921	1923	S, A

University of Southern Maine students smoke outside their dorm at the Gorham, Maine, campus on Friday 6 December 2002. The University of Southern Maine in September 2002 banned smoking in its dorms, forcing smokers to walk at least 50 feet away from the buildings to light up.
AP/WIDE WORLD PHOTOS

Enforcement was lax. "Tobacco manufacturers sent cigarette papers through the mails; retail dealers sold matches for twenty cents or so and gave cigarettes away" (Warfield 1930). The prohibitions certainly did not stop the rise in cigarettes consumption, as can be seen from the following table (from Doron 1979), which shows average annual cigarette consumption:

Years	Average Consumption (Billion Units)
1900–1909	4.2
1910–1919	24.3
1920–1929	80.0

Even if concerns with health were not totally absent (cigarettes were called "coffin nails"), the main driving force was morality. For the reform and religious groups who pressured the state to eliminate it, cigarette smoking was an evil, destructive to the moral and physical fiber.

Cigarette prohibition was an element of the broader social reform movement of the Progressive era. The catalysts behind regulation were temperance organizations such as the Woman's Christian Temperance Union. In order to achieve a social order based on Christian and family values, they condemned and fought frivolous activities such as dancing, drinking, smoking, and gambling.

During the war, billions of cigarettes were distributed by organizations such as the Young Men's Christian Association, the Salvation Army, the Red Cross, and the federal government to soldiers fighting in Europe. Patriotic organizations in Kansas sent cartons of cigarettes to the front lines, even though their sale was illegal in that state. Anyone who questioned these shipments was deemed unpatriotic. Soldiers returning from World War I made cigarette smoking common and more respectable. By 1927, all prohibitory laws, except those regarding minors, had been repealed.

This second prohibitionist wave was no more successful than the first one. Cigarette smoking became generalized among men after World War I and among women after World War II. As Cassandra Tate noted, "Back then, the world was one big smoking section" (Tate 1999). Even though smokers were actually never the majority (42% of adults in 1965), smoking was embedded in the cultural landscape.

THE THIRD WAVE: 1970 TO THE PRESENT. From the end of the 1960s, following the two landmark reports linking smoking to cancer by the Royal College of Physicians in the United Kingdom (1962) and U.S. Surgeon General (1964), the wind turned for smokers. From being a social norm, smoking became an antisocial behavior and smokers became outcasts. Since then, the habit has been denounced, discouraged, banned, and taxed. North America led the crusade, followed by Europe. In the 1990s, the antismoking movement spread to some extent to the rest of the world. With the exception of Africa, all countries have some restrictions on smoking in public places. However, their severity and coverage vary widely, tending to be much milder and much less respected outside North America.

The prohibition battle was fought on three fronts: advertising, smoking in public places, and among the youth. The progression in the United States has been as follows: In 1971, 8 percent of the American population lived in states with some restrictions; fifteen years later, 80 percent of the American population lived in states with some restrictions; in the 2000s, the figure is 100 percent.

The earliest prohibitions around tobacco focused on advertising. Cigarette advertisements were banned on radio and television in 1971 in the United States and Canada (in the latter by voluntary agreement rather than legislation). Some European countries had already done so several years earlier (Italy, 1962; the United Kingdom, 1965). In the 1990s, advertising bans were extended to print media and to sports-events promotions in many countries, raising much controversy. Probably the most heated example centers on the Grand Prix Formula 1 because of the international character of the competition.

There have also been successful efforts to prohibit smoking in public places. The first regulations in the 1970s established separate smoking and nonsmoking sections in various public places: airplanes, trains, buses, restaurants, halls, and workplaces. Over time, the antismoking movement continued to press for more drastic action, arguing that the segregation did not eliminate the health risk for nonsmokers. Clean Indoor Air acts and regulations have moved to ban indoor smoking by steps: first in flights less than two hours, then all domestic flights, then all flights; in governmental buildings then private enterprise workplaces; in restaurants and finally in bars. Since the 1990s, comprehensive smoking bans in all workplaces, including bars and restaurants, have been enacted in California, New York, Boston, Toronto, and other cities and states.

C. Everett Koop

Dr. C. Everett Koop was appointed surgeon general by President Ronald Reagan in 1981 and he immediately became an outspoken foe of tobacco by advocating "a smoke-free environment by the year 2000." His efforts went beyond medical advisory reports and cigarette package labeling; he became the first surgeon general to use his position to speak out resolutely to the public about the dangers of tobacco use. His 1986 Surgeon General's Report on the dangers of passive smoke became an important tool in the fight to eliminate smoking in public buildings, transportation, and eventually the workplace—including workplaces commonly associated with smoking such as bars and nightclubs. In appreciation of his tireless antismoking efforts and his work on many other public health issues, President Bill Clinton presented Koop with the Presidential Medal of Freedom, the nation's highest civilian award.

❙ DONALD LOWE

PROHIBITIONS

Former U.S. Surgeon General Dr. C. Everett Koop testifies in Concord, N.H., Tuesday 12 February 2002 that cigarette smoking is more addicting than heroin or cocaine. Koop was speaking to a bill that uses a new distribution formula for millions of dollars in tobacco settlement funds. AP/WIDE WORLD PHOTOS

paternalistic fatherly. Although paternalism presumes an obligation for the stronger to provide for the weaker, it implies superiority and dominance over them as well. For example, masters often had paternalistic feelings for their slaves, whom they considered child-like.

A crucial target for both the tobacco companies and the antismoking movement are children and young adults, since most smokers start smoking in their teens. In the mid-twentieth century, youngsters viewed smoking as a ritual to adulthood; in the 2000s they consider smoking a rebellious gesture against adults. Legislations banning cigarette sale to minors were thus adopted or reactivated everywhere in the 1990s. Public health officials note that they are the most difficult of the antismoking measures to enforce. Legislation was also adopted to restrict automatic machine cigarette selling that make it possible for children to purchase cigarettes. At the federal and state levels, so-called "Pro-Children" acts are banning smoking in and around state funded facilities providing children's services (for instance, school grounds in New York State).

Previous prohibition movements against alcohol or illicit drugs were generally driven by moral factors. However, the current efforts against tobacco focus on health as the primary concern. However, opponents charge that antitobacco activists often seem to be trying to impose their values on smokers "for their own good." Indeed, in the 1960s and 1970s, the focus of antitobacco efforts was on the harm to smokers from their own smoking. But over time the focus has shifted to address the harm inflicted to nonsmokers by environmental tobacco smoke.

As public health researchers Ronald Bayer and James Colgrove argue, the strong emphasis on nonsmokers' welfare was an astute strategy in the American cultural context of hostility to overtly **paternalistic** public policies. The prohibition path was not inevitable: Critics argue that private arrangements between smokers and nonsmokers could

464 Tobacco in History and Culture
AN ENCYCLOPEDIA

Beverly Mathis-Swanson smokes a cigar at the door of her bar, the One Double Oh Seven Club and Smoking Parlor, in Santa Cruz, California, 21 August 1997, after state health authorities announced that they would be enforcing California's Smoke-Free Workplace Act. The act, which prohibited smoking in all "enclosed spaces at a place of employment," came into force 1 January 1995 but gave a two-year exemption to bars and casinos. Mathis-Swanson was spokesperson for a group called Tavern Owners United for Fairness. AP/WIDE WORLD PHOTOS

have been devised. However, advocates of restrictions on public smoking argue that the rights of employees, even those who work in bars, to a safe workplace free of hazardous exposures is fundamental.

The Future: A Smoke-Free World?

Prohibition is not a dead issue. Advocacy organizations like the Foundation for a Smokefree America work toward a goal of preventing young people from starting to smoke and helping adults to quit. They can point to the sharp decline in the proportion of smokers in the U.S. population: from 42 percent in 1965 to 25 percent in 1990, and 28 percent in 2000. Other countries with similar policies like Canada, Australia, Britain, and Sweden are in the same range of below 30 percent. The fact that smokers tend to be concentrated among people of lower socioeconomic status may have facilitated their stigmatization.

However, the rest of the world is still far behind North America. Smoking rates of the male population are still above 60 percent in countries like China, Russia, and Japan, and approximately 40 percent in India, Brazil, Mexico, and European countries like France and Spain. The World Health Organization predicts that the tobacco "epidemic" will get worse, shifting from developed to developing nations and touching an increasing number of women.

Moreover, the fact that one of out four people continues to smoke in the United States despite an incredibly hostile environment suggests that smoking will not vanish. Social reprobation may even have the unintended effect of making it more attractive to young people as symbols of rebellion. While the addictive properties of nicotine and the financial strength of the tobacco industry are major contributors to the continuing use of tobacco, cigarettes are also deeply entrenched in American history and culture, so much so that their prohibition remains uncertain.

See Also Advertising Restrictions; Antismoking Movement Before 1950; Antismoking Movement From 1950; Regulation of Tobacco Products in the United States; Smoking Clubs and Rooms; Smuggling and Contraband.

▮ RUTH DUPRÉ

BIBLIOGRAPHY

Alston, Lee, Ruth Dupré, and Tom Nonnenmacher. "Social Reformers and Regulation: The Prohibition of Cigarettes in the United States and Canada." *Explorations in Economic History* 39 (2002): 425–445.

Bayer, Ronald, and James Colgrove. "Science, Politics, and Ideology in the Campaign Against Environmental Tobacco Smoke." *American Journal of Public Health* 92, no. 6 (2002): 949–963.

Borio, Gene. The Tobacco History Timeline. Online. Available: <http://www.tobacco.org/resources/history>.

Brooks, Jerome. *The Mighty Leaf: Tobacco through the Centuries.* Boston: Little, Brown, 1952.

Corti, Egon Caesar, Conte. *A History of Smoking.* London: George G. Harrap, 1931.

Doron, Gideon. *The Smoking Paradox: Public Regulation in the Cigarette Industry.* Cambridge, Mass.: Abt Books, 1979.

Gabb, Sean. *Smoking and Its Enemies: A Short History of 500 Years of the Use and Prohibition of Tobacco.* London: FOREST, 1990. Available: <http://www.seangabb.co.uk/pamphlet/faghist.htm>.

Gottsegen, Jack. *Tobacco: A Study of Its Consumption in the United States.* New York: Pitman, 1940.

James I. *A Counterblaste to Tobacco.* 1604. Reprint, London: Rodale Press, 1954.

Kiernan, Victor Gordon. *Tobacco: A History.* London: Hutchinson Radius, 1991.

Mackay, Judith, and Michael. Eriksen. *The Tobacco Atlas.* Geneva: World Health Organization, 2002.

Mill, John Stuart. *On Liberty.* 1859. Reprint, New York: The Liberal Arts Press, 1956.

Mises, Ludwig Von. *Human Action: A Treatise on Economics.* New Haven, Conn.: Yale University Press, 1949.

Parker-Pope, Tara. *Cigarettes: Anatomy of an Industry from Seed to Smoke.* New York: The New Press., 2001.

Rogozinski, Jan. *Smokeless Tobacco in the Western World 1550–1950.* New York: Praeger, 1990.

Sobel, Robert. *They Satisfy: The Cigarette in American Life.* Garden City, N.Y.: Anchor, 1978.

Spinoza, Benedict. "Tractatus Politicus/Treatise on Politics." In *The Political Works of Benedict de Spinoza.* Translated and edited by A. G. Wernham. Oxford: Clarendon Press, 1958.

Sullum, Jacob. *For Your Own Good: The Anti-Smoking Crusade and the Tyranny of Public Health.* New York: The Free Press, 1998.

Tate, Cassandra. *Cigarette Wars: The Triumph of "the Little White Slaver."* New York: Oxford University Press, 1999.

Tennant, Richard. B. *The American Tobacco Industry.* New Haven, Conn.: Yale University Press, 1950.

Thornton, Mark. *The Economics of Prohibition.* Salt Lake City: University of Utah Press, 1991.

Tollison, Robert D. *Smoking and Society: Toward a More Balanced Assessment.* Lexington, Mass.: Lexington Books, 1986.

Wagner, Susan. *Cigarette Country: Tobacco in American History and Politics.* New York: Praeger, 1971.

Psychology and Smoking Behavior

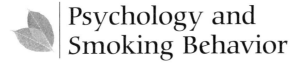

Cigarette smoking causes more premature disease and death worldwide than any other known human behavior. Although the damage caused by smoking is the subject of biology and medicine, smoking is a behavior, and thus has also been the subject of much behavioral research.

The Behavior of Smoking

Modern psychological science concerns itself primarily with the study of behavior, attempting to answer questions such as: Why do individuals engage in a given behavior? What factors lead individuals to stop or continue an undesired behavior? And finally, can we use this knowledge to develop more effective treatments to help people cease the unwanted behavior?

With respect to cigarette smoking, distinctions between "psychological" and "biological" factors have given way in the late twentieth century to perspectives that view smoking as a biobehavioral, or, even more broadly, a biopsychosocial, disorder. Thus, smoking is best viewed as a behavior that is governed by multiple, intertwined factors, including **physiological** (biological), social, and psychological ones. Researchers have clearly demonstrated that nicotine (the ingredient of cigarette smoke most responsible for affecting mood or thought) exerts influence on multiple brain systems, all of which, in turn, affect behavior, thinking, and feeling. It follows that in the study of smoking behavior, attempts to disentangle psychological from biological factors ultimately create a false and unnecessary distinction. Instead, modern psychological science attempts to understand the interplay between various social, individual, and biological influences that, together, promote smoking and tobacco use.

physiology the study of the functions and processes of the body.

Tobacco Smoking as an Addictive Behavior

The once controversial question of whether cigarette smoking may constitute an actual addiction has been universally answered with a resounding "yes." Indeed, an accumulation of well over 3,000 scientific papers has led to the unequivocal conclusion that cigarettes and other forms of tobacco use are addicting, that nicotine is the drug in tobacco most responsible for addiction, and that the pharmacological and behavioral processes that cause addiction to tobacco are similar to those responsible for addiction to other drugs. These facts do not necessarily imply that every smoker is dependent on nicotine (see sidebar). However, the vast majority of smokers who smoke with any degree of regularity ultimately progress to nicotine addiction. To fully appreciate this conceptualization of smoking, a brief overview of precisely how drug dependence (or addiction) is defined by research scientists is necessary.

Simply stated, addiction arises when, for a given individual in a given set of circumstances, drug use results in a powerful rewarding (reinforcing) experience and abstinence from the drug (even for relatively short time periods) causes unpleasant physical and emotional experiences that are alleviated by taking the drug once again. When this leads to compulsive drug use that seems to take over the person's behavior and is difficult to stop, addiction has taken hold. Does this pattern of addictive behavior hold for cigarette smoking? Indeed, it does. For example, research has demonstrated the existence of a reliable withdrawal syndrome that results when the smoker goes without smoking for a given period of time. These withdrawal symptoms include: (1) **dysphoric** or depressed mood; (2) insomnia; (3) irritability, frustration, or anger; (4) difficulty concentrating; (5) restlessness; (6) decreased heart rate; and (7) increased appetite or weight gain. Craving,—an intense, often uncontrollable desire—for the drug is also frequently reported by smokers who are deprived of nicotine.

Another hallmark of addiction is difficulty in stopping the behavior. That is, people who are addicted to a drug often report that, whereas they may sincerely want to quit, they are unable to do so. Do smokers have a difficult time quitting? Yes. Most U.S. smokers say they want to quit, but only 3 percent are actually able to stop permanently each year. According to a literature review by John Hughes and colleagues (2004), smokers who try to quit without treatment have as high as 97 percent failure rate. Even people facing imminent life-threatening consequences often are unable to quit: Most smokers who have had heart attacks ultimately return to smoking. In sum, then, tobacco smoking is a behavioral disorder typified by persistent desires and unsuccessful efforts to quit, thus resulting in resuming smoking.

Cigarette smokers also meet other criteria for being considered addicted. These include development of tolerance (that is, a need for increased amounts of the drug to achieve desired effects), a great deal of time spent in activities necessary to obtain or use the substance (for example, chain-smoking), willingness to give up other things in favor of smoking (for example, avoiding events in nonsmoking venues, risking their health), and use of the drug despite knowledge of having a physical problem (for example, lung disease) that is likely to have been caused by the substance. Relative to the users of other drugs, a higher percentage of smokers are considered addicted. Interestingly, many drug abusers who also smoke say that it would be harder to stop smoking than to stop using their other drugs (even though they find other drugs like alcohol or cocaine more pleasurable). In sum, tobacco smoking can be a highly addicting behavior, comparable to, or even exceeding, the addictive potential of other, "harder" drugs of abuse, such as cocaine or heroin.

Why Do Smokers Smoke?

Research has clearly revealed that nicotine is reinforcing in both animals and humans. Even among addicted smokers, however, not all cigarettes are smoked solely in response to nicotine withdrawal. Indeed, when asked, cigarette smokers themselves consistently attribute their smoking

dysphoria a feeling of unhappiness and discomfort; being ill-at-ease. Cigarette smokers can experience dysphoria when deprived of cigarettes.

to a variety of other motives. These motives are governed by both negative reinforcement (for example, smoking to reduce stress) and positive reinforcement (for example, smoking to celebrate when already feeling good) processes.

The most commonly cited reason for smoking (among both novice and nicotine-dependent smokers) is smoking's alleged ability to reduce subjective stress and anxiety. Smokers often report that they smoke more when angry, depressed, or anxious, and that smoking helps to alleviate these negative mood states. It is not clear that either part of this statement is true, however. Some field studies have shown that negative feelings do not make smokers more likely to smoke. Laboratory studies assessing smoking's effect on anxiety have yielded inconsistent results. Thus, although most smokers clearly believe that smoking reduces negative emotions, this effect has been difficult to reliably produce under controlled, laboratory conditions. There is one exception: When negative emotions are due to nicotine withdrawal, nicotine provides quick relief.

Another interesting aspect of smoking's reputed relaxing properties is that nicotine is a central nervous system stimulant. Thus, smoking a cigarette actually increases autonomic nervous system arousal (for example, heart rate), generating something resembling the "stress response." But how can a drug that produces a "stress response" be perceived as relaxing? More research will clearly be needed in order to adequately answer this question. Of course, some smokers also attribute their smoking to nicotine's stimulant (arousing) properties.

Researchers believe that some of the pleasurable experiences associated with smoking are not solely attributable to nicotine. For instance, research suggests that the sensorimotor aspects of smoking (for example, the taste, the smell, the handling of the cigarette) can become reinforcing in and of themselves, largely as a result of their association with smoking. Through repeated pairing, the act of smoking likely becomes "conditioned" to a variety of emotional states (such as anxiety) and situations (such as after eating). In other words, the smoker associates a particular situation with the act of smoking a cigarette. Consider for a moment a typical pack-a-day smoker, who smokes 20 cigarettes a day. At 10 puffs per cigarette, this adds up to 200 administrations of nicotine a day, or over 72,800 "hits" a year. (No other drug of abuse is self-administered at such a high rate.) As a result of such frequent administrations of nicotine across a variety of situations, smoking invariably becomes linked to specific cues, causing the smoker to smoke some cigarettes "out of habit" rather than out of a craving for nicotine.

Finally, it is important to note that the majority of research on smoking motives (and other aspects of smoking behavior) has been conducted in developed Western countries, primarily in the United States, Europe, and Australia. Researchers do not know the extent to which smoking to reduce stress, for example, is a potent motive for smoking among smokers in developing countries. Moreover, well-validated measures of nicotine dependence that are suitable for use in the United States, for example, may be unsuitable in other countries, where smoking practices and beliefs differ. The lack of information about smoking behavior in other countries is another serious research gap that warrants attention.

The Mystery of Tobacco "Chippers"

In the 1990s, Saul Shiffman and colleagues described a group of smokers, called "chippers," characterized by their apparent invulnerability to developing nicotine dependence. These smokers smoked regularly for years, yet rarely smoked more than five cigarettes a day and did not appear to suffer from nicotine withdrawal when they went without smoking. How did they do it? Although research is still attempting to answer this question, Shiffman and his colleagues made the following observations about chippers:

They typically smoke their first cigarette of the day hours after waking (whereas most addicted smokers smoke much sooner).

They metabolize nicotine at the same rate as regular smokers.

They report frequent casual abstinence (for example, not smoking for several days) from smoking (unlike addicted smokers).

Based on self-report questionnaires, chippers evidence more self-control, and are less impulsive (more able to resist temptation), compared to regular smokers.

Whereas regular smokers show marked changes in mood, craving, sleep disturbance, and cognitive performance when deprived of nicotine, chippers show none of these changes.

Smoking and Psychopathology

Research shows that smokers suffer more mental illness than non-smokers. Smokers are more likely to suffer from depression, anxiety disorders, substance abuse, conduct disorder, and schizophrenia, to name but a few. As an example, whereas approximately 23 percent of the United States population smoke regularly, as many as 90 percent of schizophrenics are heavy smokers. According to one analysis, persons with a psychiatric diagnosis smoke the majority of cigarettes consumed in the United States. Of course, questions arise as to what these associations mean and whether they inform scholastic understanding of smoking behavior. Given that virtually all of these psychological disorders are accompanied by negative mood states, the most common interpretation of the mental illness relationships is that smokers smoke in order to regulate their mood (self-medicate). However, the empirical evidence that smoking genuinely alleviates unpleasant mood is scant. It is important to note, however, that several longitudinal studies have suggested that some forms of psychopathology (for example, depression and delinquency problems) significantly increase the chances that someone will go on to become a smoker.

Whereas these studies suggest that suffering mental illness increases the risk of becoming a smoker, investigations conducted in the 1990s suggest this relationship goes the other way, too: Smoking itself can predict the onset of anxiety and depressive disorders. This fact suggests that the link between smoking and psychopathology may be attributable, at least in part, to other factors (such as genetic variations) that render individuals vulnerable to both smoking and psychopathology. Several biologically based personality variables, particularly neuroticism (anxiety) and psychoticism (distorted thinking), are associated with both smoking and various psychological disorders, including depression and anxiety. Thus, it is conceivable that genetically transmitted vulnerabilities may predispose people to both smoking and to psychopathology.

What Factors Promote Smoking Initiation?

The factors that promote smoking among regular, adult smokers likely differ from those associated with smoking initiation (which typically occurs during adolescence). So, why do individuals begin smoking in the first place? No one is born addicted to smoking, so addiction-related motives can be ruled out as an explanation for smoking onset. Research suggests other factors: (1) peer influence, which is arguably the most important predictor of who becomes a smoker; (2) sibling (and parental) smoking; (3) beliefs that smoking confers advantages in social life; (4) perception that tobacco use is the norm (at least in one's own social circles); and (5) prior experimentation with cigarettes, which is a strong predictor of subsequent smoking. Finally, some of the smoking motives expressed by adults are probably applicable to understanding smoking uptake among adolescents as well.

According to the leading scientific theories, smokers typically proceed through stages of smoking on their way to becoming nicotine-dependent (see figure). Broadly stated, during the early stage, smokers smoke for psychosocial motives, prompted by friends and social situations in which smoking is viewed as normal behavior. Most smokers are believed

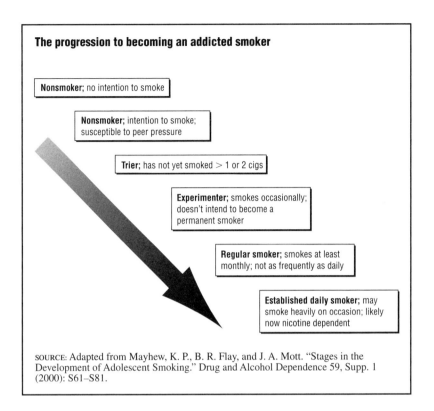

SOURCE: Adapted from Mayhew, K. P., B. R. Flay, and J. A. Mott. "Stages in the Development of Adolescent Smoking." Drug and Alcohol Dependence 59, Supp. 1 (2000): S61–S81.

to progress rapidly to the next stage, in which their smoking is driven by the positively reinforcing pharmacological effects of nicotine. At this stage, smokers appear to seek both the relaxing and stimulating effects of the drug. Some smokers then progress to the final stage where their smoking is primarily governed by the need to stave off or escape from withdrawal symptoms; that is, addiction-related motives.

Most research has focused on understanding factors that make it more likely that a person will begin smoking. Less effort has gone into understanding factors that act to protect individuals from smoking in the first place. Identifying these factors may ultimately improve smoking prevention and intervention programs.

The Changing Landscape of Smoking

One landmark in the history of smoking behavior was the 1964 publication of the United States Surgeon General's report, *Smoking and Health*, wherein the link between smoking and cancer was first widely disseminated. As a result of this groundbreaking health information, many people began to quit smoking. Indeed, since the 1960s, the public's recognition of the health dangers attributable to smoking has grown significantly. One need only look at social policy change since the 1990s to see profound societal and legal shifts in attitudes toward smoking. And society has been witness to a gradual, yet steady, decline in overall smoking prevalence rates, at least in the United States. Analyses suggest that the decline in smoking prevalence is due to multiple factors: increased awareness of risk; rising cigarette prices; restrictions placed on smoking in public (and some private) places; promotion of

Children are influenced by exposure to cigarette products and the smoking behaviors of adults and young adults. In low-income countries, particularly where distribution of tobacco products is uncontrolled, children often use colorful cigarette packs as toys. Tobacco companies consider cigarette packs to be the primary vehicles for advertising their brands, and free merchandise puts tobacco products within easy reach of children. PHOTO BY ANNA WHITE

quitting; and help in quitting. Together, these factors have made smoking a far less appealing and affordable behavior.

Conclusions

Cigarette smoking is a destructive, complex behavior that is governed by multiple, interrelated factors. As such, understanding the psychology of smoking demands a multidisciplinary approach that considers biological, psychological, and social factors. As a field, psychology has made tremendous strides in the understanding of the processes that promote and maintain tobacco smoking. However, psychologists and behavioral scientists have more work to do. Millions of people die every year from diseases directly attributable to smoking. Whereas state-of-the-art smoking cessation treatments are available (including nicotine replacement therapy and behavior therapy), far more research into the mechanisms underlying smoking initiation, maintenance, and cessation is still needed.

See Also Addiction; Bad Habits in America; Consumption (Demographics); Nicotine; Quitting; Youth Tobacco Use.

▮ JON D. KASSEL
▮ SAUL SHIFFMAN

BIBLIOGRAPHY

Benowitz, Neal L., ed. *Nicotine Safety and Toxicity*. New York: Oxford University Press, 1998.

Bolliger, Christoph T., and Karl-Olav Fagerström, eds. *The Tobacco Epidemic*, Vol. 28. New York: Karger, 1997.

Fiore, M. C., et al. *Treating Tobacco Use and Dependence: Clinical Practice Guideline*. Rockville, Md.: U.S. Department of Health and Human Services, Public Health Service, 2000.

Gilbert, David G. *Smoking: Individual Differences, Psychopathology, and Emotion*. Washington, D.C.: Taylor and Francis, 1995.

Kassel, J. D., L. R. Stroud, and C. A. Paronis. "Smoking, Stress, and Negative Affect: Correlation, Causation, and Context Across Stages of Smoking." *Psychological Bulletin* 129 (2003): 270–304.

Mayhew, K. P., B. R. Flay, and J. A. Mott. "Stages in the Development of Adolescent Smoking." *Drug and Alcohol Dependence* 59 (2000): Supp. 1: S61–S81.

Orleans, C. Tracy, and John Slade, eds. *Nicotine Addiction: Principles and Management*. New York: Oxford University Press, 1993.

Shiffman, S., et al. "Smoking Behavior and Smoking History of Tobacco Chippers." *Experimental and Clinical Psychopharmacology* 2 (1994): 126–142.

Shiffman, Saul, and Thomas A. Wills, eds. *Coping and Substance Use*. Orlando, Fla.: Academic Press, 1985.

Slovic, Paul, ed. *Smoking: Risk, Perception, and Policy*. Thousand Oaks, Calif.: Sage, 2000.

U.S. Department of Health and Human Services. *Reducing Tobacco Use: A Report of the Surgeon General—Executive Summary*. Atlanta, Ga.: U.S. Department of Health and Human Services, Centers for Disease Control and Prevention, National Center for Chronic Disease Prevention and Health Promotion, Office on Smoking and Health, 2000.

Wagner, Eric F., ed. *Nicotine Addiction Among Adolescents*. New York: Haworth Press, 2000.

Public Relations

Public relations (PR) is an important element of a firm's promotional mix or strategy, whereby the firm communicates information to the public with a goal of influencing their attitudes and behavior. The primary purposes of promotion are to inform, persuade, and remind. Many PR activities may be regarded as persuasion-based because the communications are

PUBLIC RELATIONS

linking the firm with desirable attributes and images. Scholars and social historians often consider PR synonymous with "image management," or less positively with "spin" or "media manipulation."

Although marketing and sales activities commonly have an objective of selling a firm's products, PR efforts are typically focused on enhancing the image of the firm or the entire institution. PR involves a very broad range of activities; thus it can be challenging to find agreement about how the term is best defined. Nevertheless, Denny Griswold, founder of *Public Relations News*, a leading PR newsletter for publicists and professionals in the PR field, offers a widely accepted definition of PR: "The management function which evaluates public attitudes, identifies the policies and procedures of an individual or an organization with the public interest, and plans and executes a program of action to earn public understanding and acceptance" (Seitel 2001). PR can involve communication through both paid and unpaid means, with PR practitioners communicating on either a personal or non-personal basis with several publics including customers, competitors, the academic community, the government, regulatory authorities, trade associations, special interest groups, the investment community, suppliers and distributors, employees, and the press. PR practitioners perform several functions, including press relations, product publicity, corporate communication, lobbying, employee and investor relations, and crisis management.

Early PR Efforts by the Tobacco Industry

Ivy Lee and Edward Bernays are considered key pioneers of modern public relations. Lee was a former Wall Street reporter who became involved in publicity work in 1903 and formed a PR agency with George Parker in 1904. Lee's clients included the hard coal industry, the Rockefeller family (one of the wealthiest families in the United States, who owned the Colorado Fuel and Iron Company), and the American Tobacco Company. During the mid-1920s, George Washington Hill, president of American Tobacco, sought Lee to direct the firm's PR activities, which included guidance and critical comment about various advertising campaigns. One of Lee's assignments was to improve relations with those possessing candy and sugar interests. Relations had become strained due to an effective Lucky Strike advertising campaign that encouraged female consumers to "Reach for a Lucky instead of a sweet." Lee's firm was paid a retainer of $40,000 per year for its efforts, which was a considerable amount at the time.

In 1928, the American Tobacco Company also became a key client of Bernays (he previously worked for Liggett and Myers, the producers of Chesterfields), who was the mastermind of a PR strategy that encouraged women to start smoking cigarettes in public places. A promotional campaign was introduced at the 1929 Easter parade in New York City, in which ten young women, including Ruth Hale (a leading feminist), lit "torches of freedom" as a protest against women's inequality. This freedom march gained front-page exposure in newspapers.

Bernays was assigned the formidable task of altering consumers' perceptions about the color green because internal market research indicated that many women did not hold a favorable view of Lucky Strike's green packaging. In 1934, Bernays was noted for facilitating green as the fashion color of the year by coordinating both a "Green Ball" with New York's socialites and a "Green Fashions Fall Luncheon" with leading

PUBLIC RELATIONS

Edward Bernays, a pioneer of modern public relations, died in 1995 at the age of 103. He continued to attend PR-related meetings until shortly before his death.
© BETTMANN/CORBIS

fashion editors. The color was omnipresent at both events and supplemented with press releases indicating that green was a symbol of hope, victory (over depression), solitude, and peace. Presentations were given by both a noted art academic on the subject of "green in the work of great artists" and a renowned psychologist on green's psychological connotations. A Color Fashion Bureau was established, and letters and announcements were sent to influential interior decorators, department store managers, home-furnishings buyers, art industry groups, and women's clubs to communicate that green was a color with several virtues. These events were organized by Bernays without other participants knowing the identity of his client.

Hill recognized the importance of PR. Lee and Bernays were simultaneously on the American Tobacco payroll at great expense for a considerable number of years. At first, neither Lee nor Bernays was aware of the duplicity and duplication of many of their efforts for the same company. Interestingly, once Lee and Bernays discovered their concurrent employment, Hill explained, "If I have both of you, my competitors can't get either of you" (Pollay 1990).

The Tobacco Industry Research Committee and the Tobacco Institute

During the early 1950s, scientific and popular articles began to more commonly associate lung cancer with smoking, and smokers became increasingly "health concerned." Several epidemiologic studies linked smoking with lung cancer in the early 1950s and *Reader's Digest* articles in 1952 and 1953 highlighted the relationship of smoking with cancer. Tobacco firms became increasingly uneasy about the negative publicity, which prompted the industry to hire Hill and Knowlton, a renowned PR firm, in 1953. Recommendations by Hill and Knowlton led to the formation of the New York–based Tobacco Industry Research Committee (TIRC) in 1954. On 4 January 1954, a full-page PR advertisement, using the headline "A Frank Statement to Cigarette Smokers," circulated in 448 newspapers in 258 U.S. cities reaching an estimated readership of more than 43 million, to announce that the TIRC was being established with a mandate of supporting scientific research related to the health effects of tobacco use. The advertisement cast doubt on unfavorable research findings and included the statements: "We [the tobacco industry] accept an interest in people's health as a basic responsibility, paramount to every other consideration in our business. We believe the products we make are not injurious to health. We always have and always will cooperate closely with those whose task it is to safeguard the public health" (Glantz et al.).

The TIRC, renamed the Council for Tobacco Research in 1964, continued to function with the purpose of maintaining uncertainty and "controversy" over the health effects of smoking by maintaining files on experts, carefully monitoring media stories, arranging meetings with key press editors and writers, generating favorable publicity through news releases, and providing grants for scientific research that was reportedly independent. The TIRC put forward arguments that more research was needed before a conclusive link could be made between smoking and cancer, placed an emphasis on people's genetic susceptibility to cancer, speculated which other factors were attributable to lung cancer, claimed that tests on mice were not applicable to humans, and attempted to discredit the existing studies that reached unfavorable conclusions.

The Washington-based Tobacco Institute was established in 1958 to take over lobbying and PR activities, representing its members on matters of common interest pertaining to litigation, politics, and public opinion. By the late 1980s, the annual budget of the Tobacco Institute was estimated to be more than $20 million. The tobacco industry increasingly focused its efforts toward resisting increases in excise taxes, restrictions on indoor smoking, and proposals to ban advertising. The tobacco industry also continued to deny that nicotine was addictive—on 14 April 1997, seven U.S. tobacco executives made an infamous testimony to this effect in U.S. Congress—even though internal research indicated otherwise. The Council for Tobacco Research and the Tobacco Institute were both abolished in accordance with the 1998 U.S. Master Settlement Agreement.

Key PR Tools

Tools commonly used by PR practitioners include new product publicity, product placement, Internet websites, and event or "issue" sponsorship. New product publicity involves efforts to generate positive media attention about specific products or services being introduced to the marketplace. During the 1950s, cigarette promotions frequently portrayed newly introduced filtered products as technological breakthroughs and made assertions that were meant to reassure smokers

---

Tobacco PR Consultants Try to Create "Controversy" About Secondhand Smoke

The tobacco industry continues to use science as a PR tool to resist or forestall proposed bylaws that prohibit smoking in indoor, public settings. Internal tobacco industry documents, which are publicly accessible through various court proceedings, reveal that tobacco firms and their PR consultants have aggressively discredited, undermined, and refuted scientific research findings relating to the health consequences of environmental tobacco smoke exposure. When assessing Philip Morris' PR activities, researchers Elisa Ong and Stanton Glantz said that the firm has "gone beyond 'creating doubt' and 'controversy' about the scientific evidence that demonstrates that active and passive smoking cause disease, to attempting to change the scientific standards of proof" (2001).

PUBLIC RELATIONS

with health concerns. In 1958, for example, Philip Morris organized a press conference at New York's Plaza Hotel to unveil its new "high filtration" Parliament brand. At the conference, a Philip Morris executive described the new filter—coined "Hi-Fi"—as "hospital white" and explained that this was an event of "irrevocable significance." Meanwhile, test tubes bubbled in the hotel foyers and personnel, wearing long white laboratory coats, responded to questions.

In 1964, consumer health concerns were reawakened with the release of the first U.S. Surgeon General's report on smoking and health. The tobacco industry responded by launching several low (machine-measured) yield products, which were supported by promotions implying that they were healthier or less hazardous. During the 1970s, several line extensions of familiar products were introduced, making use of the "Light" product descriptor. In the 2000s, corporate websites of tobacco firms emphasize their efforts to introduce new products that purportedly deliver lower levels of toxins to the smoker, again portraying these products as technological breakthroughs.

PR practitioners may also garner publicity through product placement, which entails efforts to have product exposure during special events or in films, television programming, stage-theater, concerts, and computer games. Philip Morris, for example, had product placement arrangements for the motion pictures *Superman II* (1980) and *License to Kill* (1989), and allegedly spent $200,000 to have actor Martin Sheen smoke Marlboro throughout *Apocalypse Now* (1979). Internal tobacco industry documents also reveal that, in 1983, the Brown & Williamson

James J. Morgan, left, president and CEO of Philip Morris USA, and Michael E. Sxymanczyk, executive vice president of sales and marketing, unveil their company's Action Against Access program in New York, 27 June 1995. Although promoted as a program to prevent underage tobacco sales, internal tobacco industry documents reveal that the purpose of this and similar programs was to avert or delay regulatory measures and to improve public opinion about tobacco industry marketing practices. AP/WORLD WIDE PHOTOS

Tobacco Corporation agreed to pay $500,000 to actor Sylvester Stallone in exchange for him smoking the firm's brands in a minimum of five feature films. Cigarette product placement payments are prohibited in accordance with the 1998 U.S. Master Settlement Agreement, but many tobacco control groups and health practitioners remain concerned about how tobacco use is portrayed in television programming and film.

Using Internet websites, PR initiatives have attempted to "reposition" tobacco corporations such that they are perceived as responsible firms within a controversial industry. One employed strategy involves tobacco firms communicating their support of youth prevention campaigns and youth access laws. The Tobacco Institute's first prominent program related to youth access included distribution of both a booklet titled "Helping Youth Say No" and signs for posting at retail that stated, "It's the Law: We Do Not Sell Tobacco Products to Persons Under 18." The R.J. Reynolds Tobacco Company initiated the "Right Decisions, Right Now" program in 1991, procuring actor Danny Glover as a spokesperson. "We Card" is an active industry-initiated program that was established in 1995 under the auspices of the Coalition for Responsible Tobacco Retailing. The self-described mandate of this coalition is to prevent underage tobacco sales in the United States by providing training and education opportunities to owners, managers, and front-line employees of retailers that sell tobacco. Member organizations of this coalition include tobacco firms, retailers, and law enforcement agencies. Philip Morris began publicizing its "Action Against Access" program in 1995 as a complement to "We Card," and in 1998, the firm formed a Youth Smoking Prevention department that has an annual budget of more than $100 million.

Internal tobacco industry documents reveal, however, that the purpose of these programs is to avert or delay regulatory measures and to improve public opinion about tobacco industry marketing practices. These industry-sponsored programs have been ineffective at diminishing youth tobacco use, largely because they portray smoking as an adult activity (thus, framing cigarettes as "forbidden fruit") and emphasize the influential role of peer pressure and parents as role models (meanwhile, omitting discussion about the role of industry marketing practices).

Finally, tobacco firms often become involved in event or "issue" sponsorships to improve the image of both the company and its products through being associated with useful works. Sponsorship objectives are typically distinguishable as either corporation related or product related. In addition to enhancing the company's image, common corporation-related objectives include: 1) increasing public awareness of the company and its services; 2) altering public perception; 3) being involved in the community and having local relevance; 4) building business relations and goodwill; and 5) enhancing employee relations and motivation. Tobacco industry representatives commonly maintain that their sponsorships allow events to be staged that might otherwise be denied, and view sponsorships as an opportunity to be regarded as good corporate citizens. With respect to "issues" sponsorship, Philip Morris ran prominent $100 million promotional campaigns during the late 1990s that communicated the philanthropic and social activism activities of the firm, relating to topics such as feeding the hungry and supporting domestic violence crisis centers. The campaigns did not specify, however, that the money spent toward promoting these efforts considerably exceeded what was actually contributed to those in need.

See Also Advertising; American Tobacco Company; British American Tobacco; Lobbying; Marketing; Philip Morris; Regulation of Tobacco Products in the United States; Sponsorship.

■ TIMOTHY DEWHIRST

BIBLIOGRAPHY

Cosco, Joe. "Tobacco Wars: The Battle Between Titans Heats Up." *Public Relations Journal* 44 (1988): 14–19, 38.

Drope, Jacqui, and Simon Chapman. "Tobacco Industry Efforts at Discrediting Scientific Knowledge of Environmental Tobacco Smoke: A Review of Internal Industry Documents." *Journal of Epidemiology and Community Health* 55 (2001): 588–594.

Ewen, Stuart. *Captains of Consciousness: Advertising and the Social Roots of the Consumer Culture.* New York: McGraw-Hill, 1976.

Glantz, Stanton A., et al. *The Cigarette Papers.* Berkeley: University of California Press, 1996.

Irwin, Richard L., and Makis K. Asimakopoulos. "An Approach to the Evaluation and Selection of Sport Sponsorship Proposals." *Sport Marketing Quarterly* 1 (1992): 43–51.

Lamb, Charles W., Jr., Joseph F. Hair, Jr., and Carl McDaniel. *Marketing,* 5th ed. Cincinnati, Ohio: South-Western College Publishing, 2000.

Landman, Anne, Pamela M. Ling, and Stanton A. Glantz. "Tobacco Industry Youth Smoking Prevention Programs: Protecting the Industry and Hurting Tobacco Control." *American Journal of Public Health* 92 (2002): 917–930.

Muggli, Monique E., et al. "The Smoke You Don't See: Uncovering Tobacco Industry Scientific Strategies Aimed Against Environmental Tobacco Smoke Policies." *American Journal of Public Health* 91 (2001): 1,419–1,423.

Ong, Elisa K., and Stanton A. Glantz. "Constructing 'Sound Science' and 'Good Epidemiology': Tobacco, Lawyers, and Public Relations Firms." *American Journal of Public Health* 91 (2001): 1749–1757.

Pollay, Richard W. "Propaganda, Puffing and the Public Interest." *Public Relations Review* 16 (1990): 39–54.

Pollay, Richard W., and Timothy Dewhirst. "The Dark Side of Marketing Seemingly 'Light' Cigarettes: Successful Images and Failed Fact." *Tobacco Control* 11 (2002): i18–i31.

Pratt, Cornelius B. "The 40-Year Tobacco Wars: Giving Public Relations a Black Eye?" *Public Relations Quarterly* 42 (1997–1998): 5–10.

Seitel, Fraser P. *The Practice of Public Relations,* 8th ed. Upper Saddle River, N.J.: Prentice Hall, 2001.

Tye, Larry. *The Father of Spin: Edward L. Bernays and the Birth of Public Relations.* New York: Crown, 1998.

Yach, Derek, and Stella A. Bialous. "Junking Science to Promote Tobacco." *American Journal of Public Health* 91 (2001): 1745–1748.

Quitting

European observers noted as early as the sixteenth century that tobacco users found it difficult to quit their practice once they had adopted the habit. The Spanish archbishop Bartolomé de Las Casas Cuzco observed in 1527 that Spanish soldiers on Hispaniola seemed unable to stop using the plant. In 1604 King James I of England in his *Counterblaste to Tobacco* wrote that smokers became "bewitched" to tobacco and overcome by "lust" for the "vile custome." Sir Francis Bacon observed in 1622, "The use of tobacco . . . conquers men with a certain secret pleasure so that those who have once become accustomed thereto can hardly be refrained therefrom" (Slade 1998).

Nineteenth-century American authors frequently warned readers that tobacco *enslaved* smokers, **snuff** users, and tobacco chewers. In 1852 one author warned boys that tobacco users were bound "in chains not easily broken" and compared tobacco with **opium** (Trask 1852).

Industrialist Henry Ford pointed out the addictive quality of cigarette smoking in 1914 in his popular book *The Case Against the Little White Slaver*. By the 1930s many medical writers saw the tobacco habit as "a form of drug addiction" (Dorsey 1936).

Thus, from an early time doctors and laypersons generally understood that quitting tobacco use was very difficult. However, a scientific consensus did not emerge until 1988, with the publication of *The Health Consequences of Smoking: Nicotine Addiction: A Report of the Surgeon General*, which said that the smoking habit was a biologically based addiction like that associated with cocaine or heroin. How then might one quit, given the great difficulty in doing so?

snuff a form of powdered tobacco, usually flavored, either sniffed into the nose or "dipped," packed between cheek and gum. Snuff was popular in the eighteenth century but had faded to obscurity by the twentieth century.

opium an addictive narcotic drug produced from poppies. Derivatives include heroin, morphine, and codeine.

Three Twentieth-Century Models

In advice typical for the nineteenth century, one author recommended that users focus their "stern, resistless will" on breaking their dependency (Trask). He added that prayers, staying busy, signing a pledge, drinking copious amounts of pure water, and hydrotherapy could aid the slave to tobacco.

QUITTING

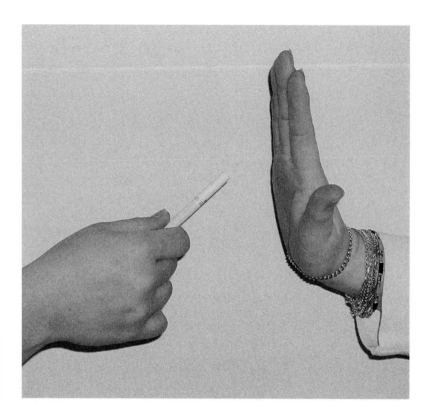

If only quitting smoking was this easy. Unfortunately, mere willpower Is seldom power enough to quit the addictive habit. © KELLY A. QUIN. REPRODUCED BY PERMISSION

In the first half of the twentieth century writers continued to recommend the nineteenth-century smoking cessation measures and added recommendations such as drinking copious amounts of fruit juices, deep breathing, and psychoanalysis. A few physicians advised patients to use amphetamines, tranquilizers, or lobeline sulphate as aids to cessation. Taking the alkaloid lobeline ($C_{22}H_{27}NO_2$) and tobacco simultaneously caused smokers to become nauseous because of a cross-tolerance between lobeline and nicotine. Some physicians believed it also helped reduce cravings for nicotine because of similar pharmacological effects on the nervous system. By the 1930s scientists understood that lobeline "caused a brief stimulation of the motor centers in the spinal cord and medulla. This stimulation is soon followed by depression, and later paralysis with large doses. The feature of the action of the drug is the stimulation of the motor nerve endings in the involuntary muscles" (Dorsey 1936).

As strong scientific evidence linking smoking with lung cancer emerged in the 1950s, health scientists began to design formal programs to assist those smokers who were unable to quit on their own. The clinical treatment programs were generally based on techniques and ideas about self-control or pharmacological interventions.

WAYNE MCFARLAND. Dr. Wayne McFarland was a pioneer in the field of clinical treatment. In the 1950s he developed the Five Day Plan to quit smoking, and began to conduct smoking withdrawal clinics on a large scale in 1962. Although McFarland was associated with the Seventh Day Adventist Church, his Five Day Plan was a nonreligious smoking cessation program often cosponsored by local hospitals and

The Five Day Plan

The Five Day Plan, developed in the 1950s by Dr. Wayne McFarland, sought to strengthen smokers' willpower and to weaken the physical craving for nicotine experienced during withdrawal. Participants met for five evenings at various convenient places such as school auditoriums, hotels, civic halls, or hospitals for one and a half to two hours. To motivate the smokers, participants were given frightening lectures on the hazards of smoking and shown graphic films of smokers having cancerous lungs removed. Ex-smokers gave hopeful testimonials and participants were encouraged to select a buddy for mutual support.

Along with the evening meetings, participants were given behavioral and psychological tools to use while on their own. The plan advocated complete and immediate cessation. To sustain abstinence, a three-pronged assault, with mental, physical, and spiritual components, was made on the addiction.

The plan's author advised taking large quantities of fruit juices and copious amounts of water with the hope that this would reduce craving sensations. Rhythmic breathing was also recommended to increase the oxygen supply to the brain and to fortify willpower.

Various behavioral interventions were suggested. One might take walks after meals or help his or her spouse with the dishes. Long hot showers in the morning and evenings, as a form of hydrotherapy, were recommended for their calming effect. The participants were taught to avoid spicy foods, sugar, coffee, and alcohol. The belief was that these stimulating foods would induce craving.

To fortify the will, the plan recommended repeating the mantra: "I choose not to smoke." Though the plan was nonreligious, it did incorporate a generic spirituality into its cessation armamentarium. Participants were instructed to ask for divine help at moments of crisis in accord with their own beliefs. In substance, the plan's recommendations echoed early-twentieth- and late-nineteenth-century recommendations. The main difference was its formal organization and system of social support.

local voluntary health groups such as the National Tuberculosis and Respiratory Disease Association (now the American Lung Association, or ALA). The clinics were free or only a small nominal fee was charged. Between 1961 and 1964, 50,000 Americans completed the Five Day Plan. Often more than 100 people attended a single Five Day Program at one location for one week. In the greater Los Angeles area alone, 8,000 to 10,000 people had completed 300 clinics by 1970. Because of the difficulty of following ex-smokers over time and the lack of prospective studies on the attendees, researchers do not know how successful the program was. Evidence from the time suggests that significant numbers of people were able to quit for a short time, but many returned to smoking later.

BORJE E.V. EJRUP. Another early programmatic attempt to aid smokers in quitting their addiction was made by the Swedish physician Borje E.V. Ejrup. He began his work in Stockholm in 1955 and continued it at the New York Hospital/Cornell Medical Center in the 1960s. By 1967 he had treated 7,000 patients with his method.

Ejrup was especially interested in the "hard core" smokers with a strong **physiological** dependence on nicotine. He gave patients lobeline hydrochloride in injections and in oral form in order to support them in breaking their physiological dependence. In addition, Ejrup prescribed

physiology the study of the functions and processes of the body.

the tranquilizer meprobamate to allay anxiety. Because many patients were concerned about weight gain, he also gave them an amphetamine to reduce hunger.

Patients came to Ejrup's Tobacco Withdrawal Clinic every weekday for the first two weeks for injections and tablets. They also received individual counseling from a physician. Ejrup advised physicians to be dramatic in their counseling sessions as they attempted to warn, persuade, and cajole would-be ex-smokers.

DONALD T. FREDRICKSON. Donald T. Fredrickson, M.D., director of the Smoking Control Program at the New York City Department of Health, planned and directed the first smoking control program of the New York City Department of Health from 1964 to 1967. Nearly 100 volunteer ex-smokers, drawn mainly from the upper middle class, comprised most of the staff of the program. Fredrickson's program had three phases, beginning with motivational lectures, and progressing through group sessions involving discussion, question and answer, and mutual encouragement.

The core of Fredrickson's program was based on a lay self-help model derived from Alcoholics Anonymous, Gamblers Anonymous, and Weight Watchers. In addition, he derived elements of his model from conversations with ex-smokers, reports of other clinics, and studies of a handful of behavioral scientists.

Fredrickson believed that habituation to smoking was, in part, learned behavior, and the smoker needed to learn to manage emotional and psychological states without cigarettes. He instructed smokers that they needed to be highly motivated and to faithfully exercise the virtues of patience and persistence in order to alter their behavior through psychological retraining. Hopefully the smoker would experience the program as a positive exercise in self-mastery while achieving a new dimension of self-control.

ADAPTATIONS. The three early programs described above became the models on which many later smoking withdrawal clinics in North America and Europe were based. McFarland's Five Day Plan was transplanted to the United Kingdom and Canada and was often co-sponsored by various health agencies and hospitals in the United States. Adaptations of Ejrup's pharmacological and intensive counseling treatment regimen were deployed by clinics in the United States, the United Kingdom, Denmark, and Germany. Fredrickson's self-help, self-control program became a model for other stop smoking clinics by 1970 including those of the American Cancer Society (ACS), the American Lung Association (ALA), and the Los Angeles County Department of Health.

Programs in the 1960s to 1970s

The ACS, ALA, local health departments, local hospitals, and local voluntary health associations began offering free or low-cost smoking cessation clinics in small numbers in 1964. By 1974, 13,000 smokers in California alone had participated in ACS stop smoking clinics. In the

1970s other voluntary agencies like the Young Men's Christian Association (YMCA) and the American Heart Association (AHA) began to offer smoking cessation clinics. During the period from 1977 to 1981, the ACS held 18,000 stop smoking clinics across the United States. All of these agencies continued to offer stop smoking clinics in the 1980s, notable among these were the American Lung Association's Freedom from Smoking Program and the ACS's Fresh Start smoking cessation program.

These clinics, generally based on a self-control rationale, usually included lectures, question-and-answer sessions, self-evaluation tests, the buddy system, some form of individual therapy or group therapy or group support in which participants shared experiences and stories, and handbooks that gave advice about changing behaviors. The noncommercial clinics usually had from four sessions to twelve or more sessions lasting from one to eight weeks and sometimes as long as six months.

In addition to these noncommercial stop-smoking clinics, by 1970 there was a $50 million per year industry of for-profit smoking cessation programs. Some of these programs included Smokewatchers, Quit Now, Smokenders, and Schick Centers. By 1977 Smokenders alone reported that it had 150,000 graduates of its eight-week program. The commercial clinics sometimes used adaptations of the Fredrickson model, hypnosis, and aversive conditioning.

Due to high drop-out rates and the difficulties of following patients over time, researchers are not certain of how effective these programs were. Impressions from the time and current data indicate that clinics had some initial success but over time many smokers returned to tobacco use.

Besides these clinical interventions, health agencies, among whom the ACS was the largest and most active, also attempted to induce cessation through educational campaigns directed at the population level. The ACS waged their "Who Me? . . . Quit Smoking!" and "The Time to Stop Is Now" campaigns beginning in 1965. In 1968 the ACS began its "I Quit" or "IQ" smoking cessation campaign. Other notable population level interventions included the ACS's "Target Five" campaign from 1977 to 1981, in which 20 million adults were reached with antismoking messages and, during which, the ACS sponsored 18,000 Quit Smoking! clinics through local affiliates. The Great American Smokeout, held annually since 1977, has been another prominent attempt at intervening at the population level. For example, in 1983, 19 million Americans participated in the Great American Smokeout. During this event the ACS, through a national publicity campaign, attempts to persuade smokers to try to quit for one day hoping that a fraction of them will quit permanently.

EXPERIMENTAL STUDIES. Clinical delivery of smoking cessation treatment preceded the large increase in formal, experimental studies of smoking cessation that began in the mid-1960s, a field that one study described as still in its infancy in 1968. During the 1970s there was a great deal of wide-ranging research into smoking cessation methods. Most of the research was based on behavioral strategies of aversion or self-control. In aversion strategies the idea was to associate unpleasant

stimuli with smoking so that smoking would no longer be experienced as pleasurable. Among the aversion methods studied were giving electric shocks to people while they smoked, having people rapidly smoke cigarette after cigarette or smoke so much that they became ill, blowing smoke in the face of a cigarette user as he or she smoked creating irritation and discomfort, and having smokers concentrate on negative, disgusting, or unpleasant images in their minds while they smoked. It was hoped these negative associations would deter smoking.

In other studies researchers hoped to help participants resist the idea or craving to smoke; in essence, to help them increase their ability to control their smoking behavior. Researchers did this by making contingency contracts wherein smokers would receive some reward, such as money, if they avoided smoking, and wherein they would have to pay money if they smoked. They tried social contracts, like the buddy system, wherein smokers attempted to quit with a partner. It was hoped that through increased social support smokers could resist the temptation to smoke. In the 1980s experimental research began to focus on physician advice models, work site interventions, and community wide approaches. In these approaches it was hoped that less intensive interventions directed at a much larger population would end up, on balance, creating more ex-smokers than intensive interventions directed at individuals and small groups. In addition, researchers increasingly studied nicotine replacement strategies such as the nicotine patch with and without behavioral components. These interventions continued to demonstrate modest effects with relatively low quit rates because of the strength of "multiple societal, psychosocial, biobehavioral, and biological processes that maintain smoking behavior" (Lichtenstein and Glasgow 1992).

The 1990s to the Present

Based on research from the 1990s into the new millennium, scholars and medical professionals understand that nicotine is the addicting drug that has the poorest success rate for cessation when compared to alcohol, cocaine, and opioids. Withdrawal symptoms might include craving sensations, irritability, anger, anxiety, difficulty concentrating, restlessness, decreased heart rate, or weight gain.

Approximately 87 percent of those who successfully quit smoking do so on their own, while only 13 percent quit with the help of a formal program or drug therapy. Smokers usually have to make several attempts at quitting before achieving success.

Since the 1990s some progress had been made with medical interventions to help with smoking cessation. Specifically, nicotine replacement therapies in the form of gum, patch, and nasal spray, and the use of bupropion, a non-nicotine-based quitting medication, have shown promise. Researchers have shown that the concomitant use of drug therapy (including nicotine replacement) and receiving counseling of some kind give the addicted smoker the best chance at quitting.

See Also Addiction; Nicotine; Quitting Medications.

▌COLIN TALLEY

BIBLIOGRAPHY

Bernstein, Douglas A. "Modification of Smoking Behavior: An Evaluative Review." *Psychological Bulletin* 71 (1969): 418–440.

Brandt, Allan. "The Cigarette, Risk, and American Culture." *Daedalus* 119 (Fall 1990): 155–176.

———. "Blow Some My Way: Passive Smoking, Risk, and American Culture." In *Ashes to Ashes: The History of Smoking and Health*. Edited by S. Lock, L. A. Reynolds, and E. M. Tansey. Amsterdam and Atlanta, Ga.: Editions Rodopi B.V., 1998.

Diehl, Harold S. *Tobacco and Your Health: The Smoking Controversy*. New York: McGraw Hill, 1969.

Dorsey, John L. "Control of the Tobacco Habit." *Annals of Internal Medicine* 10 (November 1936): 628–631.

Gehman, Jesse Mercer. *Smoke Over America*. Paterson, N.J.: Beoma Publishing House, 1943.

Goodman, Henry A., ed. *World Conference on Smoking and Health: Summary of the Proceedings*. New York: National Interagency Council on Smoking and Health, 1967.

Koslowski, Lynn T., Jack E. Henningfield, and Janet Brigham. *Cigarettes, Nicotine, and Health: A Biobehavioral Approach*. Thousand Oaks, Calif.: Sage Publications, 2001.

Lichtenstein, Edward, and Carolin S. Keutzer. "Modification of Smoking Behavior: A Review." *Psychological Bulletin* 70 (1968): 520–533.

Lichtenstein, Edward, and Brian G. Danaher. "Modification of Smoking Behavior: A Critical Analysis of Theory, Research, and Practice." In *Progress in Behavior Modification*, Vol. 3. Edited by Michel Hersen, Richard M. Eisler, and Peter M. Miller. New York: Academic Press, 1976.

Lichtenstein, Edward, and Russell E. Glasgow. "Smoking Cessation: What Have We Learned Over the Past Decade?" *Journal of Consulting and Clinical Psychology* 60 (1992): 518–527.

McFarland, J. Wayne, and Elman J. Folkenberg. *How to Stop Smoking in Five Days*. Englewood Cliffs, N.J.: Prentice Hall, 1964.

Richardson, Robert G., ed. *The Second World Conference on Smoking and Health*. New York: Health Education Council, 1971.

Ross, Walter S. *Crusade: The Official History of the American Cancer Society*. New York: Arbor House, 1987.

Rustin, Terry A. "Management of Nicotine Withdrawal." In *Principles of Addiction Medicine*, 2nd ed. Edited by Allan W. Graham and Terry K. Schultz. Chevy Chase, Md.: American Society of Addiction Medicine, 1998.

Schwartz, Jerome L. "A Critical Review and Evaluation of Smoking Control Methods." *Public Health Reports* 84 (June 1969): 483–506.

Slade, John. "The Pharmacology of Nicotine." In *Principles of Addiction Medicine*, 2nd ed. Edited by Allan W. Graham and Terry K. Schultz. Chevy Chase, Md.: American Society of Addiction Medicine, 1998.

Trask, George. *Thoughts and Stories on Tobacco for American Lads*. Boston: G. C. Rand, 1852.

Wilkins, Jeffery N., David G. Gorelick, and Bradley T. Conner. "Pharmacologic Therapies for Nicotine Dependence." In *Principles of Addiction Medicine*, 2nd ed. Edited by Allan W. Graham and Terry K. Schultz. Chevy Chase, Md.: American Society of Addiction Medicine, 1998.

Quitting Medications

Up to the end of the 1970s behavioral treatments were the only procedures with some efficacy available. Particularly the aversive method "rapid smoking" had a reasonably good efficacy record.

Early Development of Nicotine Replacement Therapy

The first scientifically evaluated drug treatment for tobacco dependence or smoking cessation, nicotine replacement therapy (NRT), was conceptualized in 1967 and developed in Sweden during the 1970s. Two physicians at the University of Lund, Stefan Lichtneckert and Claes Lundgren, approached the nearby pharmaceutical company AB Leo with the idea of using nicotine for smoking cessation after they had observed crew members in a submarine use smokeless tobacco. The notion that tobacco use was driven by nicotine was not widespread in the late 1960s, and AB Leo's research director, Ove Fernö, himself a heavy smoker, agreed to fund Drs. Lichtneckert and Lundgren's research.

After trying several administration forms (including aerosol) the researchers chose gum, mainly for safety reasons. The first gum was abandoned because it released its nicotine too quickly. In order to slow down the release, Dr. Fernö introduced the use of an ion exchanger into which the nicotine could be incorporated, a complex binding the nicotine to the gum until it comes into contact with saliva when it is released. However, using the **ion exchanger complex** slowed down the release of nicotine; in order to improve absorption a buffer was added to the gum.

Around 1973 Professor Michael Russell at the Department of Psychiatry at Denmark Hill in London became interested in the idea of using nicotine in smoking cessation after having used behavioral methods without much success. In the United States, Murray Jarvik and Nina Schneider were the first researchers to experiment with nicotine gum and they became great ambassadors of the product. At the Medical School at the University of Lund, a Smoking Cessation Clinic established and headed by Professor Håkan Westling came to use gum liberally for its patients beginning in 1970. The experience from the uncontrolled clinical use by Drs. Lichtneckert and Lundgren was important for testing out various reformulations of the gum that were used later in the centers above.

Initial Marketing Authorizations

In Sweden there was a discussion about whether the food or the medicine agencies should regulate nicotine gum. Gum was considered food, but nicotine was not an approved food additive. After several years, during which the Swedish tobacco monopoly showed an intention to market the gum, the government decided that the gum should be regulated as a medicine.

Nicotine gum was first approved in Switzerland in 1978. In a U.S. regulatory agency advisory committee meeting in 1983, there was a lot

ion exchanger complex a method of controlling the nicotine levels of cigarettes by adding ion exchangers, usually resins, to the tobacco or filter.

QUITTING MEDICATIONS

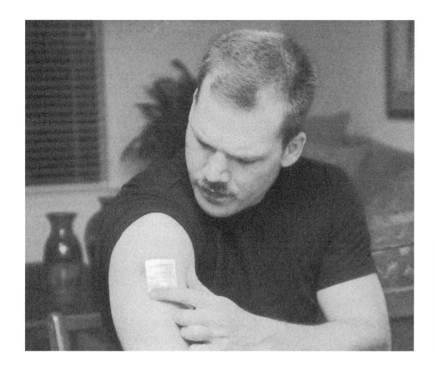

People who are trying to quit smoking may use skin patches (shown here) or chewing gums that contain nicotine. The medications help relieve nicotine withdrawal symptoms and can make it easier to break the addictive patterns of tobacco use. PHOTOGRAPH BY ROBERT J. HUFFMAN/FIELD MARK PUBLICATIONS

of controversy surrounding the gum's possible approval. Fear of abuse of the gum was cited as one of the FDA's concerns.

Nicotine Nasal Spray

It was evident that a gum could not deliver nicotine with the same speed and to the same extent that a cigarette could. As a way to remedy the relatively slow absorption of nicotine from the gum, in 1979 Professor Russell and Dr. Fernö agreed to investigate the absorption of nicotine through the nose. Russell's group headed most of the clinical development with the nicotine nasal spray and advocated that in a smoker's clinic with heavy dependent smokers it was the most effective tool. In the United States in the 2000s, nicotine nasal spray and nicotine inhalers are available by prescription.

The Nicotine Patch

In the early 1990s it became clear that not every cigarette smoker could or liked to chew gum, and many experienced side effects like irritation in the oral cavity and some indigestion. This led to underdosing and thoughts on how to remedy the problem. In 1984, Drs. Jed Rose and Murray Jarvik from the University of California Los Angeles (UCLA) began to experiment with nicotine given transdermally (through the skin). They applied for a patent for transdermal delivery of nicotine, which was later approved. Around the same time, U.S.-based ALZA Corporation and the German-based Lohman Therapie Systeme started to develop nicotine patches, which are similar to adhesive bandages and are available in different shapes and sizes. The nicotine patch releases a constant amount of nicotine in the body; the nicotine dissolves through the skin and enters the body, thus providing relief from some of the withdrawal symptoms people experience when they quit smoking.

pharmacokinetics the branch of medicine that deals with the action of drugs within the body, specifically absorption, distribution, and elimination.

Other Nicotine-Delivery Products

In the 1990s and early 2000s, a number of nicotine-delivery products were developed by different pharmaceutical companies. In order to provide consumer options, Pfizer developed an oral inhaler. This preparation delivers nicotine with approximately the same characteristics as a 2-milligram gum and the efficacy is also the same. The Swiss company, Novartis, was the first to develop a lozenge with a 1-milligram dose, later followed by a 2-milligram dose. UK-based GlaxoSmithKline developed a 2- and 4-milligram lozenge that roughly mimics the **pharmacokinetics** of the Pfizer and Novartis 2- and 4- products.

Nicotine Replacement Therapy in the Future

Since the mid-1980s consumers have witnessed the medicine regulatory authorities outside the United States relaxing their attitude toward nicotine and the expressed safety concerns. Since this time, many contraindications have been lifted, better availability—from prescription-only, over the counter, and general sale—has been allowed, and wider use with new indications like temporary abstinence and reduced smoking has been granted. In the 2000s, patches are only used for complete cessation while the other products are used for both cessation and relapse prevention, reducing smoking, and recreational use. The gum formulation is the best-selling product both in the United States and worldwide, followed by the patch.

"It is argued that it is not so much the efficacy of new nicotine delivery systems as temporary aids to cessation, but their potential as long-term alternatives to tobacco, that makes the virtual elimination of tobacco a realistic future target," Michael Russell wrote in *The Lancet* (1991). How nicotine will be consumed in the future has to do with how it will be regulated. Public health officials have maintained that a regulatory framework—where one agency has the power to regulate all nicotine-containing products—would be instrumental in determining the best public health impact of cessation products and abstinence from nicotine. In the United Kingdom, organizations such as the British Royal College of Physicians and the House of Commons have petitioned the British Department of Health to move in that direction.

See Also Addiction; Nicotine; Quitting.

▌ KARL FAGERSTRÖM

BIBLIOGRAPHY

Bolliger, Christoph T., and Karl-Olov Fagerström, eds. *The Tobacco Epidemic.* Vol. 28, *Progress in Respiratory Research.* New York: Karger, 1997.

Haustein, Knut-Olaf. *Tobacco or Health? Physiological and Social Damages Caused by Tobacco Smoking.* New York: Springer, 2003.

Schneider, N. *How to Use Nicotine Gum.* New York: Simon & Schuster, 1988.

 Regulation of Tobacco Products in the United States

The U.S. federal government regulates all consumer products except cigarettes and other tobacco products. Products from foods and drugs to cars and car seats are all subject to rigorous regulation to ensure that they are safe and that they work as consumers expect. But with the exception of a program to ensure the collection of taxes from their sale, there is virtually no regulation of tobacco products. In fact, tobacco is expressly exempted from regulation under a number of consumer product statutes, such as the Federal Consumer Product Safety Act.

The most logical agency in the federal government to regulate tobacco products is the U.S. Food and Drug Administration (FDA). The FDA regulates food, drugs, cosmetics, and medical devices like X-ray machines and surgical instruments. Indeed, about 25 cents worth of every dollar spent on consumer products goes to a product regulated by FDA. In the 1990s the FDA did attempt to regulate tobacco products but, as will be explained, the agency's efforts were overturned in court.

Other federal agencies have tobacco-related responsibilities that do not involve product regulation. Some are involved in studying the health effects of tobacco products, while others inform the public of the risks of tobacco products and how to quit. The work of these other agencies also will be described.

Why Regulate Tobacco Products?

A logical first question many people ask is why tobacco products should be regulated at all. Some people believe that the public is already aware of the risks associated with tobacco products and that regulation is not needed. Others question why they are permitted to be sold at all. If tobacco products are so dangerous, so this argument goes, why not ban tobacco products entirely?

Many experts believe that the answer to this very sensible question is that prohibition would not work. Almost 50 million adults smoke. Most of these tobacco users are addicted to the nicotine in tobacco. These people would still be dependent on nicotine if tobacco sales were made illegal. Experts fear that, as was the case when alcohol sales were made illegal in the early twentieth century, a black market for tobacco products would quickly come into existence.

A black market is a system of illegal sales of prohibited products. One concern is that black market tobacco products could be more dangerous than the products available today because of even more questionable ingredient quality and product purity. The tens of millions of smokers who might seek out cigarettes in a black market could thus be exposing themselves to even greater risks than they would under a system in which sales are lawful.

What Would Tobacco Product Regulation Include?

Traditional regulation of consumer products is designed to ensure that ingredients or components are safe and that the products will work as promised. FDA regulation of health claims for food and drug products is a good example of a regulatory approach that might work for tobacco products.

Food and drug manufacturers have to first demonstrate to the FDA that there is scientific support for the claims they make about their products. Most importantly, they have to submit their evidence to the FDA before they can make a claim on a product label or package. The FDA then decides if there is adequate scientific evidence to support the claim. For example, a breakfast cereal company that wants to claim that its new high-fiber cereal will reduce the risk of cancer must first prove it to the FDA. If the FDA is not satisfied that there is sufficient clinical or **epidemiological** data from the scientific studies to support the claim, the company cannot make the claim. This important consumer protection system prevents the public from being exposed to unproven claims.

epidemiological pertaining to epidemiology, that is, to seeking the causes of disease.

By contrast, in the current unregulated marketplace for tobacco products, cigarette manufacturers are free to make any claims about their products. Smokers, especially those concerned about their health and interested in quitting, have no way of knowing whether claims promising to reduce exposure to cancer-causing chemicals in smoke are actually true (see sidebar, p. 493).

Other features from the food and drug regulatory system could be applicable to the regulation of tobacco products. One would be to evaluate the safety of new ingredients before they can be added to a tobacco product already on the market. Another would be to reduce the risk of tobacco products by restricting the level of harmful compounds to which tobacco users are exposed. Yet another would be to monitor the marketplace to make sure products are being used as intended. This task is particularly important to ensure that children and adolescents are not using tobacco products.

Unregulated Health Claims for Tobacco Products

A new generation of tobacco products has entered the marketplace in the last decade. These products offer promises of reduced exposure to dangerous chemicals in tobacco smoke, and even make claims to reduce the risk of cancer and other diseases.

The products take various forms. Some burn tobacco, or use special methods to burn or heat tobacco. Others are tobacco-based but do not burn. The products that burn tobacco include Omni and Advance. Omni, manufactured by a company called Vector, invested $40 million in an advertising campaign prominently featuring the claim "Reduced Carcinogens. Premium Taste." Two-page ads for Omni bearing that claim appeared regularly in *Parade* magazine in 2002. The products that use novel methods to burn or heat tobacco include Eclipse and Accord. Their claims are similar to Omni's. The non-combusting products include Ariva, Revel, and Exalt; these promise tobacco satisfaction in situations where smoking is not possible (e.g., at work or at home).

Whether they burn or not, all of these products are aimed squarely at the health-concerned smoker. They have entered the marketplace in the absence of any independent scientific evaluation of their claims, and without any governmental scrutiny of the products or their claims.

From a public health perspective, these products may pose a significant threat to efforts to help smokers quit. Health-concerned smokers who see these products may now think that a safer cigarette genuinely exists. This may make them less inclined to try to quit.

There is the added concern that former smokers may start smoking again, thinking they can now safely consume tobacco products. Likewise, those who never smoked may light up for the first time, using one of these new products under the assumption that a safe cigarette exists.

In the absence of public health–based regulation of these products, there is no way to know whether this new generation of products will actually reduce exposure and risk. The great fear held by some public health experts is that these new products may be nothing more than a scientifically sophisticated version of the "light" cigarette. We now know, many decades too late to help smokers who switched to "light" cigarettes over the last 30 years, that "lights" were deliberately designed so as not to reduce tar and nicotine deliveries when smoked by human beings. Back then, well-intentioned public health officials encouraged health-concerned smokers to switch to "lights." Experts urge that we avoid repeating the same mistakes with today's products.

The FDA's Attempt to Regulate Tobacco Products in the 1990s

In 1994, under then-Commissioner David Kessler, the FDA announced that it would investigate the role of nicotine in the design and manufacture of tobacco products. If there was sufficient evidence that tobacco companies deliberately designed their products to create and sustain an addiction to nicotine, the FDA claimed that it should assert jurisdiction and begin to regulate those products.

From 1994 to 1996, the FDA gathered evidence from public health experts, current and former tobacco industry scientists, and tobacco industry documents. Some of the most important evidence proving what the tobacco industry knew about nicotine's role in causing addiction came from the industry itself in admissions contained within internal documents (see sidebar p. 494).

In 1996 the FDA gathered all the evidence from its nicotine investigation and made a two-part determination under the Federal Food, Drug, and Cosmetic Act: 1) that the nicotine in tobacco products was a

> ### What the Tobacco Industry Said Privately About Nicotine and Addiction
>
> Throughout the mid-1990s, thousands of previously secret internal tobacco industry documents were made available to the public. These documents were particularly helpful to the FDA during its investigation of the role of nicotine in the design and manufacture of cigarettes. Here are a few of the most revealing statements about nicotine and addiction from the industry's own documents, which helped the FDA to determine that tobacco products are drug delivery devices.
>
> > Nicotine is addictive. We are, then, in the business of selling nicotine—an addictive drug effective in the release of stress mechanisms.
> >
> > ■ (ADDISION YEAMAN, BROWN & WILLIAMSON, 1963)
>
> > In a sense, the tobacco industry may be thought of as being a specialized, highly ritualized and stylized segment of the pharmaceutical industry. Tobacco products, uniquely, contain and deliver nicotine, a potent drug with a variety of physiological effects.
> >
> > ■ (CLAUDE TEAGUE JR., R.J. REYNOLDS, 1972)
>
> > The cigarette should be conceived not as a product but as a package. The product is nicotine. . . . Think of the cigarette pack as a storage container for a day's supply of nicotine. . . . Think of a cigarette as a dispenser for a dose unit of nicotine. Think of a puff of smoke as the vehicle of nicotine. . . . Smoke is beyond question the most optimized vehicle of nicotine and the cigarette the most optimized dispenser of smoke.
> >
> > ■ (WILLIAM L. DUNN, PHILIP MORRIS, 1972)

drug; and 2) that the products (i.e., cigarettes and other tobacco products) were devices for the delivery of the drug nicotine.

Simultaneously, the FDA issued a final rule designed to reduce the numbers of children and adolescents who start smoking. The 1996 regulation made it illegal for retailers to sell cigarettes to minors. Other provisions were designed to make tobacco advertising less appealing to young people. For example, ads that children might see in magazines, other publications, or in stores would have been limited to a black-and-white, text-only format. This would have preserved the industry's ability to advertise to adults, but in a format that experts said would have been less attractive to youngsters.

Shortly thereafter, the FDA was sued by tobacco manufacturers, growers, retailers, and advertisers. They claimed that the agency's actions were illegal. The case made it all the way to the U.S. Supreme Court. While the case was being heard in the federal courts, the FDA began enforcing a few provisions of the 1996 final rule. From 1997 to 2000, the FDA worked with the states to conduct over 200,000 inspections of retailers to enforce the rule prohibiting the sale of tobacco products to minors. Over $1 million in fines was collected from retailers who illegally sold cigarettes to minors more than once.

In 2000 the U.S. Supreme Court issued a 5–4 decision that stripped the FDA of legal authority over tobacco products. The Court ruled that Congress never intended for the FDA to have regulatory powers over these products under federal law. In order for the FDA to regain these powers, Congress would have to pass a new law granting the FDA that authority.

Other Federal Agencies

Other federal agencies are involved in tobacco-related work that is not directly tied to product regulation. Here is a brief description of what some of them do.

- The **National Cancer Institute** (NCI) funds important research into such questions as how tobacco causes cancer and what can be done to make tobacco products less harmful. The NCI issues reports that summarize research findings.

- The Office on Smoking and Health within the **Centers for Disease Control and Prevention** (CDC) provides critical guidance to the states on which programs work best to prevent young people from starting to use tobacco and on how to help addicted smokers who want to quit. The CDC also works with governments around the world to share advances in prevention and treatment efforts.

- The **Office of the Surgeon General** releases regular reports on smoking and health issues. The Surgeon General reports are important summaries of what is known about tobacco-related disease and what can be done to reduce the death and disease toll caused by tobacco. The first-ever Surgeon General's report on smoking and health, published in 1964, was a landmark publication in the history of public health.

The Federal Trade Commission also compiles and releases an annual report on the **tar** and nicotine levels for all marketed cigarettes in the United States, as well as a report on the annual marketing expenditures of the tobacco industry.

tar a residue of tobacco smoke, composed of many chemical substances that are collectively known by this term.

The Future of U.S. Tobacco Regulation

The World Health Organization is leading a global effort to enact a treaty known as the Framework Convention on Tobacco Control. One of the key provisions of this treaty calls for all countries to begin to regulate tobacco products as drug delivery devices. Such regulation could consist of advertising and marketing restrictions, as well as limits on permissible levels of toxins in tobacco smoke. At the time of this writing, it is unclear whether the United States will ratify this treaty. Public health experts continue to hope that the U.S. Congress will enact new legislation granting the FDA regulatory authority over tobacco products.

See Also Additives; Advertising Restrictions; Politics; Product Design; Prohibitions; State Tobacco Monopolies; Taxation; Warning Labels; Youth Marketing; Youth Tobacco Use.

■ MITCHELL ZELLER

BIBLIOGRAPHY

Dunn, William L. "Motives and Incentives in Cigarette Smoking: Summary of CTR-Sponsored Conference in St. Martin." Richmond, Va.: Philip Morris Research Center, 1972. Cited in *Regulation of Cigarettes and Smokeless Tobacco Under the Federal Food, Drug, and Cosmetic Act*. Washington, D.C.: Department of Health and Human Services, U.S. Food and Drug Administration, 1996.

Glantz, Stanton A., et al. "Looking Through a Keyhole at the Tobacco Industry: The Brown & Williamson Documents." *Journal of the American Medical Association* 274, no. 3 (19 July 1995): 219–224.

Hilts, Philip J. *Smokescreen: The Truth behind the Tobacco Industry Cover-Up.* Reading, Mass.: Addison-Wesley, 1996.

Kessler, David A. *A Question of Intent: A Great American Battle with a Deadly Industry.* New York: Public Affairs, 2001.

Kessler, David A., et al. "The Food and Drug Administration's Regulation of Tobacco Products." *New England Journal of Medicine* 335 (26 September 1996): 988–994.

Kessler, David A., et al. "The Legal and Scientific Basis for FDA's Assertion of Jurisdiction over Cigarettes and Smokeless Tobacco." *Journal of the American Medical Association* 277, no. 5 (5 February 1997): 405–409.

Kluger, Richard. *Ashes to Ashes: America's Hundred-Year Cigarette War, the Public Health, and the Unabashed Triumph of Philip Morris.* New York: Alfred A. Knopf, 1996.

Pertschuk, Michael. *Smoke in Their Eyes: Lessons in Movement Leadership from the Tobacco Wars.* Nashville, Tenn.: Vanderbilt University Press, 2001.

Teague, Claude, Jr. "Research Planning Memorandum on the Nature of the Tobacco Business and the Crucial Role of Nicotine Therein" (14 April 1972). Available: <http://legacy.library.ucsf.edu/tid/nnv59d00>.

Yeaman, Addision. "Implications of Battelle Hippo I & II and the Griffith Filter" (17 July 1963). Available: <http://legacy.library.ucsf.edu/tid/xrc72d00>.

Retailing

The retailer is a critical link in the supply chain of tobacco. Retailers serve as the bridge between the tobacco manufacturers and the consumers of tobacco products. Retailing practices have changed substantially since the introduction of commercially produced tobacco products. However, the importance of the retailer to the viability and prosperity of the tobacco companies has remained constant.

History

In the middle of the nineteenth century, Americans were the heaviest per capita consumers of tobacco. Most Americans lived in rural areas and the "typical citizen was a native-born outdoorsman, short on cash and uneager to spend what he earned on things he could grow himself or swap with a neighbor or a traveling peddler. He tended to take his tobacco Indian-style, either chewing it or smoking it in a pipe, in an age when brand-name goods hardly existed" (Kluger). At that time, the most popular commercial tobacco products were chew and **plug** tobacco, which was sold by the town's old-time tobacconist and rural crossroads storekeepers.

These tobacco products featured imaginative names such as My Wife's Hat, Wiggletail Twist, and Sweet Buy & Buy. Outside the front

plug a small, compressed cake of flavored tobacco usually cut into pieces for chewing.

Smokers like this one frequently smoke 20 or more cigarettes a day (one pack). Instead of purchasing them by the carton, which is cheaper, most smokers purchase their cigarettes by the pack at some form of retail outlet. PHOTOGRAPH BY KELLY A. QUIN

door, a carved figure of a Native American often signaled the location of the local tobacconist shop. These shops were popular in America between the 1870s and the 1930s. In fact, the U.S. Census Bureau estimated that there were 580,000 cigar and tobacco shops in the United States in 1917.

Tobacco shops were a place of leisure and refuge and "possessed special qualities—warmth, camaraderie and congeniality—that appealed strongly to male senses. It became a pleasantly informal neighborhood forum with back rooms for pinochle, stud poker or just plain conversation. . . . The bastion of togetherness, a well-frequented social club, took on an atmosphere of exclusivity and privacy only surpassed by the local saloon. The price of admission was only a five-cent cigar" (Petrone 1996).

Eventually sales of cigarettes and smoking tobacco exceeded sales of chewing tobacco after concerns that chewing tobacco and spitting was a messy habit, socially inappropriate in a more crowded urban America, and a cause of tuberculosis and other diseases. Moreover, cigarettes were being mass manufactured very cheaply by Durham, North Carolina, entrepreneur named Buck Duke.

Another Durham resident, a farmer named John R. Green, pioneered tobacco merchandising after he purchased a smoking tobacco company in 1862. He launched a smoking tobacco under the name Bull Durham and widely advertised it, gave gifts to frequent buyers, and gave special premiums to dealers. Buck Duke also excelled in promoting his tobacco products by heavily advertising and giving under-the-table payment to tobacco retailers who most aggressively pushed his brands. He also gave them special premiums such as floor mops and imitation diamond stickpins. Other companies followed suit. In its early days before it became a major powerhouse company, underling Philip Morris even gave company stock to select retailers for helping launch its brands and giving them preferential display treatment.

RETAILING

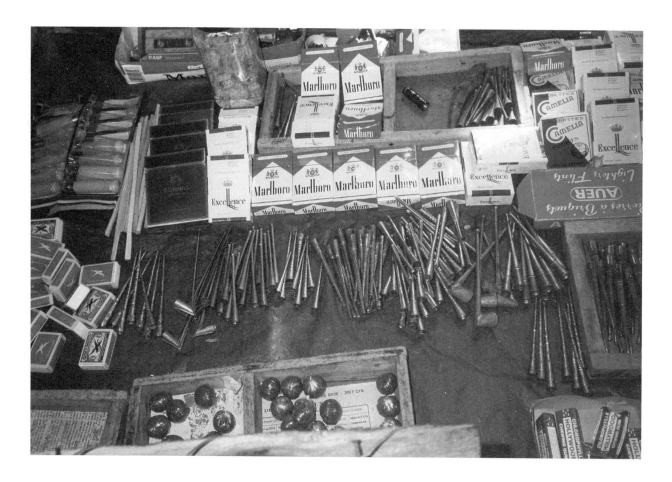

Retailing in low-income countries includes a preponderance of street vendors not commonly seen in developed countries. Here a street vendor in Senegal displays his wares on a table: Marlboros, cola nuts, gum, hard candy, and stick toothbrushes.
PHOTO BY ANNA WHITE

bidis thin, hand-rolled cigarettes produced in India. Bidis are often flavored with strawberry or other fruits and are popular with teenagers.

flue-cured tobacco also called Bright Leaf, a variety of leaf tobacco dried (or cured) in air-tight barns using artificial heat. Heat is distributed through a network of pipes, or flues, near the barn floor.

Current Retailing

In the twenty-first century, tobacco products are sold in a wide variety of stores, including convenience stores, gas stations, liquor stores, supermarkets, and pharmacies. Cigarettes are sold in bowling alleys, donut shops, bars, and smoking paraphernalia shops known as "head shops." Some bars and restaurants sell cigarettes by the pack from the bartender or a vending machine. Of these different locations, the highest sales volume of tobacco occurs at convenience stores, where more than half of all cigarettes are sold in the United States.

Most contemporary tobacco retailers offer several types of tobacco products, such as cigarettes, cigars, **bidis** (a type of cigarette imported from India), smokeless tobacco, and loose-leaf tobacco. A handful of specialty stores offer a selection of loose-leaf tobaccos (for example, **flue-cured** Bright leaf, thick-cut Virginia blend, and Turkish blend) and allow customers to roll their own cigarettes by loading their blends into premanufactured filter tubes.

Importance of Retail Outlets

The retail outlet is a critical venue for the tobacco industry. Stores are the primary location where tobacco products are sold to consumers. Cigarettes are an $80-billion industry each year in America. Scholars do not know the exact number of retail outlets that sell tobacco products in the United States, but estimates range from 534,000 to more than

A tobacco retailer in the United States displays his products. Note that tobacco products are behind the counter and are not sold via self-service. Cigarettes, both domestics and imported, are placed on shelving units behind the clerk and smokeless tobacco products are located above the shelving units. The shelving units, which are often provided by the tobacco companies, feature colorful branded advertising for popular R.J. Reynolds Brands, such as Camel and Winston. Even though this store sells a variety of consumer products, the shelving units and advertising for cigarettes are prominent and dominate the coveted retail space behind the point of purchase at the cash register. AP/WIDE WORLD PHOTOS

1.3 million. Having cigarettes available in so many locations keeps them readily accessible twenty-four hours a day to the country's 50 million smokers. Even though it is cheaper for smokers to buy their cigarettes by the carton than the pack, most smokers still purchase them by the pack. Economists have suggested that this is probably because many smokers do not want an excess quantity of cigarettes at their disposal, which may tempt them to smoke more. Many smokers also want to quit smoking and want to feel that each pack is their last.

In addition to providing the supply of tobacco, retail outlets also feature advertising and promotional materials that convey the image of cigarette enjoyment to customers in the store. This helps stimulate demand for the product. Although cigarette advertising has been banned on radio, television, and some print channels in many industrialized nations, there are few restrictions on cigarette advertising at stores. U.S. stores that sell cigarettes contain approximately ten to twenty distinct cigarette advertising and promotional items. Branded cigarette advertising appears everywhere on posters, window decals, lighted signs, display racks, clocks, and gas pumps. All customers, both youth and adult, are exposed to these advertisements. Some of these advertisements are displayed near candy shelves or video games, or at 3 feet or below, the eye level of a small child. Point-of-sale advertising may encourage youth smokers to experiment with a particular brand. These advertisements are also tempting to former smokers who can experience cigarette cravings when seeing the imagery of their former brand. According to point-of-sale marketing experts, well-designed cigarette advertising and point-of-sale displays can boost product sales by up to 10 to 20 percent.

Virtually all companies want their products to be displayed in the prime locations. They also want strategically placed point-of-sale advertising that promotes their brand imagery. How are the tobacco companies so successful in securing prime placement of their products and advertising in the most coveted locations inside of stores? The short answer is that they pay retailers. According to a 2001 report on cigarette industry advertising and promotional expenditures filed with the U.S. Federal Trade Commission, the major cigarettes companies spend over 85 percent of their $11.2 billion promotional budget at the retail outlet. Some of this money is for promotional allowances, which are paid to retailers to ensure that tobacco products get the best and most visible shelf space in the store. Cigarette companies also give payments to retailers and special discounts on cigarettes for prime placement of cigarette advertising and promotional materials. Finally, the companies also spend money on value-added promotions, which include offering multipack specials whereby customers can "buy two packs, and get one free."

Regulation

PREVENTING YOUTH ACCESS. Although some states have had laws in place since the early 1900s making it illegal for merchants to sell tobacco products to minors, this issue began to receive heightened attention in the late 1980s. In 1987, a researcher published a study showing that an eleven-year-old girl was successful in purchasing cigarettes in 75 of 100 attempts at stores (Difranza 1987). Subsequent studies confirmed the finding that underage youth had easy access to cigarettes. The most effective solution is a law that bans tobacco sales to minors and is actively enforced by penalizing storeowners or clerks who sell to underage youth. In the 2000s, as required by a 1992 federal law known as the Synar Amendment, all U.S. states prohibit the sale of tobacco products to individuals under age eighteen (a few states have nineteen as their minimum age of sale) and must show evidence that they are enforcing these restrictions. When the U.S. Food and Drug Administration (FDA) claimed jurisdiction over tobacco products in 1996, it created a federal policy banning tobacco sales to minors and it created a nationwide enforcement system. However, a coalition of major tobacco companies and retailers, including the National Association of Convenience Stores, challenged the FDA's legal authority over tobacco products and prevailed in a Supreme Court ruling delivered in March 2000. Even though the federal level enforcement system was disbanded, all state laws banning sales to minors remain in place.

Some regulations govern the manner of tobacco sales. For instance, some communities and states ban self-service of tobacco products, which requires that the product be kept behind the counter or in an overhead bin accessible only to the clerk. This prevents customers, especially teen tobacco users, from stealing the product from shelves. Communities in countries such as Australia, Canada, Iceland, and Ireland have banned tobacco product displays. That is, tobacco products must be kept under the counter or in some other location that is not visible to consumers.

Some laws prohibit minors from purchasing, using, or possessing tobacco products. Florida has some of the strictest laws banning youth possession of tobacco products, whereby minors are subject to having their driver's license revoked after multiple violations.

The exact impact on youth tobacco use of these laws is hotly contested. Several studies in the early 1990s demonstrated large decreases in youth smoking when enforcement actions lowered the rate of cigarette sales to minors. These studies did not have control groups, and controlled studies of the effect of enforcement on youth smoking rates showed that the impact was either modest or nonexistent. Restrictions on retail sales to youth may simply drive them to find other means of obtaining cigarettes. However, studies do indicate that comprehensive programs aimed at preventing youth initiation, using a combination of media campaigns, cessation programs, and changes in the retail environment, may help reduce youth initiation. Additionally, a newer research area is examining the impact of policies that impose penalties on youth for purchasing, using, or possessing the product. This area is also controversial with some evidence suggesting that these policies may have an impact on reducing youth smoking. However, many tobacco control advocates criticize this punitive approach for focusing on youth rather than on the adults who manufacture, distribute, advertise, and sell tobacco products to youth.

TOBACCO PRICING. Sales and excise taxes are commonly applied to tobacco products. Most excise taxes are paid prior to their distribution to retailers, but retailers are responsible for charging sales taxes if they are levied in that area. Half of the states in the United States are "fair trade" states, which means that they have laws that establish a minimum price for cigarettes. The minimum price is a set percentage markup applied to the manufacturer's invoice price at the wholesaler and retailer level. Despite the fact that tobacco products are addictive, increased tobacco prices reduce consumption by cutting down on both the number of people using the product as well as the amount that they consume. Youth are especially sensitive to prices, so higher prices have a greater impact on reducing their consumption than they do on the behavior of adults.

TOBACCO ADVERTISING AND MARKETING. In the United States, cigarette advertising is preempted by federal law, which means that states are blocked by the federal government from regulating tobacco advertising. In 1999, Massachusetts attempted to ban all outdoor cigarette advertising within 1,000 feet of schools and playgrounds and prohibited advertisements placed lower than 5 feet. This would have curtailed advertising at stores in close proximity to schools. However, the tobacco companies challenged the legality of the policy and won in a U.S. Supreme Court decision handed down in June 2001. This case highlighted the many challenges in regulating tobacco advertising in countries such as the United States that have constitutional provisions protecting freedom of speech.

The Master Settlement Agreement (MSA) between the major cigarette manufacturers and forty-six state Attorneys General contains some restrictions on tobacco advertising, but the only restriction on retailers is that they cannot display tobacco advertisements that are larger than 14 square feet.

Several countries have comprehensive bans on tobacco advertising, including advertising at retail locations. These countries include Canada, Finland, Norway, France, Italy, New Zealand, Portugal, Jordan, Singapore, and Thailand. Moreover, the European Union has agreed to phase out all tobacco advertising by 2006.

Internet Tobacco Sales

Websites selling tobacco products started appearing in the mid-1990s. Scholars do not have reliable data on the number of vendors over time, but one study identified 88 Internet cigarette vendors in January 2000 and more than 800 in January 2004. Some industry analysts predict

duty a tax, usually on certain products by type or origin; a tariff.

that Internet cigarette vendors will sell more than $5 billion worth of cigarettes by 2005.

Although the majority of English-language websites are located in the United States, Internet vendors are located all over the world. Many of these international vendors sell **duty**-free cigarettes at prices that are far cheaper than cigarettes sold in retail outlets because taxes and duties are not collected on these products. Duty-free vendors are located in the British Virgin Islands, Netherlands Antilles, Panama, and Portugal, but most are in Switzerland. Nearly half of the Internet vendors located in America are located on Native American reservations. Members of the Seneca tribe in western New York State have more than 100 websites selling cigarettes and this area has been called the "Internet cigarette capital of the world by one reporter for the *Buffalo News* named Michael Beebe who writes about Internet cigarette sales." Tribal vendors tout on their websites that they sell from sovereign land and that their treaties with the United States allow them to sell cigarettes tax free. The U.S. government disputes this position and the U.S. Supreme Court has ruled that tribal entities can sell products tax free only to tribal members, which means that they cannot sell tax-free cigarettes to non-tribal members either on or off of tribal lands. The other location where many Internet vendors are located is in tobacco-producing states in the southeastern United States. This is mainly because these states have very low cigarette excise taxes and the website owners can purchase cigarettes very cheaply for resale online.

In the 2000s, few regulations affect these Internet vendors, but several laws have been proposed. These policies affect both youth access and tax collection. One study published in 2003 by researchers at the University of North Carolina at Chapel Hill (Ribisl et al.) showed that youth aged eleven to fifteen were successful in purchasing cigarettes from Internet vendors in 92 percent of purchase attempts. This fueled interest in a federal law banning Internet tobacco sales to minors. Several states, such as California, Maine, and Rhode Island already have such laws, but have experienced difficulty enforcing them against out-of-state vendors, which suggests that a federal law may be needed. Some proposed laws would require Internet vendors to collect the excise taxes at the level that they are charged in the customer's state. Traditional (non-online) tobacco retailers regularly support these regulations. They feel the market is not a fair one because Internet vendors can sell tobacco products more cheaply. Finally, given the global reach of the Internet, the World Health Organization has been interested in regulating Internet tobacco marketing. Cigarette advertising on the Internet is prohibited under the terms of the Framework Convention on Tobacco Control, an international tobacco control treaty created by the World Health Assembly, but countries must ratify the treaty for it to go into effect.

See Also Marketing; State Tobacco Monopolies; Trade.

▮ KURT RIBISL

BIBLIOGRAPHY

DiFranza, Joseph. R., et al. "Legislative Efforts to Protect Children from Tobacco." *Journal of the American Medical Association* 257, no. 24 (1987): 3387–3389.

Feighery, Ellen. C., et al. "Cigarette Advertising and Promotional Strategies in Retail Outlets: Results of a Statewide Survey in California." *Tobacco Control* 10 (2001): 184–188.

Feighery, Ellen. C., et al. "How the Tobacco Companies Ensure Prime Placement of Their Advertising and Products in Stores: Interviews with Retailers about Tobacco Company Incentive Programs." *Tobacco Control* 12 (2003): 184–188.

Jacobson, Peter D., et al. *Combating Teen Smoking: Research and Policy Strategies.* Ann Arbor: University of Michigan Press, 2001.

Kluger, Richard. *Ashes to Ashes: America's Hundred-Year Cigarette War, the Public Health, and the Unabashed Triumph of Philip Morris* New York: Vintage, 1997.

Petrone, Gerard S. *Tobacco Advertising: The Great Seduction.* Atglen, Pa.: Schiffer Publishing, 1996.

Ribisl, Kurt M., "The Potential of the Internet as a Medium to Encourage and Discourage Youth Tobacco Use." *Tobacco Control* 12 (2003): Supp. 1: i48–i59.

Ribisl, Kurt M., et al. (2003) "Internet Cigarette Sales to Minors." *Journal of the American Medical Association* 290 (2003): 1356–1359.

Taylor, Allyn. L., and Douglas. W. Bettcher. "Who Framework Convention on Tobacco Control: A Global 'Good' for Public Health." *Bulletin of the World Health Organization* 78, no. 7 (2000): 920–929.

Tofler, A., and Simon Chapman. "'Some Convincing Arguments to Pass Back to Nervous Customers': The Role of the Tobacco Retailer in the Australian Tobacco Industry's Smoker Reassurance Campaign 1950–1978." *Tobacco Control* 12 (2003): Supp. 3: iii7–iii12.

U.S. Department of Health and Human Services. "Preventing Tobacco Use among Young People: A Report of the Surgeon General." Atlanta, Ga.: U.S. Department of Health and Human Services, Public Health Service, Centers for Disease Control and Prevention, National Center for Chronic Disease Prevention and Health Promotion, Office on Smoking and Health, 1994.

———. "Reducing Tobacco Use: A Report of the Surgeon General." Atlanta, Ga.: U.S. Department of Health and Human Services, Centers for Disease Control and Prevention, National Center for Chronic Disease Prevention and Health Promotion, Office on Smoking and Health, 2000.

 # "Safer" Cigarettes

Since the beginnings of the modern cigarette in the late nineteenth century, public health officials have attributed a variety of toxic effects to cigarette smoking. At the same time, manufacturers, scientists, entrepreneurs, and public health leaders have, at various points, promoted or recommended product changes that would allegedly make cigarette smoking less harmful, though not entirely harmless or "safe" for use.

For example, in the 1880s tobacco smoke was known to contain nicotine, which both physicians and the public widely believed to be poisonous. Entrepreneurs developed novel products that allegedly blocked nicotine and other constituents, such as Dr. Scott's Electric Cigarettes, containing a cotton filter which, the manufacturer claimed, "strains and eliminates the injurious qualities from the smoke" (Tate 1999). In the 1930s and 1940s, cigarette advertisements for major brands, such as Lucky Strike, Chesterfield, and Camel, routinely included health-related statements and testimonials from physicians. For example, Camel cigarette ads promised "28% less nicotine," while Philip Morris promised reduced "throat irritation."

Filters and Tar

As studies linking cigarette smoking and lung cancer became widely publicized in the early 1950s, tobacco manufacturers predicted that there would be an increase in consumer demand for cigarettes with filter tips, especially among "health conscious" consumers. P. Lorillard launched Kent cigarettes in 1952 with its "micronite" filter, which contained fibers that the company claimed trapped dust particles in the smoke. Kent advertisements claimed the filter removed "7 times more nicotine and **tars**" than other filter cigarettes and offered "the greatest health protection in cigarette history." Kent sales received a substantial boost in 1957 when *Reader's Digest* highlighted the brand in an article titled "Wanted—and available—filter tips that really filter," which reported that Kent yielded 14 to 40 percent less tar than other leading filter brands. Kent sales shot up from 3.5 billion cigarettes in 1956 to

tar a residue of tobacco smoke, composed of many chemical substances that are collectively known by this term."

"SAFER" CIGARETTES

epidemiological pertaining to epidemiology, that is, to seeking the causes of disease.

37.5 billion in 1958, making it the fifth most popular and fastest growing of any cigarette brand ("With Filters" 1958).

By 1962 over half (54.6%) of all cigarettes produced in the United States had filters, compared with only 1.4 percent in 1952. The landmark report of the Advisory Committee to the Surgeon General on Smoking and Health, released on 11 January 1964, concluded that the hazards of cigarette smoking were substantial enough to warrant "appropriate remedial action." However, the committee concluded the available evidence was insufficient to draw any conclusions about the possible benefits of filters.

Nevertheless, **epidemiological** studies described in the Surgeon General's report demonstrated that there was a clear dose-response relationship between the number of cigarettes a person smoked and his or her risk of lung cancer. Additionally, animal studies showed that tobacco "tar," the particles in tobacco smoke, caused tumors when painted on laboratory animals. Based on this evidence, another expert committee organized by the Surgeon General in 1966 concluded, "The preponderance of scientific evidence strongly suggests that the lower the 'tar' and nicotine content of cigarette smoke, the less harmful are the effects."

In 1967, at the first World Conference on Smoking and Health, U.S. Surgeon General William H. Stewart warned that a "stalemate" had been reached in smoking prevention and cessation efforts. While some people had quit smoking because of health warnings, young people continued to take up the habit (Stewart 1967). Thus, Stewart and other public health leaders believed that they were obligated to do something to help prevent disease in people who would not, or could not, quit smoking. For example, the National Clearinghouse for Smoking and Health, a government office, began an educational campaign in 1968 titled "If You Must Smoke . . ." aimed at people who wanted to reduce their risk but did not want to quit smoking. The pamphlet gave five suggestions: Choose a cigarette with less tar and nicotine; don't smoke the cigarette all the way down (the last few puffs have more tar and nicotine); take fewer draws; reduce inhaling; and smoke fewer cigarettes.

Consumer advocates in Congress, including Senators Maurine Neuberger and Warren G. Magnuson, proposed legislation to require cigarette makers to disclose the average amount of tar and nicotine in cigarettes of each brand. This information would allow consumers to compare brands objectively and to choose brands with lower tar and nicotine content. In the long run, consumer advocates hoped, legitimate competition between manufacturers would lead to changes in cigarette design that would make cigarettes less hazardous, ultimately benefiting consumers. Cigarette manufacturers opposed these proposals. But when the Federal Trade Commission proposed to require cigarette companies to disclose tar and nicotine information to consumers, the major companies agreed to voluntarily provide this information on cigarette packages and advertising.

Scientific Research on "Less Hazardous Cigarettes"

In 1968, President Lyndon Johnson ordered the creation of a federal task force to address the growing incidence of lung cancer. The group's first recommendation was for an organized research program aimed at developing a "less hazardous cigarette." At the time, public health leaders and scientists were optimistic that an organized, collaborative research program, bringing together government, academia, and industry, could develop techniques to identify and remove hazardous ingredients in cigarette smoke.

Over the following decade, the National Cancer Institute (NCI), the federal government's cancer research arm, spent more than $50 million on research to develop "less hazardous cigarettes." The majority of funds were spent on developing animal tests, including exposing dogs to cigarette smoke. By the mid-1970s, the research program had identified some potential design changes they believed would make cigarettes less harmful, including use of reconstituted tobacco sheet, inert filler, and high-porosity paper. Additionally, public health leaders predicted that new "light" (low-tar) cigarettes entering the market would substantially reduce lung cancer rates. NCI Director Frank Rauscher, speaking before Congress, predicted of the new low-tar products: "If these cigarettes are acceptable to the public taste wise, we should see a diminution of the increasing curve of lung cancer incidence in the next years" (Rauscher 1976).

However, by the late 1970s, attitudes toward this strategy began to change. Government officials and voluntary agencies took a tougher stance against tobacco with a renewed commitment toward helping smokers to quit and preventing young people from starting. The focus of scientific research shifted toward studies of strategies for smoking prevention and cessation, the addictive nature of nicotine, and the effects of smoking on nonsmokers. The 1981 Surgeon General's report, *The Changing Cigarette*, took a far more cautious approach than earlier reports to claims about the health benefits of switching to lower tar cigarettes, acknowledging that there is no safe level of smoking and that switching to low-tar cigarettes may reduce lung cancer risk but "the benefits are minimal."

Similar government-led efforts to promote the development of reduced risk products were pursued in the United Kingdom. Starting in 1973, a government laboratory began monitoring tar and nicotine yields from brands of cigarettes, and the government published public information posters and leaflets classifying familiar cigarette brands into "Low," "Medium," and "High" tar categories. Additionally, an Independent Scientific Committee on Smoking and Health (ISCSH), made up of scientific experts in biology and medicine, provided advice to the government and the tobacco industry. The ISCSH developed guidelines for the testing and approval of additives and synthetic materials used in cigarettes. Some scientists believed that synthetic materials could be developed to replace tobacco that would be less harmful than tobacco when burned. Two synthetic tobacco substitutes were approved for commercial use in 1977, Cytrel and NSM (new smoking material), both using modified cellulose. In July 1977, twelve cigarette brands were launched containing at least 25 percent synthetic material in place of tobacco. However, these products never gained popularity among smokers and were eventually taken off the market.

Novel Cigarettes and New Claims

Tobacco companies did experiment with other types of technological innovations to develop cigarettes that could potentially be marketed as less harmful. Documents and testimony of former tobacco company employees introduced in lawsuits over the past ten years suggest that some tobacco companies had developed innovative technologies but did not pursue them because of fear of legal actions. For example, in the 1970s, Liggett Group, Inc. began a research effort called the XA Project, which focused on blending additives to tobacco to neutralize cancer-causing compounds. However, the company abandoned the project reportedly because of company lawyers' concerns that marketing the product would require admitting that conventional cigarettes were hazardous, thereby making the company vulnerable to lawsuits from smokers who used the company's conventional products.

In 1988, the R.J. Reynolds Tobacco Company introduced a high-tech cigarette called Premier, which was touted as a virtually smokeless cigarette. It contained aluminum capsules with tobacco pellets inside, which were heated instead of burned. The product required its own instruction booklet showing consumers how to light it. While R.J.R. reportedly spent more than $800 million developing the brand, smokers who tried it said it left an unpleasant charcoal taste in their mouths. Additionally, public health officials argued that the Food and Drug Administration should regulate it as a drug-delivery device. Reynolds abandoned the brand less than a year after it was introduced.

But despite previous setbacks in the marketplace, manufacturers have continued to develop and market high-tech cigarettes with claims that they reduce exposure to toxic ingredients in tobacco smoke or reduce secondhand smoke. For example, some new cigarettes employ tobacco that has been genetically modified to produce lower levels of some cancer-causing agents. Additionally, tobacco lozenges containing powdered tobacco are being marketed to smokers for situations where they cannot smoke. However, in 2001 an expert committee convened by the Institute of Medicine, a nongovernmental U.S. scientific organization, determined that these products have not yet been evaluated sufficiently to determine whether they are in fact less harmful.

In the 2000s, scientific and public health experts urge the need for government regulation of tobacco products as a crucial step toward reducing tobacco-related harm. An effective regulatory plan could provide the U.S. government with the authority to require changes in products to reduce their toxicity, to evaluate ingredients in new products as they enter the market, and to oversee advertising claims made by manufacturers about potential reduced risk products. But even if innovative high-tech products can reduce health risks for smokers in the United States and other developed countries, they are unlikely to make an impact on the rapidly expanding cigarette markets in developing countries, where government oversight and public concern about the health effects of smoking are substantially weaker.

See Also Cigarettes; Menthol Cigarettes; Product Design; Toxins.

■ MARK PARASCANDOLA

BIBLIOGRAPHY

The Changing Cigarette: A Report of the Surgeon General. Washington, D.C.: U.S. Government Printing Office, 1981.

Congressional Record—Senate. 89th Cong., 2nd Sess., 27 July 1966, volume 94: 16468–16475.

National Commission on Smoking and Public Policy. *A National Dilemma: Cigarette Smoking or the Health of Americans.* New York: National Cancer Society, 1978.

Rauscher, F.J. Testimony. Cigarette Smoking and Disease, 1976. Hearings before the Subcommittee on Health of the Committee on Labor and Public Welfare. U.S. Senate, 94th Cong., 2d sess., on S.2902, 19 February, 24 March, and 27 May 1976. Washington, D.C.: U.S. Government Printing Office, 1976.

Schwartz, J. "Safer Smoke." *Washington Post Magazine* (31 January 1999): 9–24.

Smoking and Health. Public Report of the Advisory Committee to the Surgeon General of the Public Health Service. Public Health Service Publication No. 1103. Washington, D.C.: U.S. Government Printing Office, 1964.

Status Report. Smoking and Health Program, National Cancer Institute. Bethesda, M.D.; National Cancer Institute, 1979.

Stewart, William H. *Influencing Smoking Behavior. World Conference On Smoking and Health (Summary of the Proceeding), September 11–13, 1967.* New York: National Interagency Council on Smoking and Health, 1967.

Stratton, Kathleen et al., eds. *Clearing the Smoke: Assessing the Science Base for Tobacco Harm Reduction.* Washington, D.C.: National Academy Press; 2001.

Swann, Cheryl, and Sir Peter Froggatt. *The Tobacco Products Research Trust, 1982–1996.* London: Royal Society of Medicine Press, 1996.

Tate, Cassandra. *Cigarette Wars: The Triumph of "the Little White Slaver."* New York: Oxford University Press, 1999.

U.S. Congress. Reviewing Progress Made Toward the Development and Marketing of a Less Hazardous Cigarette. Hearings before the Consumer Subcommittee of the Committee on Commerce. U.S. Senate, 90th Cong., 1st sess., 23–25 August 1967. Washington, D.C.: U.S. Government Printing Office, 1968.

"With Filters, Record Cigarette Sales." *Business Week* (27 December 1958): 49–54.

Sailors

Sailors, or more broadly speaking mariners of all types, were vital to the transmission of tobacco from America to Europe. Sixteenth- and seventeenth-century witnesses attested that people involved in the maritime trade (ship captains, sailors, and slaves) were among the first to use tobacco in the Old World. In 1571 the Dutch herbalist Matthias de l'Obel described "many sailors, all of whom have returned from [the Indies] carrying small tubes . . . [which] they light with fire" (Goodman 1993). In 1619, a Spanish writer observed that sailors and "all of the people who travel by sea" inaugurated tobacco use in Spain, and that initially tobacco was "thought of as something vile and low, and a thing of slaves and tavern drinkers, and people of low consideration" (Norton 2000). Throughout the seventeenth and eighteenth centuries, depictions of sailors often showed them with a pipe in hand. (However, sailors were restricted to using **snuff** or chewing tobacco while on ship, since open fires were a shipboard hazard.)

Demographic figures also suggest that sailors served as agents of tobacco diffusion. The rapid increase in transatlantic commerce in the

snuff a form of powdered tobacco, usually flavored, either sniffed into the nose or "dipped," packed between cheek and gum. Snuff was popular in the eighteenth century but had faded to obscurity by the twentieth century.

SAILORS

Satirical English cartoon of veteran sailors exchanging tales in London, 1801.
© HULTON-DEUTSCH COLLECTION/CORBIS

second half of the sixteenth century meant that mariners became an ever more important and conspicuous segment of society. During one of the peak years, in 1594, at least 150 ships sailed between Seville and the Indies, which required more than 7,000 men to crew according to estimates. These 7,000-plus men would have been a visible presence in Seville, which had a population of about 130,000 in those years. The increasing demographic weight of sailors in the last decades of the sixteenth century meant that there was a critical and visible mass of tobacco consumers from whom the custom could spread to other groups in society. Chronology bears this out, for it is in the last decade of the sixteenth century that tobacco began to be systematically exported to Europe.

Not only were sailors a conspicuous group of tobacco aficionados, but they also helped transform tobacco from an exotic good erratically imported to a readily available commodity by developing a nascent distribution system. One way that sailors supplemented their pathetic income was to bring over small quantities of goods to sell. A ship manifest from 1602 reveals that a ship captain brought back about 181 kilograms of tobacco into Seville on his own account; this was a time period in which the important and wealthy merchants did not yet take an interest in importing tobacco. Critics of the plan to make tobacco's sale the exclusive prerogative of the Crown in 1636 argued that it would hurt the marginal members of society who depended on the tobacco trade. They evoked pilots, sailors, and passengers who, returning from the Indies, relied on sales of meager amounts of tobacco in order to pay off their boats fares. The petitioners described tobacco as so thoroughly entrenched in the local petty economy that if its free trade were prohibited by the monopoly, many subjects would not be able to pay for their upkeep, they would default on their debts, and a wave of bankruptcies would wreck the economy in Seville.

Why were humble mariners early agents for tobacco's diffusion to the Old World? In the first place, sailors were a group of Europeans who had enduring contact with Native Americans, particularly in the sixteenth century. Ships sailing from Europe to the Americas often anchored on islands that were unconquered—such as many of the Lesser Antilles until the seventeenth century—and crews traded with the local Indians to get provisions. In such interactions, sailors were initiated into tobacco rites of Native Americans. For instance, during Francis Drake's 1585–1586 expedition to the West Indies, the crew stopped on Dominica to procure food and potable water, and also traded for tobacco with the Island Caribs.

The fact that sailors occupied a marginal social position also likely contributed to their precocious adoption of the Native American custom of consuming tobacco. Sailors existed on the lower rungs of the very stratified societies of early modern Europe, in terms of both pay and status. Tobacco promised relief to the overworked and undernourished: It was said to ease fatigue and suppress pangs of hunger and thirst. Such effects would have been attractive to poor sailors in precarious economic circumstances.

Another reason that humble mariners may have had a class-related propensity to be on the vanguard of tobacco users was that they were less constrained by the status concerns that inhibited their higher ranking peers from bringing home the tobacco habit. While those of a superior rank might have felt free in the frontier ambience of the Indies to experiment with native practices, once back home they would have been reluctant to maintain a practice associated with New World "savagery." Already hovering near the bottom of the social hierarchy, sailors could do little damage to honor or status they did not possess.

Sailors were ideal agents of transmission because they functioned as an intact mobile community. When they returned to the Old World they did not simply disperse, but often continued to maintain links with each other. In Seville, sailors tended to live in certain neighborhoods such as the Triana neighborhood across the river from the main part of town. Seamen who came from elsewhere congregated in inns and taverns that catered to sailors. Because tobacco was learned as a social habit, linked to rituals of sociability, makes sense that the practice would be easier to maintain if one had a community with which to share the habit.

■ MARCY NORTON

BIBLIOGRAPHY

Castro y Medinilla, Juan. *Historia de las virtudes y propriedades del tabaco.* Cordoba: Salvador de Cea Tesa, 1620.

Goodman, Jordan. *Tobacco in History: The Cultures of Dependence.* London: Routledge, 1993.

Linebaugh, Peter, and Marcus Rediker. *The Many-Headed Hydra: Sailors, Slaves, Commoners, and the Hidden History of the Revolutionary Atlantic.* Boston: Beacon Press, 2000.

Norton, Marcy. "New World of Goods: A History of Tobacco and Chocolate in the Spanish Empire, 1492–1700." Ph.D. diss., University of California at Berkeley, 2000.

Ortiz, Fernando. *Cuban Counterpoint: Tobacco and Sugar.* Translated by Harriet de Onís. New York: Alfred A. Knopf, 1947. Reprint, Durham, N.C.: Duke University Press, 1995.

Pérez-Mallaína, Pablo E. *Spain's Men of the Sea: Daily Life on the Indies Fleets in the Sixteenth Century.* Translated by Carla Rahn Phillips. Baltimore: Johns Hopkins University Press, 1998.

Secondhand Smoke

sidestream smoke the smoke that rises from a burning cigarette.

mainstream smoke the cigarette smoke actually inhaled by the smoker.

Secondhand smoke (SHS) is the mixture of gases and particles from a burning cigarette (or other tobacco product) that end up in the surrounding air. The sources include smoke that comes off the lit end of a cigarette, called **"sidestream smoke,"** as well as smoke exhaled by the smoker. The smoke delivered directly to the smoker from the cigarette is called **"mainstream smoke."** Thus, exhaled smoke is sometimes referred to as "exhaled mainstream smoke." Secondhand smoke has a number of names, including "environmental tobacco smoke" (ETS) and "tobacco smoke pollution." The act of breathing in secondhand smoke has been called "involuntary smoking" and "passive smoking" (in contrast to the "active smoking" of the cigarette smoker).

In the United States, public health advocates have argued for the use of the term "secondhand smoke" as a matter of policy because it focuses attention on the nonsmoker who breathes in other people's smoke. Tobacco industry researchers introduced the term ETS in the early 1970s, believing it to be a more precise description of the earlier term "passive smoke." The term "tobacco smoke pollution" is used infrequently in public discussions but is notable for its formal use in indexing scientific literature by the U.S. National Library of Medicine.

Composition and Effects

In the 2000s, researchers know much about secondhand smoke's composition and effects. More than half of the smoke (by weight) from a burning cigarette is sidestream smoke. It is qualitatively similar to mainstream smoke: Both are produced by the combustion of tobacco and contain more than forty known or suspected human carcinogens, such as benzo(a)pyrene, 4-aminobiphenyl, and formaldehyde; irritants such as ammonia and nitrogen oxides; and compounds that affect cardiovascular function, such as smoke particles and 1,3-butadiene. The exact concentrations of compounds in mainstream and sidestream smoke differ quantitatively and change over time. Sidestream smoke, which is produced while the cigarette sits idle, actually contains more harmful compounds than mainstream smoke, because it is generated at a lower, "dirtier" burning temperature. An additional reason exhaled smoke and sidestream smoke differ is because exhaled mainstream smoke has been filtered through the smoker's lungs. The health effects of active and passive smoking are not necessarily identical—although they are similar for cardiovascular effects—because it matters how the smoke is breathed, and specifically where the smoke lands in the lungs.

Based on scientific research since the 1970s, experts have reached a number of conclusions about the effects of secondhand smoke on human health. In children and infants, it is a cause of respiratory symptoms and infections (for example, bronchitis), fluid in the middle ear, asthma, reduced lung function (difficulty breathing), sudden infant death syndrome (SIDS), and low birth weight. In adults, it is a cause of lung cancer, nasal sinus cancer, asthma, and cardiovascular disease. Evidence since the early 1990s suggests that secondhand smoke exposure also increases the risk of breast cancer. In policy as well as medical discussions of the health effects of secondhand smoke, lung cancer is generally the focus; however, mortality from cardiovascular disease is much greater. In the United States alone, every year the exposure of nonsmokers to secondhand smoke is estimated to cause 3,000 lung cancer deaths and between 35,000 and 62,000 cardiovascular disease deaths. The short-term effects on the cardiovascular system are also significant: In as little as thirty minutes, a nonsmoker's heart, blood, and blood vessels can be adversely effected similar to a pack-a-day smoker. The accumulation of research findings has established secondhand smoke as an important toxic air contaminant, and exposures should be prevented.

Rise of Secondhand Smoke as Public Health Concern

Secondhand smoke received relatively little attention from the public health community until the 1970s, when the environmental consciousness of the decade offered a new way to think about it. Cigarette smoke was characterized as one more form of pollution to which the public was involuntarily subjected. Thus, smokers were not just harming themselves but also those around them, implying a different sense of individual responsibility and a rationale for smoking restrictions. The U.S. tobacco industry was aware of the changing political climate as early as 1978, when a study for the Tobacco Institute, the industry's trade organization, described the antismoking movement's focus on passive smoking as "the most dangerous development to the viability of the tobacco industry that has yet occurred" (Roper Organization 1978). Although the tobacco companies had begun basic chemical research on secondhand smoke as early as the 1930s, their research efforts expanded greatly in the 1980s, following the increased scientific and public attention to the topic.

When the first **epidemiological** studies of secondhand smoke exposure were published in the early 1980s, an energized antismoking movement was quick to embrace results that indicated that secondhand smoke was not just a nuisance, but also a toxic air pollutant with serious health consequences. In one of the most influential studies, published in 1981, Japanese epidemiologist Takeshi Hirayama followed the health status of a group of 91,540 nonsmoking wives in Japan from 1966 to 1979 and concluded that those who had husbands that were heavy smokers showed a twofold increased risk of lung cancer compared to those who had nonsmoking husbands. This study and several others published that same year set off an international debate about the health consequences of secondhand smoke exposure. On one side were

epidemiological pertaining to epidemiology, that is, to seeking the causes of disease.

the tobacco industry and its allies who argued that risks were small and inconsequential; on the other side were public health advocates who argued that even small risks would produce a large amount of disease in a population that was widely exposed.

Exposure data and additional scientific evidence about the health effects of secondhand smoke accumulated, and in 1986 two scientific reviews were released that were a watershed for public policy: The U.S. Surgeon General's report, *The Health Consequences of Involuntary Smoking*, and the National Academy of Science's report, *Environmental Tobacco Smoke: Measuring Exposures and Assessing Health Effects*. The reports concluded that secondhand smoke caused lung cancer in healthy adult nonsmokers and respiratory symptoms in children. By the late 1980s the U.S. Environmental Protection Agency (EPA) had begun a risk assessment of the respiratory effects of secondhand smoke, including as evidence thirty epidemiological studies of women who had never smoked living with smoking and nonsmoking husbands. The 1992 final report, which took four years to complete, confirmed earlier findings and took the additional step of classifying secondhand smoke as a known human **carcinogen.** Specifically, it was categorized as a Group A carcinogen, which means that the weight of the evidence conclusively demonstrates that the substance causes cancer in humans.

carcinogen a substance or activity that can cause cancer. Cigarette smoking has been proven to be carcinogenic, that is, cancer causing.

As of 2004, the most comprehensive risk assessment of the health effects of secondhand smoke is the 1997 report by the California Environmental Protection Agency (CalEPA), *Health Effects of Exposure to Environmental Tobacco Smoke—Final Report*. In addition to examining respiratory effects, the CalEPA report is notable for assessing the effects on the cardiovascular system among other diseases. Other important summary statements, all of which have reached similar conclusions, include the 1998 report of the British Scientific Committee on Tobacco and Health and the 2002 World Health Organization's (WHO) International Agency for Research on Cancer (IARC) Monograph on Tobacco Smoking.

Sparking and Fueling Controversy

The tobacco industry was aggressive in its criticism of the scientific research and the risk assessments on secondhand smoke because the results threatened cigarette company profits by increasing the likelihood of regulation and litigation. Researchers know much about industry efforts to confuse the public and prevent meaningful regulation of secondhand smoke. The release of previously confidential internal company documents, which were produced during litigation, reveals that the tobacco industry challenged scientific research in two ways: with its own research and with public relations campaigns that portrayed unwelcome scientific findings as controversial and inconclusive and risk assessments as faulty and biased. Beginning in the 1980s, the industry organized teams of experts worldwide to promote the position that other pollutants (for example, mold or gases given off by carpets) were the cause of indoor air quality problems and that expensive building ventilation systems provided the best resolution, a position it still actively promotes. In short, the tobacco industry sought to avoid any legislative action on banning smoking indoors in the United States and internationally through an organized and constant effort to generate public doubt about mainstream scientific conclusions of secondhand smoke's health consequences.

All scientific research has limitations, and the tobacco industry was quick to use experts who could exploit weaknesses in the epidemiological studies or present invalid counter-results to muddy the conclusions. Many of the hired experts, who did not always disclose their financial ties to the industry, argued the studies were methodologically flawed. For instance, they claimed that researchers did not adequately account for the possibility that a current nonsmoker might have once smoked ("misclassification bias"). Critics also suggested numerous potential confounders, other factors that might explain why some people developed lung cancer, particularly that living in a household with smokers was associated with other lifestyle factors that could contribute to cancer, such as poor diet, lack of exercise, or hazardous occupation. Additionally, critics argued that studies did not adequately account for exposure to secondhand smoke; that is, the estimates of how much smoke people in the study were exposed to might be very uncertain. Peoples' exposures can be measured indirectly by measuring indoor air concentrations or by questionnaires and more directly by personal monitors or by the biochemical analysis of salvia, urine, and blood, which contain traces of tobacco smoke chemicals.

One of the first instances of the tobacco industry turning a weakness to its public relations advantage came in 1981. In the same year that the landmark Hirayama research was published, Lawrence Garfinkel of the American Cancer Society also published a study that found an increased risk of lung cancer from exposure to secondhand smoke. The results of the quantitative analysis, however, did not achieve what scientists call "statistical significance" (the probability that the result was a chance or random finding was greater than the conventional).

A powerful argument against smoking is the deadly effects of secondhand smoke on people who inhabit the same relative space as smokers. In this U.S. ad, showing a bride with a cigarette being extinguished on her chest, the caption reads, "Women married to a smoker have a 91% greater risk of heart disease. Secondhand smoke. Still want to breathe it?" AP/WIDE WORLD

The tobacco industry took the opportunity to misrepresent the study in a series of advertisements in major U.S. newspapers, making the blanket claim that Garfinkel's results were "insignificant." When the industry followed a similar public relations strategy in Australia, a successful lawsuit declared that the advertisements were false and misleading. Almost two decades later, when the World Health Organization released findings from a ten-year epidemiological study in 1998, the tobacco industry misrepresented the statistical significance of this study in the media. The tobacco industry had already worked behind the scenes to undercut the study using their network of scientific consultants in Europe, known as Project Whitecoat. This strategy included "seeding" the medical literature with letters to the editor from pro-tobacco scientists. In addition, industry development of a network of influential scientists and experts successfully hindered regulation of secondhand smoke in Latin America in 1990s.

The tobacco industry also sought to generate research that would be useful for defending against regulation and litigation. The research results were often published with minimal scientific peer review, as proceedings of symposia or sponsored publications rather than in medical journals. In the case of the landmark Hirayama study, the tobacco industry sought to refute Hirayama's results, nearly a decade later, by enlisting a different group of Japanese scientists to help produce a different "Japanese spousal study," but the company kept its involvement with Japanese researchers hidden. In another development, three of the United States tobacco companies—Philip Morris, R.J. Reynolds, and Lorillard—created a nonprofit research organization called the Center for Indoor Air Research in 1988, which funded around 244 published studies over the next decade. The organization was used by the tobacco industry to fund research to deflect attention from secondhand smoke as a significant indoor air pollutant and to produce data to challenge the findings on the health effects of secondhand smoke for political and legal purposes. The organization was disbanded in 1998, as part of the terms of the Master Settlement Agreement between state governments and the major tobacco companies. As of 2004, Philip Morris is funding scientific research through an openly sponsored external grants program it created in 2000.

The risk assessments of secondhand smoke published in the 1990s received a great deal of attention because of their implications for smoking restrictions policy. In an attempt to undercut the conclusions the of the 1992 EPA risk assessment, the tobacco industry successfully sued the EPA in federal court. The industry won a judgment in July 1998, on the grounds that the EPA had exceeded its authority established by Congress under the 1986 Radon Gas and Indoor Quality Research Act and did not follow proper administrative procedures, such as not properly representing industry interests in the review process. The judicial decision blocked the full implementation of the report as it related to lung cancer until it was overturned on appeal and formally dismissed by the U.S. District Court for the Middle District of North Carolina on 23 March 2003.

Secondhand smoke is one of the world's most significant air toxins and has been recognized as a global public health concern. Policy goals in the early twenty-first century focus on protecting children and workers, especially those in bars and restaurants where exposures are the highest. As of 2003, the American Lung Association considered nine U.S. states (Delaware, California, New York, Connecticut, Maine, Florida, Maryland, Utah, and Vermont) as having strong smoke-free air laws, and the number is expected to grow. The World Health Organization has made tobacco control and smoke-free environments one of its priorities, and legislatures are actively working on the issue worldwide.

See Also Antismoking Movement Before 1950; Antismoking Movement From 1950; Litigation; Lobbying; Medical Evidence (Cause and Effect); Product Design; Smoking Clubs and Room; Toxins.

▌JOSHUA DUNSBY

BIBLIOGRAPHY

Bero, Lisa. "Implications of the Tobacco Industry Documents for Public Health and Policy." *Annual Review of Public Health* 24 (2003): 267–288.

Brandt, Allan M. "Blow Some My Way: Passive Smoking, Risk, and American Culture." *Clio Medica* 46 (1998): 164–187.

Glantz, Stanton. Tobacco Scam: How Big Tobacco Uses and Abuses the Restaurant Industry. Online. 2002. Regents of the University of California. Available: <http://www.tobaccoscam.ucsf.edu>.

Glantz, Stanton A. et al., eds. *The Cigarette Papers*. Berkeley: University of California Press, 1996.

National Cancer Institute. *Health Effects of Exposure to Environmental Tobacco Smoke: The Report of the California Environmental Protection Agency.* Smoking and Tobacco Control Monograph no. 10. Bethesda, Maryland: U.S. Department of Health and Human Services, National Institutes of Health, National Cancer Institute. NIH Pub. No. 99-4645. Online. 1999. Available: <http://cancercontrol.cancer.gov/tcrb/monographs>.

Roper Organization. "A Study of Public Attitudes toward Cigarette Smoking and the Tobacco Industry. Volume 1." May 1978. Tobacco Institute Collection. Bates No.: TIMN0210766/0210819. Available: <http://legacy.library.ucsf.edu/tid/ebr72f00>.

U.S. Environmental Protection Agency. Environmental Tobacco Smoke Page. Online. Available: <http://www.epa.gov/iaq/ets/>.

Sexual Politics *See* Women.

Shamanism

The literature on shamanism and shamans has grown exponentially in recent years, and with it the uncritical tendency to call almost any religious specialist and curer of illness in traditional or tribal societies a "shaman."

The term derives from a language of tribal Siberia but has become widely adopted into many of the major languages. This is because of numerous perceived correspondences in different parts of the world in the shaman's role in his or her society and the belief systems in which shamans function. These include supernatural "election," initiatory ordeals, theories of illness and techniques of curing, the nature of the human soul, ecstatic trance, relations between the living and the dead, and relations between human beings and the spirits.

The differences are often overlooked. But so is one constant of shamanic practice in the traditional world: the ecstatic trance in which the shaman believes himself, with the concurrence of his social group, to project his soul on out-of-body journeys to the spirit world to seek knowledge and advice for the benefit of his society. The techniques by which shamans attain the desired visionary state vary. In Indian South and Middle America they most often involve plants with intoxicating qualities. Tobacco, *Nicotiana rustica*, is one of the most ancient and widely distributed of these "plants of the shaman."

Tobacco Shamanism in Native Cultures

One of the clearest indications of the importance of tobacco is seen in South American shamanism. The Matsigenka of eastern Peru, whose name means "people," call their shamans *seripi'gari*, "the one who is intoxicated by tobacco." This fact seems to set this Arawakan-speaking people apart from other Amazonian Indians who call their shamans by names that reflect their identity with the jaguar. But the difference is only apparent: It is intoxication with *Nicotiana rustica*, tobacco that facilitates the Matsigenka shaman's initiation, his recruitment of the jaguar and other animals as spirit helpers, and his ultimate transformation into the powerful predatory feline after death.

Throughout his lifetime, the Matsigenka shaman maintains a close relationship with spirits that reside in the mountains, where he subsists exclusively on tobacco, either in solid or liquid form, and where he keeps jaguars and pumas as pets or "dogs." Matsigenka shamans also own "jaguar stones," sacred rocks they receive during initiation and which they are obliged to feed regular rations of tobacco. These stones help them transform into helping jaguar spirits; conversely, if neglected that can cause their owner's death.

Tobacco is thus indispensable to Matsigenka shamanic ideology and practice. However, as is the case with tobacco shamanism among other indigenous groups, it is accompanied by other visionary or "hallucinogenic" plants, in this case *ayahuasca* (*Banisteriopsis* spp.), a Quechua term meaning "vine of souls," and the solanaceous *Brugmansia*, formerly known as *Datura arborea*, or tree datura. Scholars believe that the botanicals' effects on consciousness are enhanced or heightened by nicotine, which is known to activate norepinephrine and serotonin, hormones that occur naturally in the brain and that share the same structure with and are thus closely related to several plant "hallucinogens," including psilocybine and mescaline. *Ayahuasca* contains tryptamines; *Brugmansia* contains scopolamine, hyoscyamine, and noratropine as its principal **alkaloids** and belongs to the same nightshade family as the Nicotianas.

alkaloid an alkaloid is an organic compound made out of carbon, hydrogen, nitrogen, and sometimes oxygen. Alkaloids have potent effects on the human body. The primary alkaloid in tobacco is nicotine.

Full-scale tobacco shamanism, in which tobacco is the sole ecstatic intoxicant, to the exclusion of whatever other psychotropic species are available in the environment, is in fact rare. The Warao of the Orinoco Delta in Venezuela provide the best studied example. To trigger the ecstatic trance that is one of the cornerstones of shamanism everywhere, the religious specialists of the Warao inhale extraordinary quantities of smoke perfumed with powdered or crushed *caraña* (*Protium heptaphyllum*), a resinous gum, from cigars as long as 3 feet. Warao shamans share with the spirits irresistible hunger for tobacco as their essential spirit food and feel ill when it is not available—this despite the fact that the swampy environment of the delta precludes tobacco cultivation and necessitates its importation from the island of Trinidad and areas adjacent to Warao territory.

So deeply embedded in the intellectual culture is tobacco and its effects that the Warao have constructed a complex and highly sophisticated universe with "houses" and "bridges" of tobacco smoke ringed by sacred mountains of the world directions, whose ruling spirits the shaman keeps contended and favorably disposed toward the people with gifts of tobacco.

Origin and Diffusion of Tobacco

The genus *Nicotiana* consists of some sixty-four species, the great majority native to the Americas. Only about a dozen have ever been used for tobacco. Of these, only two, *Nicotiana rustica*, and its much milder sister species, *N. tabacum*, have achieved cultural importance and wide distribution—the former as a shamanic intoxicant and "spirit food," and the latter, also widely dispersed through the Indian Americas from the 1700s onward, mainly for recreational smoking. *N. tabacum* is the progenitor of modern commercial tobacco blends. The word

"Rustica" means "wild," but in fact both species are cultivated hybrids. Students of the genus *Nicotiana* believe them to have been in existence as long ago as 8,000 years, and that *Nicotiana rustica* and *N. tabacum* are among the first fruits of South American tropical agriculture.

Scholars also suggest that the high content of the alkaloid nicotine in *N. rustica*—as high as just under 19 percent in the leaf and several times that in the stems, compared with a low of 0.6 percent and a high of 9 percent in *N. tabacum*—and the resultant **physiological** and psychological effects to the point of addiction, account for its rapid and wide dispersal, and its quick adoption into visionary shamanism. To this discussion Johannes Wilbert adds the concept of "natural modeling," the close functional relationship between the botany and pharmacology of tobacco and the physical and mental effects of its principal alkaloid, nicotine, on the human organism. Thus, certain shamanic beliefs, behaviors, physical and mental effects, and "otherworldly" experiences such as death and resurrection may all be to some degree attributable to the actual experience of nicotine intoxication. For example, *N. rustica* is often found growing in disturbed soil, such as burial sites, which conforms to the widespread belief that tobacco is a gift of ancestors and ancestor spirits.

physiology the study of the functions and processes of the body.

Belief in tobacco as a life-giving force is confirmed by the proven effectiveness of fumigation with tobacco smoke, the application to the skin of tobacco poultices, and enemas against a variety of external and internal pathogens. The indigenous cultures believe the shaman's breath itself has healing powers; made visible by tobacco smoke it is doubly efficacious. The early European literature on Indian South America contains numerous eyewitness accounts of shamans repeatedly blowing clouds of tobacco smoke over the bodies of patients. European travelers might have interpreted this as "superstition," but there may in fact be a true biological effect of absorption of nicotine through the skin. In addition, nicotine may rid the skin of external pathogens and microorganisms, since botanists consider tobacco to be a powerful natural insecticide.

The Death and Resurrection Continuum

Shamans in various parts of the world undergo initiatory ecstasy and symbolic death and resurrection and repeat these traumatic experiences throughout their lifetime, often with the aid of a variety of "hallucinogenic" plants. This is evidenced by the name the Aztecs gave the very potent intoxicating seeds of the morning glory *Turbina corymbosa*. Tobacco is particularly well suited to dramatize the deathlike catatonic state and return to life that, thanks to the rapid biotranformation of nicotine in the body, is experienced by some candidate shamans in their initiatory rituals.

The Warao initiation ritual provides an instructive example of the death and resurrection continuum During initiation, the master shaman feeds the candidate enormous quantities of "spirit food"—tobacco—and this continues as he falls into a deathlike trance and commences an out-of-body journey through a series of obstacles where, to pass safely across an abyss filled with hungry jaguars, snapping alligators, and blood-thirsty sharks, he consumes more tobacco smoke. Demons wait to slash him with sharp-bladed spears and knives until finally he reaches a great tree that has a hole through its center with rapidly opening and closing doors. This is the threshold between life and death through which he must pass at precisely the right moment, lest he be crushed

SHAMANISM

By the time the early European explorers first arrived in the Americas, the many native cultures already had long traditions of chewing, smoking, and snuffing tobacco and other psychotropic plants ritually and socially in their pursuit of the supernatural. This colored engraving, printed in 1592 by Theodor de Bry, depicts a shamanic tobacco dance among the Tupinamba Indians of Brazil. RARE BOOKS DIVISION, THE NEW YORK PUBLIC LIBRARY, ASTOR, LENOX, AND TILDEN FOUNDATIONS

to death. When he fails to find his own bones among those of less fortunate predecessors, he returns and is restored to new life.

In another ritual, the Warao shaman, after smoking incessantly for an entire month, embarks on a frightful initiatory journey in which he is repeatedly "killed" by spirits and buried in coffins in foul-smelling swampy soil and beneath stone slabs. At last, he escapes and is restored to life. Initiation rituals like this one are virtually endless, not only in South America but wherever shamanism continues to be practiced throughout the world.

Transformation of Sight and Voice

Through constant use of tobacco, shamans are marked by bodily transformations of voice and sight. Initiatory trauma in tobacco shamanism often includes the tearing out of the vocal cords and the voice box. And, indeed, a dark-timbered and guttural singing and speaking voice is a mark of the tobacco shaman (as it is often also of the habitual smoker in the West). But nothing distinguishes the shaman more than "the paranormal sight which permits him to see the hidden and to foresee the future" (Wilbert 1986). In fact, tobacco shamans experience profound changes in their eyesight, including better near vision during the day and, conversely, better eyesight under advanced nicotine intoxication in the evening and at night, this being facilitated by the release of **epinephrine** or glycogen or both. Fully initiated tobacco shamans may actually experience more or less acute amblyopia, or dimness of vision, due to the action of nicotine on the pupil, but with improved night vision have no difficulty seeing in a world that is primarily black and white.

Full recovery of vision usually occurs several weeks after nicotine intoxication. "Once fully initiated and endowed with the appropriate voice and sight," concludes Wilbert, "the tobacco shaman displays other characteristics that give evidence of his position apart from normal human beings: He eats little, he suffers no pain, he cures the sick, and he is very combative." Like other characteristics this last is a universal of shamanism, regardless of the presence or absence of tobacco or any other visionary plant. Shamans have to be combative because they do battle against evil spirits, demons, sorcerers, witches, and predatory animals that threaten their clients.

In the American tropics shamans identify themselves with the jaguar, the most powerful of all the animals and also the one that, like the shaman himself, is not bound to a single **ecological** niche. Jaguars and jaguar transformation are widely associated with tobacco and nicotine intoxication. One kind of tobacco called *kumeli*, literally means "tiger (jaguar) tobacco," which the shamans of the Carib-speaking Akawaio in Guiana smoke to achieve jaguar-like combativeness in order to drive away and destroy evil spirits.

epinephrine also called adrenaline, a chemical secretion of the adrenal gland. Epinephrine speeds the heart rate and respiration.

ecology the interrelationships of a natural environment. For example, the ecology of a forest includes animals, plants, water, atmosphere, weather, and land forms.

Tobacco as Sacramental Food

Tobacco is an appetite suppressant, and tobacco shamans eat little. At least in part this must explain why indigenous peoples generally classify tobacco as "food," regardless of the method of ingestion, and attribute to the spirits the same hunger for it as experienced by humans, and, conversely, the same feelings of satisfaction. The idea that in making a gift of tobacco to humanity, the spirits somehow forgot to keep some for themselves, thus making themselves dependent on human beings for their essential nourishment, must have originated close to its initial cultivation, experimental use, and subsequent dispersal, perhaps in north central Peru or the valleys between Peru and Ecuador. Similar versions of this story are shared over wide areas, including North America. In these beliefs, not just the shaman's own helping spirits—many of which inhabit his own body—but a vast company of spirits and deities scattered throughout the environment and the upper- and underworld are wholly dependent on tobacco as nourishment. According to shamanic tradition, these spirits thus must rely on humans as the only producers of the sacred sustenance, in exchange for which they bestow health, rain, fertility, and other benefits.

Tobacco is also one of the most toxic "foods" known: One or two drops of an extract of the nicotine in a single cigar placed on the skin or the tongue is sufficient to kill an adult. That shamans do not share this dire fate presumably has to do with the rate and method of absorption into the gastrointestinal tract; in addition, the shaman culture holds the firm conviction that tobacco is a beneficial and very sacred bounty originating with the gods and spirits, which may play a role in the effect of nicotine on their bodies.

South American shamans absorb nicotine through smoking, sucking, drinking, licking, smoking, and snuffing, the first being by far the most common across the continent and northward across Central America into Mexico and North America, as far north as the sub-Arctic and Arctic. Tubular pipes dating to the second and first millennium B.C.E. have been excavated in Mexico as well as in California, but smoking is likely to be much older than that. At some point tobacco made its appearance among the indigenous peoples of the Northwest Coast and Alaska, where by the early nineteenth-century pipes of wood or stone had evolved into true works of art. The method of diffusion that far north, whether overland from tribe to tribe, or by trade through Russian traders and colonists, is unknown. What scholars do know is that both ritual and recreational smoking was widely practiced in Siberia by the eighteenth century, and that at some point, both *N. rustica* and *tabacum* reached Nepal and the Himalayas. In Nepal, *N. rustica* became assimilated into shamanism as a ritual intoxicant consecrated to the Hindu god Shiva, while *N. tabacum* joined the company of indigenous and introduced recreational "drug" plants.

See Also Hallucinogens; Mayas; Native Americans; Social and Cultural Uses.

▪ PETER T. FURST

BIBLIOGRAPHY

Baer, Gerhard. "The Role of Tobacco in the Shamanism of the Matsigenka, Eastern Peru." *Acta Americana* 3, no. 2 (1995): 101–116.

Butt Colson, Audrey. "The Akawaio Shaman." In *Carib-speaking Indians: Culture, Society, and Language*. Edited by Ellen B. Basso. Tucson: University of Arizona Press, 1977.

Müller-Ebeling, Claudia, Christian Rätsch, and Surendra Bahadur Shahi. *Shamanism and Tantra in the Himalayas*. Rochester, Vt.: Inner Traditions, 2000.

Wilbert, Johannes. "Tobacco and Shamanistic Ecstasy among the Warao Indians of Venezuela." In *Flesh of the Gods: The Ritual Use of Hallucinogens*. Edited by Peter T. Furst. New York: Praeger, 1972.

———. *Tobacco and Shamanism in South America*. New Haven, Conn.: Yale University Press, 1986.

Sharecroppers

sharecropping a form of agricultural labor that gained popularity after the Civil War. Laborers, usually families, lived and worked on land belonging to a proprietor. They grew staple crops like tobacco and cotton. Rather than regular cash wages, they were paid with shares of the crop at harvest time.

In the 2000s, tobacco production is fully mechanized. Herbicides destroy weeds before they are able to affect the plant, and the metal claws of mechanical harvesters collect ripened tobacco leaves as they roll down fields row by row. But for almost a century, sharecroppers predominated in the arduous and unrelenting cultivation of Bright leaf tobacco in the American South, from planting and weeding to harvesting and curing.

A system of labor in which workers received a share of the crop as compensation, **sharecropping** emerged after the Civil War (1861–1865)

In the 2000s, tobacco production is fully mechanized. But for almost a century after the Civil War, tobacco cultivation in the American South, especially Virginia and North Carolina, was handled mainly by sharecroppers. This photograph was taken in 1939 in North Carolina. LIBRARY OF CONGRESS

in the Bright leaf tobacco belts of Virginia and North Carolina. Favored by manufacturers for its mild taste, Bright leaf, or **flue-cured tobacco,** turned a golden hue from the intense heat of flues in the curing barn. Sharecropping represented a new way for landowners to control the labor of former slaves. Devastated by defeat and stripped of their most valuable investment, former slave owners were suffering financially. Lacking cash to pay wages, they adopted the practice of financing tobacco production by engaging in a lien on the prospective crop. With their aspirations for land unfulfilled, former slaves had no choice but to accept the new form of labor.

flue-cured tobacco also called Bright Leaf, a variety of leaf tobacco dried (or cured) in air-tight barns using artificial heat. Heat is distributed through a network of pipes, or flues, near the barn floor.

The Sharecropping System

Sharecropping contracts varied state by state and farm by farm. However, certain common features characterized them. Little more than a nod sealed most agreements between sharecroppers and landlords. Each year in December the head of a sharecropping family typically committed the labor of the entire family to cultivate about three to six acres of tobacco. Sharecroppers usually worked for no more than a half share of the crop, which they collected only after cultivating and harvesting. In effect, they were advancing a season's worth of labor to the landlord before receiving pay. Landlords furnished sharecroppers with housing, mules, seed, fertilizer, and tools, and extended them credit for food and necessities, usually through country stores. Sharecroppers also could keep gardens. At the end of the season, the landlord paid sharecroppers after deducting what they owed for living expenses, with interest. Often, the season's labor offered little actual profit and the possibility of debt.

SHARECROPPERS

tenant farmers landless farmers who rented acreage from landowners. The tenant family usually moved to a house on the rented land where they lived and worked. The rental was payable in cash or sometimes a specified amount of produce. The tenant often owned draft animals and implements and had established credit. Tenants were typically more independent than sharecroppers and occupied a higher place in the hierarchy of rural America.

During the late nineteenth century, agricultural depression contributed to an increase in the number of sharecroppers and **tenant farmers** as many white farmers lost their land. In 1890, more than 40 percent of farmers in North Carolina and 30 percent of farmers in Virginia occupied these two farming classes. Tenancy also had become common in the flue-cured culture after the Civil War. Unlike sharecroppers, tenant farmers paid the landowner for renting and working a plot of land with either cash or a share from the proceeds of the year's crop.

Despite differences in their tenure status and location, the way that farmers raised tobacco changed very little across time and space. Whether sharecroppers, tenant farmers, or small landowners, they relied on the labor of all family members and neighbors in all the states that produced Bright leaf tobacco, which by the twentieth century included South Carolina, Georgia, and Florida. Between January and December, they planted seeds, transplanted tobacco seedlings, and eliminated weeds and pests as the plants matured in the fields. Beginning in July, they started priming, the task of removing leaves individually as they ripened. Each day, women performed the delicate task of stringing the harvested tobacco to sticks before it was hung in the curing barn. After filling the barn, the men of the family spent several nights at the barn to control the temperature of the heated air as it cured the tobacco. After harvesting and curing, they prepared the leaves for market by grading them.

Changes to the System

During the 1930s, sharecroppers felt the brunt of low prices brought about by the Great Depression. As tobacco prices plummeted, landlords often absorbed their loss in income by passing debt onto their sharecroppers or dismissing them. The Agricultural Adjustment Act (AAA), a New Deal policy under President Franklin D. Roosevelt, revived Bright tobacco farming through a system of price supports and acreage reduction. While the AAA cotton program contributed to the displacement of sharecroppers and tenant farmers, the tobacco program helped stabilize sharecropping and tenancy. The number of sharecroppers and tenant farmers reached its height in the late 1930s, comprising an estimated 48 percent of all farmers in tobacco regions in 1937, according to the President's Committee on Farm Tenancy.

The 1950s witnessed the first significant dip in the number of sharecroppers. Federal farm policies and another slump in agricultural prices encouraged landowners to cut back acreage. Between 1940 and 1959, the number of sharecroppers and tenant farmers in Wilson County, North Carolina, a major tobacco-producing area, fell by nearly 40 percent, from 3,027 to 1,840. By the 1960s, the number of sharecroppers dropped even more dramatically as flue-cured tobacco became mechanized. In 1978, only 409 sharecroppers and tenant farmers remained in Wilson County. In addition to acreage reduction, migration to urban areas and the mechanical transformation of tobacco production contributed to the decline of sharecropping. Sharecroppers, once **ubiquitous** in Wilson County and across the tobacco South, are now as rare as agricultural machinery once was.

ubiquitous being everywhere; commonplace; widespread.

■ ADRIENNE PETTY

BIBLIOGRAPHY

Badger, Anthony. *Prosperity Road: The New Deal, Tobacco, and North Carolina*. Chapel Hill: University of North Carolina Press, 1980.

Daniel, Pete. *Breaking the Land: The Transformation of Cotton, Tobacco, and Rice Cultures since 1880*. Urbana and Chicago: University of Illinois Press, 1985.

Kerr-Ritchie, Jeffrey R. *Freedpeople in the Tobacco South: Virginia, 1860–1900*. Chapel Hill: University of North Carolina Press, 1999.

Prince, Eldred E. Jr., with Robert R. Simpson. *Long Green: The Rise and Fall of Tobacco in South Carolina*. Athens: University of Georgia Press, 2000.

Tilley, Nannie May. *The Bright Tobacco Industry, 1860–1929*. Chapel Hill: University of North Carolina Press, 1948.

United States Department of Commerce. Bureau of the Census. *Sixteenth Census of the United States: 1940, Agriculture*. Volume I, State Reports, Part 3, Statistics for Counties. Washington: U.S. Government Printing Office, 1942.

———. *United States Census of Agriculture, 1959*. Volume I, Counties, Part 26, North Carolina. Washington: U.S. Department of Commerce, 1961.

———. *1978 Census of Agriculture*. Volume I, State and County Data, Part 33, North Carolina. Washington: U.S. Department of Commerce, 1981.

Slavery and Slave Trade

According to the most recent estimates of the Atlantic slave trade, over 10 million Africans were forcibly exported to the Americas between the 1440s and 1860s (Klein 1999). The major reason for bringing all of these Africans to the Americas was the production of cash crops to make profits and satisfy European tastes. Tobacco was the first exotic luxury in the Americas to become an item of mass consumption.

From Servants to Slaves

At the beginning of the seventeenth century, the typical tobacco-field worker was an English or Irish **indentured servant.** By the century's end, it was an imported African slave. The early story of tobacco is one of this shift in labor relations.

Although tobacco was a familiar, even sacred, plant to Native Americans, the combination of European colonialism and changing consumer choices transformed it into one of the most lucrative commodities in the New World. In the seventeenth century the English colonies of the Chesapeake Bay (Virginia and Maryland) were the greatest tobacco-producing regions in the British Empire. The annual export of tobacco leaf increased from 65,000 pounds in the 1620s to 40 million pounds by 1700. British seamen, planters, and adventurers developed a taste for pipe smoking, chewing, and taking **snuff,** and tobacco gradually became popular among the general populace; some say this is because of tobacco's addictive qualities, while others claim it is because tobacco appeased appetites and was energizing. By the end of the century, annual tobacco consumption in England and Wales peaked at over

indentured servant a person who agreed to work for another for a specified term (usually a few years) to satisfy a financial obligation. During the American colonial period, immigrants sometimes paid their passage with indentured service.

snuff a form of powdered tobacco, usually flavored, either sniffed into the nose or "dipped," packed between cheek and gum. Snuff was popular in the eighteenth century but had faded to obscurity by the twentieth century.

SLAVERY AND SLAVE TRADE

This engraving (c. 1730) shows a relaxed Virginia planter watching his slaves work in the tobacco field. Also depicted are the planter's manor house, hogsheads containing packed tobacco, and sailing ships near the wharf and in the harbor. The Chesapeake Bay colonies were a major tobacco-producing region in the early British Empire. THE GRANGER COLLECTION

two pounds per capita (Goodman 1993). This consumer choice helped spawn the enslavement of Africans.

The development of British capitalist agriculture, together with the removal of laboring people from the countryside during the seventeenth century, created a surplus population that resulted in increased immigration to New World colonies. Some immigrants were unwilling convicts, but many were free laborers who contracted for several years of service in exchange for transportation and post contract freedom dues, such as money, land, and supplies. Between 1630 and 1680, 75,000 indentured laborers entered the English mainland colonies, around three-fourths of whom were males (Blackburn 1997).

The profitability of tobacco, together with high mortality rates and abundant lands in the Chesapeake Bay region, however, created a labor shortage for tobacco planters. As a result, they made an economic decision to follow the existing model of the sugar industry, which was dependent on slavery, elsewhere in the New World. Moreover, there already existed an operative slave trade, an African slave promised a much longer work life than an indentured servant, and Africans could be treated more harshly because they were considered **heathen** and beyond the pale of Christian civilization.

The English slave trade began in earnest with the Royal African Company in 1672. The constant demand for slave labor, however, proved too much, and, after 1689, the slave trade was managed by independent traders. About 20,000 African slaves were brought to the English mainland colonies during the seventeenth century, most of them in the 1680s and 1690s. Around one-third of the slaves were women and girls (Blackburn).

Although English and Irish servants were of a different provenance to African slaves, both groups were unfree laborers who forged a common front. The harboring of runaways, sexual unions between African men and English or Irish women, and interracial conviviality resulted in the passage of numerous laws prohibiting such actions by the Virginia legislature. These laws do not appear to have been completely effective. African slaves and Irish servants joined Nathaniel Bacon's rebellion against English colonial rule in 1676, with over one hundred of them refusing to surrender until they were guaranteed their freedom. They eventually failed. Moreover, colonial laws were passed in the 1660s and 1670s to codify African slavery in the Chesapeake region.

heathen any person or group not worshiping the God of the Old Testament, that is, anyone not a Jew, Christian, or Muslim. May also be applied to any profane, crude, or irreligious person regardless of ethnicity.

The Tobacco Revolution

The story of tobacco and slavery in the eighteenth century is one of a major regional transformation wrought by massive slave imports, a new planter regime, and the making of African American slave culture. In the Chesapeake region tobacco production increased from 30 million pounds in 1710 to over 100 million pounds by 1775 (Blackburn). Expanding consumer demand fueled the increase in production. Even though there appears to have been a decline in individual consumption, the general trend was upward. Western European consumption increased from about 70 million pounds of leaf in 1710 to 120 million pounds at the end of the century (Goodman). Furthermore, American plantation products like rum and tobacco became an acceptable means of exchange for slaves on the western and southwestern African coasts during the zenith of the Atlantic trade.

The increase in tobacco production resulted in greater demand for African slaves in the Chesapeake area. In 1700 there were 22,000 Africans in the mainland colonies, with 13,000 Africans in the Chesapeake district, but by 1760 there were 327,000 Africans in North America, with more than half laboring in the Chesapeake region (Thorton 1998). Many of the new imports to the Chesapeake came directly from the Bight of Biafra on Africa's western coast. It has been estimated that between 1710 and 1760, Igbo or Biafrans constituted around 40 percent of the total number of slaves brought to Virginia. Scholars disagree on why planters chose Igbo or Biafran slaves. Some argue for planter indifference, others for perceptions of slave resistance, and still others for physical abilities (Gomez 1998). Although some historians also have argued that slaves in Africa were chosen by planters for their intimate knowledge of crops like rice and tobacco, it is clear that physical strength was the most important criterion in the buying of slaves.

There is general agreement on the gender of slave imports; slave men outnumbered slave women by about two to one. This is because

SLAVERY AND SLAVE TRADE

plantation historically, a large agricultural estate dedicated to producing a cash crop worked by laborers living on the property. Before 1865, plantations in the American South were usually worked by slaves.

Creole originally, a person of European descent born in the Spanish colonies. Later, the term was applied to persons of mixed European and African descent. As an adjective, it can describe admixtures of European and African cultural components such as language, cookery, and religion.

males were more likely to be marketed in African societies, and planters in the Americas sought strong adult males for cash crop production. The tobacco region saw an early shift from importing male slaves to the natural reproduction of slaves. The combination of a more temperate climate, less destructive work patterns, and a more established planter society led to higher reproductive rates in the tobacco fields of the Chesapeake district than in the rice **plantations** of the Lowcountry and the sugar plantations of the Caribbean. By mid-century, it has been estimated that four-fifths of the slaves in the Chesapeake area were native-born (Berlin 1998). By the time of the American Revolution, the Chesapeake region was unique for its enslaved **Creoles** and a more equal gender ratio. In contrast, imported slave men continued to predominate in other cash crop regions—such as rice, indigo, and sugar—in the mainland and Caribbean colonies. This made the tobacco region unique among New World slave societies.

Unlike the planters in the Caribbean colonies, American planters remained in the colonies, made large fortunes from slave work, and built impressive homes. Most important, they formed an "interlocking directorate" through marriage partnerships, business interests, and political representation (Berlin 1998). These new tobacco lords fashioned a new paternal order. Small plots worked by planters, farmers, and their families were replaced by larger holdings worked by gangs of slaves supervised by overseers. There were more workdays, longer hours, closer supervision, and harsher punishments through the whip, manacle, and branding iron. Women and children filled the new gangs in the tobacco fields. Indeed, great financial rewards came from the production and reproduction of slave women. Thomas Jefferson observed: "a woman who brings a child every two years [is] more profitable than the best man on the farm [for] what she produces is an addition to the capital, while his labor disappears in mere consumption" (Berlin 1998). Many slave women and children were confined to routine agricultural, reproductive, and domestic work. In contrast, the work of some slave men became more diverse as tobacco planters switched to cereal grain production and livestock farming. Indeed, some slaves became skilled artisans in this new agricultural world. Slaves like "Jem" were reputed to "do any kind of smith's or carpenter's work" and "any kind of farming business" (Berlin 1998).

This tobacco revolution wrought resistance by slaves. During the early eighteenth century, planters uncovered several conspiracies and plots. Moreover, many slaves temporarily escaped slavery as suggested by thousands of advertisements for so-called runaways placed by slave owners in colonial newspapers. Furthermore, planters complained unceasingly to each other and in newspaper articles about productivity problems caused by slave sickness, work refusal, deliberate misunderstanding, breaking tools, and direct challenges to their authority. To them, the problem was how best to manage slaves; to the slaves, it was how best to carve out niches of freedom in a coercive slave regime.

But the most important collective expression of this resistance to the system of slavery and dehumanization in the tobacco kingdom was the development of slave culture. It was born from a complicated **synthesis** between the African past and the Chesapeake present. Moreover, it was made in the tobacco fields as well as during communal moments away from slaveholders. The building of slave quarters on

synthesis the blending of several elements into a coherent whole.

SLAVERY AND SLAVE TRADE

plantations removed the immediate supervision of the planter's family. The spread of tobacco cultivation for fresh soils entailed the constant movement of slaves into new areas. This growing slave network made connections, friendships, and kinships. One planter's complaint about the "continual concourse of Negroes on Sabboth and holy days meeting in great numbers" illustrates the extent of this broader slave community (Berlin 1998). Moreover, language, recreation, and spirituality provided the basis for a common cultural identity. The language of tobacco slaves had multilingual roots reflecting both African and British tones and idioms. Recreational activities such as music, song, dice, and athletics along with personal styles of dress, headgear, haircuts, and social interaction (for example, sucking teeth to demonstrate frustration) were crucial ingredients for cultural survival in a dehumanizing system. Even though renamed by their owners, slaves often clandestinely retained their own names. Births, marriages, and deaths reflected ancestral African customs either remembered by older generations or reintroduced by newer African imports. As many historians and anthropologists know, rural communities have long memories. Slave religion was dominated by African spiritual forms until the late eighteenth century when evangelical Christianity began to make its mark in the slave quarters.

Expansion and Contraction

The story of tobacco and slaves between the American Revolution and the Civil War is a story of contradictory pulls and tensions. Among the most important were expanded single-crop production and **diversification**, unique **manumission** rates, and new internal slave trading.

diversification in agriculture, avoidance overdependence upon one crop by producing several different crops.
manumission the act of voluntarily emancipating (freeing) a slave.

Tobacco continued to draw African slaves into the internal regions of Virginia and beyond from the late eighteenth century onward. In Pittsylvania county the slave population grew from 271 in 1767 to 4,200 by 1800 (Kerr-Ritchie 1999). Between 1720 and the Revolution, more than 15,000 Africans were transported into the internal regions. Over the next few decades, tobacco and slavery spread through southwest Kentucky and northeastern Tennessee. According to the U.S. Census, tobacco production in Kentucky amounted to over 108 million pounds in 1859, or 25 percent of the national total (Robert 1938). Much of this tobacco was worked by the 225,483 slaves listed in the state (University of Virginia 1998). This was also the major slaveholding region in the state.

Although tobacco and slaves continued to dominate the political economy of the Chesapeake, the region also experienced a marked degree of diversification from the early national period onward. This was due to a combination of factors including soil erosion, poor quality leaf, and depressed European markets due to the French revolutionary wars (1789–1815). Chesapeake planters switched from the old staple to crop mixtures of cereals and livestock according to both market prices and increasing demands from growing urban areas like Baltimore, Washington, Norfolk, and Richmond. Some slaves produced crops for these markets while others grew their own crops and marketed these themselves.

From the 1830s onward, the tobacco industry took off in Virginia and North Carolina. According to the 1860 U.S. Census, tobacco

SLAVERY AND SLAVE TRADE

capital investment spending money to make an enterprise more efficient and more profitable. For example, modern tobacco growers make capital investments in advanced machinery to lower production costs.

plug a small, compressed cake of flavored tobacco usually cut into pieces for chewing.

republicanism originally a government without a monarch; republicanism has come to mean a form of representative democracy responsible to the electorate.

manufacturing in the nation accounted for $9 million of **capital investment** worth nearly $22 million. Slaves provided the labor in these tobacco factories. The same federal returns counted 12,843 factory workers in Virginia and North Carolina, over four-fifths of whom were males (Robert). Manufacturers owned 48 percent, while slaveholders who hired out their slaves owned 52 percent. Their tasks included stemming, dipping, twisting, lumping, and prizing leaf. The chief product was **plug** or twist for chewing tobacco. One major difference between slaves engaged in rural production and those who did domestic work was a degree of quasi-freedom experienced by the latter group through work hiring, wage payment, and self-support. On 18 December 1856 the *Daily Dispatch* of Richmond observed: "For some years past our tobacco manufacturers have been compelled, in order to secure labor, first to purchase the consent of the negroes to live with them, and then to hire them of their owners." Consequently, they "have allowed the servants to dictate their own terms as to the amount of board money to be given, the extent of daily labor to be performed, and the price to be paid for such overwork as they may feel disposed to do" (Robert). But they still inhabited a slave society. The volume of newspaper and private reports on slaves shirking work, feigning illness, stealing goods, and torching warehouses points to the limitations of freedom in the tobacco factory. In February 1852 the hired slave Jordan Hatcher killed the factory overseer William Jackson with an iron bar because of an attempted whipping (Robert).

The manumission of slaves in the aftermath of the Revolution was particularly pronounced in the tobacco region. By 1810 there were over 108,000 free black people in the Upper South, most of whom lived in the urban areas of tidewater Virginia and Maryland (Berlin 1998). This regional concentration of free blacks continued. By mid-century, over 85 percent of all free blacks lived in the major tobacco region of Maryland, Virginia, North Carolina, and Delaware (Morgan 1992). The reasons for this expansion are complex. Economic decline and diversification reduced the need for slave labor, and manufacturing in Baltimore and Richmond encouraged the development of free wage labor. Furthermore, revolutionary **republicanism** and evangelical Christianity encouraged a natural rights philosophy in which all men were created equal.

At the same time, large numbers of slaves were being transported south. After the abolition of the Anglo-American transatlantic slave trade in 1808, there emerged an internal slave market. Comparisons of the U.S. Census returns for slaves by state suggest that between 1810 and 1860, nearly half a million slaves were exported from Virginia (Tadman 1996). Most slaves ended up working in the cotton states of the Lower South. Although it is difficult to determine exactly how many left the tobacco fields, it is unlikely that this region was not affected. On the one hand, tobacco slaves were needed for their labor and so were less likely to be sold. Conversely, many planters reduced their slave dependency by switching crops and selling their surplus slaves. Other planters immigrated with their slaves to the newer regions.

These contradictions had a major impact on slave culture. By the 1850s, slaves had deep roots in the tobacco region. They had Africanized the region, while the region had Americanized them. Slave culture was nurtured by manumission since the civil restrictions on people of African descent required the interaction of free people with slaves on all

After the Civil War, African Americans in tobacco-producing regions were making the transition from slave labor to wage labor, sharecropping, and tenant farming. These workers in 1899 sort tobacco at the T. B. Williams Tobacco Company in Richmond, Virginia. LIBRARY OF CONGRESS

social levels. Moreover, the diversification of slavery entailed many slaves, especially men, to work, travel, and experience the world beyond tobacco and slavery. On the other hand, the expansion of tobacco and slavery, together with the development of the internal slave trade, undermined these cultural foundations by breaking up families and disrupting kinships. Even though slaves met this challenge through "broad marriages"—a marriage between people from different plantations or environments—and extended familial relations beyond the plantation, the impact of the slave trade must have been devastating. In interviews conducted over eighty years later, virtually every surviving ex-slave recalled the times when "Dey carry you down south" (Perdue, Barden, and Phillips 1976).

Emancipation

Tobacco and slavery underwent a revolutionary transformation during the Civil War era. For over two centuries, tobacco and slaves had ruled the political economy of the Upper South. In 1859 the Virginia piedmont was the primary tobacco region, returning over 120 million pounds, or over one-fourth of the U.S. total (Kerr-Ritchie). Moreover, slaves and slave owners were concentrated in the tobacco-producing regions east of the Chesapeake tidewater region. With the advent of secession and Civil War, this old regime broke down. The western part of Virginia, with almost no tobacco production and few slaves, stayed in the Union and formed its own state in 1863. The two major tobacco-producing states of Maryland and Kentucky also stayed in the Union. Moreover, the eastern front was largely fought in the old Chesapeake region with destructive consequences for tobacco and slavery. The Confederate enlistment of many planters and farmers left slaves to their own devices. Other slaves in the tobacco belt self-emancipated themselves toward Union lines in northern and southern Virginia. Many of these slaves enlisted in the fight against slavery. Some 43,375 men of African descent from the four major tobacco states of Virginia, Kentucky, Maryland and North Carolina fought for the Union military (Berlin 1992).

sharecropping a form of agricultural labor that gained popularity after the Civil War. Laborers, usually families, lived and worked on land belonging to a proprietor. They grew staple crops like tobacco and cotton. Rather than regular cash wages, they were paid with shares of the crop at harvest time.

The official end of armed hostilities and the legal end to slavery in 1865 brought three fundamental changes to the old tobacco regime. Although tobacco planters continued to dominate state politics, they never regained their previous political and legal influence in national affairs. Furthermore, emancipation entailed a shift from the supervision of slaveholding plantations and farms to semiautonomous production through wage labor, tenant farming, and especially family-based **sharecropping**. Finally, former slaves began the slow process of carving emancipation in their own image through the establishment of visible institutions like marriage, church, school, and the ballot box. Meanwhile, the old dominion was rapidly being replaced by a new dominion of tobacco capitalists, cigarette consumers, and younger working generations in the fields and factories.

See Also Africa; Caribbean; Chesapeake Region; Christianity; Labor; Plantations.

■ JEFFREY R. KERR-RITCHIE

BIBLIOGRAPHY

Berlin, Ira. "The Tobacco Revolution in the Chesapeake." In his *Many Thousands Gone: The First Two Centuries of Slavery in North America*. Cambridge, Mass.: Belknap Press of Harvard University Press, 1998. Pp. 109–141.

Berlin, Ira, et al. *Slaves No More: Three Essays on Emancipation and the Civil War*. New York: Cambridge University Press, 1992.

Blackburn, Robin. "The Making of English Colonial Slavery." In his *The Making of New World Slavery: From the Baroque to the Modern, 1492–1800*. London: Verso, 1997. Pp. 217–276.

Gomez, Michael A. *Exchanging Our Country Marks: The Transformation of African Identities in the Colonial and Antebellum South*. Chapel Hill: University of North Carolina Press, 1998.

Goodman, Jordan. *Tobacco in History: The Cultures of Dependence*. London: Routledge, 1993.

Kerr-Ritchie, Jeffrey R. *Freedpeople in the Tobacco South: Virginia, 1860–1900*. Chapel Hill: University of North Carolina Press, 1999.

Klein, Herbert S. *The Atlantic Slave Trade*. New York: Cambridge University Press, 1999.

Morgan, Edmund S. *American Slavery, American Freedom: The Ordeal of Colonial Virginia*. New York: Norton, 1975.

Morgan, Lynda J. *Emancipation in Virginia's Tobacco Belt, 1850–1870*. Athens: University of Georgia Press, 1992.

Morgan, Philip D. *Slave Counterpoint: Black Culture in the Eighteenth-Century Chesapeake and Lowcountry*. Chapel Hill: University of North Carolina Press, 1998.

Perdue Jr., Charles L., Thomas E. Barden, and Robert K. Phillips, eds. *Weevils in the Wheat: Interviews with Virginia Ex-Slaves*. Charlottesville: University Press of Virginia, 1976.

Robert, Joseph C. *The Tobacco Kingdom: Plantation, Market, and Factory in Virginia and North Carolina, 1800–1860*. Durham, N.C.: Duke University Press, 1938. Reprint, Gloucester, Mass.: P. Smith, 1965.

Tadman, Michael. *Speculators and Slaves: Masters, Traders, and Slaves in the Old South*. Madison: University of Wisconsin Press, 1996.

Thorton, John Kelly. *Africa and Africans in the Making of the Atlantic World, 1400–1800*. New York: Cambridge University Press, 1998.

University of Virginia. Geospatial and Statistical Data Center. United States Historical Census Data Browser. Online. 1998. University of Virginia. Available: <http://fisher.lib.virginia.edu/census>.

Walsh, Lorena S. "Slave Life, Slave Society, and Tobacco Production in the Tidewater Chesapeake, 1620–1820." In *Cultivation and Culture: Labor and the Shaping of Slave Life in the Americas*. Edited by Ira Berlin and Philip D. Morgan. Charlottesville: University Press of Virginia, 1993. Pp. 170–199.

Walvin, James. "The Indian Weed: Tobacco." In *Fruits of Empire: Exotic Produce and British Taste, 1660–1800*. New York: New York University Press, 1997. Pp. 66–88.

Smoke *See* Chemistry of Tobacco and Tobacco Smoke; Cigarettes; Product Design; Tar; Toxins.

Smoking Clubs and Rooms

There is something that is quintessentially Victorian about the image of the smoking room or the club. Middle-class smokers are known in particular to have celebrated their smoking habits, their refinement of taste, and their individual discernment. Countless pamphlets, books, poems, and periodical articles acted as etiquette guides for the aspiring connoisseur. Various anecdotal "whiffs" and "pipefuls" were presented as amusing relief for busy city gentlemen who sought solace in their tobacco in the smoking rooms of clubs, hotels, and bars in the great metropolitan centers of London, New York City, and Montreal. Nowhere is this mood better encapsulated than in the English novelist Ouida's *Under Two Flags: A Story of the Household and the Desert* (1867):

> . . . *that chamber of liberty, that sanctuary of the persecuted, that temple of refuge, thrice blessed in all its forms throughout the land, that consecrated Mecca of every true believer in the divinity of the meerschaum, and the paradise of the narghilé—the smoking-room.*

Early Victorian proscriptions against smoking in the presence of women had encouraged the establishment of exclusively male rooms where men could retire after dinner, though it was the successor too of the salons and coffee houses of eighteenth-century civil society. The smoking room at the gentleman's club was a place for escape and conversation. It was a glorified masculine space (only a handful of female smoking clubs were established), mythologized as an idealized smoking utopia of rest, meditation, and sheer dedicated concentration on the joys of one's cigar.

Often smoking rooms were built in individual homes. Most famous of these was that of Queen Victoria's husband, Prince Albert, whose smoking room door was apparently the only one that did not bear both the legend "V & A," a solitary "A" sufficing. Like the club, the

SMOKING CLUBS AND ROOMS

The smoking room of an Atlantic steamer, 1940. Drawn by T. De Thulstrup. © CORBIS

smoking room was similarly a place of escape but it was also a space upon which the gentleman was expected to stamp his individuality. The smoking room was to be filled with all the trophies of an adventurous life, as well as the paraphernalia of smoking idiosyncrasy. Mirroring the attention to smoking detail found in Sir Arthur Conan Doyle's detective, Sherlock Holmes, J. M. Barrie, in *My Lady Nicotine* (1890), wrote separate chapters on each of the items found in his smoking room: his favorite blend of tobacco, his favorite pipes, his tobacco pouch, his smoking-table, and even his favorite smoking companions.

Smoking-room culture was not restricted to the English-speaking world. When Chichikov visits a minor Russian landowner, Nozdrev, in Nikolay Gogol's *Dead Souls* (1842), he is treated to a tour of his exhibition of tobacco pipes and other smoking instruments, all testaments to his individual character and masculinity.

One must not be too literal in what one considers a smoking room. Much of the importance attached to the club lays in regulated masculinity, in its isolation from the outside world and in the codes of behavior created around the culture of smoking that seek to protect a particular group. Other smokers, denied access to the salubrious surrounding of the club, created their own spatial boundaries through smoking. Voluntary associations—based around sport, trade, or mutual aid—frequently held smoking "concerts" in the late-nineteenth century to celebrate either their achievements or existence, the highly ritualized manner of their smoking serving to identify their special sets of interests. In bars and public houses all around the world men have used cigarettes to create the same fraternal circle offered to the gentlemen of the club. As anthropologists have noted, the proffering of cigarettes to friends and colleagues assists in defining the group, enclosing a community to the exclusion of others, particularly nonsmokers. Often, as mass observation found in 1930s Britain (Mass-Observation), this could be accompanied by a particular language of smoking, in this case serving to define a particularly aggressive and male working-class smoking identity:

> *Smokers tend to talk of pitching and throwing the stub, rather than, more tamely, of dropping it; and quite often it is sent flying to some distance. Their actions, moreover, even more than their language, are frequently clothed in aggressiveness. Some speak of "grinding," "crushing," even "killing" a stub, and a favourite trick is to burn it to death in the fire or to drown it in the nearest available liquid. One man said: "I cannot let a stub smoulder. I must crush it out."*

In the 1990s several attempts were made to revive the atmosphere of the smoking club with the establishment of a number of **cigar bars**, most notably in New York City and London. These, however, were only a minority interest and many proved short lived. What is more significant to social historians is the complete transformation in the meaning of the smoking room. Whereas Victorian smokers sought to create a regulated space in order to block out the outside world and protect the interests of the tobacco consumer, the twenty-first-century smoking room serves to exclude the smoker from the outside world of majority nonsmokers. The image connoted by Ouida, therefore, stands in sharp contrast to the reality of the small, congested, and uncomfortable smoking room found, for instance, in the modern international airport.

cigar bars cocktail lounges catering to cigar smokers. Cigar bars became popular in the 1990s as many restaurants and bars banned smoking.

■ MATTHEW HILTON

BIBLIOGRAPHY

Barrie, J. M. *My Lady Nicotine*. London: Hodder and Stoughton, 1890. Reprint (with Margaret Ogilvy), New York: Scribner's, 1926.

Corti, Egon C. *A History of Smoking*. New York: Harcourt, 1932.

Hilton, Matthew. *Smoking in British Popular Culture, 1800–2000: Perfect Pleasures*. Manchester and New York: Manchester University Press, 2000.

Mass-Observation, File Report 3192, 1949, Man and His Cigarette (Tom Harrisson Mass-Observation Archives, Brighton, England; Harvester Press Microform Publications, 1983) pp. 126–127.

Smoking Restrictions

The social geography of smoking underwent a profound transformation during the twentieth century. At midcentury, cigarettes were a common feature of public places, including restaurants, offices, trains, and hospitals. But from the middle to the end of the century, the range of locations where smoking was considered appropriate inexorably narrowed. The first restrictions were small-scale and incremental, as separate sections were established that could keep smokers and nonsmokers apart without requiring that smoking cease altogether. But as an antismoking movement gained momentum on the strength of both changing cultural norms and scientific evidence about the harms of secondhand smoke, an activity that had once been synonymous with sociability was redefined as unpleasant and dangerous, and smokers began to be banished from many indoor public spaces.

The Spread of Public Smoking

snuff a form of powdered tobacco, usually flavored, either sniffed into the nose or "dipped," packed between cheek and gum. Snuff was popular in the eighteenth century but had faded to obscurity by the twentieth century.

During the nineteenth century, when tobacco was most commonly consumed as **snuff** or in cigars and pipes, public attitudes varied widely about the acceptability of its use. The United States and some European nations had vigorous antitobacco movements, closely aligned with temperance and religious crusades, which argued that smoking was a form of moral degeneracy and fought for its prohibition. Several states in the United States outlawed the sale of tobacco, though these measures were later repealed. As mass production and distribution techniques made rolled cigarettes the most prevalent type of tobacco product in the twentieth century, smoking became an increasingly mainstream and popular activity and moved steadily into the public sphere. This trend was fueled in large measure by the aggressive advertising of cigarette manufacturers, who sought to connect their product with images of modernity and sophistication. Cigarette consumption rose steadily during the first half of the century, and by the 1950s smoking had become a fixture of American society, a symbol of pleasure and sociability. About one-half of men and one-third of women smoked, and few public spaces were off limits to the enjoyment of cigarettes.

epidemiological pertaining to epidemiology, that is, to seeking the causes of disease.

The first **epidemiological** studies demonstrating the link between smoking and lung cancer in the 1950s began to transform both professional and popular attitudes about the dangers of cigarettes. But the increasing attention to potential health hazards did little to change the acceptability of public smoking. Although smoking rates declined slightly in the wake of landmark reports documenting the link between smoking and lung cancer issued by the Royal College of Physicians in Great Britain in 1962 and the U.S. Surgeon General in 1964, these documents did not immediately trigger a notable shift in the places smoking was allowed. It was not until the 1970s that the modern-day movement emerged to remove cigarettes from public space.

The Emergence of Nonsmokers' Rights in the 1970s

The first nonsmokers' rights groups that were formed in the early 1970s drew explicitly on the rhetoric and discourse of the civil rights and environmental movements, claiming that everyone had a right to breathe clean air in places of public accommodation. Prominent early organizations included Group Against Smokers' Pollution (GASP), a grassroots association with chapters in several states, and Americans for Non-Smokers' Rights, based in Berkeley, California. Although such groups suggested that nonsmokers could suffer physical harm from cigarette smoke, there was scant data to support this idea, and regulations were

SMOKING RESTRICTIONS

advocated primarily on the ground that smoking was a noxious annoyance. The first limitations were imposed on public transportation. In 1971 United Airlines became the first air carrier to institute nonsmoking sections for their passengers, and in 1973 the Civil Aeronautics Board required that all U.S. airlines create such sections. Similar regulations were instituted that set aside a limited number of seats for smokers on interstate buses. Over the following decade, cities and states around the country began to enact regulations on indoor spaces. In 1973, Arizona passed ground-breaking legislation limiting places where smoking was allowed; Minnesota followed suit two years later, requiring no-smoking zones in buildings open to the public. Many regulations were enacted at the local level. In 1977, Berkeley, California, became the first city to pass an ordinance limiting smoking in restaurants.

These measures served a dual purpose. Public health advocates who were appalled at the toll of illness and death that smoking extracted saw them as a way not only to clear the air that was shared by all, but also to decrease the social legitimacy of smoking. Thus, while the bans were generally framed as a protection of innocent third parties, they also conferred a secondary benefit to smokers themselves by encouraging people to smoke less or not at all.

As a result of a smoking ban in restaurants and bars in many of Cape Cod's towns, three men stand in a snow storm to smoke their cigarettes outside the Hyannis, Massachusetts, bar/restaurant Bobby Byrne's, 6 March 2003. AP/WORLD WIDE PHOTOS

The 1980s: From Nuisance to Toxin

Although the Surgeon General's 1972 report on smoking identified secondhand smoke as a potential danger to nonsmokers, concerns about the precise nature and extent of the harm remained speculative. It was not until the following decade that scientific evidence began to accumulate that secondhand smoke was a health hazard in addition to a nuisance. In 1980 and 1981, scientific journals published epidemiological research from the United States, Greece, and Japan that suggested that those who breathed "environmental tobacco smoke"

SMOKING RESTRICTIONS

Japanese smokers gather around ashtrays in a corner of a Tokyo railway station, 18 October 1998. Restrictions on smoking have been springing up in Japan. Smokers are limited to special parts of train platforms and small no-smoking sections are common in family-style eateries. AP/WORLD WIDE PHOTOS

carcinogen a substance or activity that can cause cancer. Cigarette smoking has been proven to be carcinogenic, that is, cancer causing.

suffered from decreased lung function and increased risk of lung cancer. Since these investigations involved people who had experienced heavy exposure to smoke in the home over long periods of time, there were questions about whether and to what extent the data could be extrapolated to other enclosed public spaces. But over the next several years, additional studies gave weight to the argument that nonsmokers suffered physical harm by breathing others' cigarette smoke. Reports from a variety of scientific agencies, including the National Academy of Sciences, the Office of the U.S. Surgeon General, and the Environmental Protection Agency, lent an official imprimatur to the danger and gave a powerful impetus to a movement that already had considerable social support. The most damning statement against environmental tobacco smoke (ETS) came in 1992, when the Environmental Protection Agency declared that ETS was a Class A **carcinogen,** placing it in the same category as such known and deadly toxins as asbestos and benzene.

538 | Tobacco in History and Culture
AN ENCYCLOPEDIA

By the mid-1980s, almost all states had enacted some restrictions on where people could smoke in public; some 80 percent of the U.S. population lived in areas covered by such laws. Between 1985 and 1988, the number of communities around the country that had enacted laws restricting public smoking almost quadrupled, to more than 300. In 1986 the U.S. Congress banned all smoking on flights of less than two hours, and two years later banned smoking on all domestic flights.

As the movement to eliminate cigarettes from public spaces gained momentum, the tobacco industry recognized the grave threat that the increasing marginalization of smoking posed to their market and undertook a variety of activities to maintain its acceptability. Attempting to reframe the issue as one of manners, not health, the industry took out advertisements that urged people to resolve disputes over public smoking through polite accommodation rather than the heavy hand of legal regulation. Industry representatives aggressively lobbied politicians and business leaders in an effort to combat regulations on smoking in restaurants and workplaces, and provided covert funding to so-called smokers' rights groups that sought to portray the move to ban public smoking as intolerant zealotry. At the same time, the industry engaged in a variety of practices to undermine scientific evidence and perpetuate uncertainty about whether secondhand smoke was truly harmful. The industry created the Center for Indoor Air Research to fund studies that would refute the growing evidence. But in spite of the enormous financial resources of the industry, the movement to limit the spaces where smoking was allowed had broad-based public support, and the spheres within which smoking was legally and socially acceptable shrank steadily and dramatically in the last two decades of the twentieth century.

Restrictions Abroad

Because smoking is a behavior deeply rooted in cultural attitudes toward pleasure, risk, sociability, manners, and individual rights, there has been wide variation in the ways that countries around the world have limited smoking in public spaces. Public sentiment in many countries runs against smoking bans as an unwarranted state intrusion on a personal habit. Although the trend in most industrialized democracies since the late 1900s has been toward enacting some form of legal regulation, the scope of these laws and the extent to which they are observed varies widely. France, for example, first passed regulation limiting smoking in places "open to the public" in the 1970s and attempted to strengthen these limits through subsequent measures in the 1990s, but the laws have never been vigorously enforced, and smoking is routine in many places where it is rare in the United States, such as schools, hospitals and restaurants. Smoking restrictions in Japan have spread much more slowly than in the United States or Europe, and have generally been justified out of consideration for others rather than as a health risk. Reflecting the growing international consensus on limiting smoking, airlines in almost all countries began in the 1980s to institute some form of restriction on smoking, though many continue to maintain smoking sections.

SMOKING RESTRICTIONS

A monk sits before a golden Buddha shrine in Thailand. Smoking inside religious shrines like this one is strictly forbidden. © PAUL SEHEULT; EYE UBIGUITOUS/CORBIS

Restrictions in the Twenty-First Century: How Far to Press?

By the end of the twentieth century, there was in the United States a strong cultural norm, buttressed by science, against smoking in indoor public spaces. Nevertheless, controversies continued to swirl over whether restrictions should extend beyond enclosed environments such as restaurants and workplaces, where the health risk to nonsmokers was clear, to outdoor areas. In 1995 the city of Palo Alto, California, banned smoking within 20 feet (6 meters) of all public buildings, and other cities began to enact ordinances that prohibited smoking in places such as parks, beaches, and sports arenas. But such moves could provoke a backlash. In 2000 the community of Friendship Heights, Maryland, banned smoking in all public places, but was forced to repeal the ban a year later in the face of widespread opposition. In 2001, the town once again made headlines after city council members introduced a measure—subsequently vetoed—that would subject people to fines for smoking in their own homes if the smoke crossed over their property line into their neighbor's home.

Even bars, which were among the last bastions of indoor smoke, came under attack. In 1998 California extended its restaurant smoking ban to include pubs, and in 2003 New York City banned smoking in virtually all bars and restaurants. The laws were justified as workplace safety measures to protect waiters, bartenders and other employees who had no choice but to spend hours in smoky environments. Smoking bans have been enacted in other countries as well, sometimes at the national level. On 30 March 2004, Ireland became the first European country to ban smoking in all workplaces, including pubs and restaurants. But it remains to be seen whether these bans can be successfully enforced in the long run.

The debate over whether smoking should be banished from all public spaces was encapsulated in an exchange in 2000 in the journal *Tobacco Control*. The journal's editor, Simon Chapman, one of the leading figures in the international antitobacco movement, argued in an editorial that the increasingly restrictive stance toward smoking in public risked tainting tobacco control advocates into "the embodiment of intolerant, **paternalistic** busybodies, who not content at protecting their own health want to force smokers not to smoke, even in circumstances where the effects of their smoking on others is immeasurably small" (Chapman 2000). But another prominent antitobacco activist argued, "Even if outdoor environmental tobacco smoke were no more hazardous than dog excrement stuck to the bottom of a shoe, in many places laws require dog owners to avoid fouling public areas. Is this too much to ask of smokers?" (Repace 2000).

As this exchange suggests, the contemporary movement to restrict public smoking has not only involved questions of health but has also touched on sensitive social, cultural, and political issues. Even as the scientific evidence about the danger of environmental tobacco smoke has grown more powerful, debates have continued to rage over how far smoking restrictions should go and the role of the state in constraining individual behavior.

paternalistic fatherly; acting as a parent. Although paternalism presumes an obligation for the stronger to provide for the weaker, it implies superiority and dominance over them as well. For example, slave masters often had paternalistic feelings for their slaves, whom they considered childlike.

See Also Advertising Restrictions; Regulation of Tobacco Products in the United States; Secondhand Smoke.

■ JAMES COLGROVE

BIBLIOGRAPHY

Bayer, Ronald, and James Colgrove. "Science, Politics and Ideology in the Campaign Against Environmental Tobacco Smoke." *American Journal of Public Health* 92, no. 6 (2002): 949–954.

Brandt, Allan M. "Blow Some My Way: Passive Smoking, Risk, and American Culture." In *Ashes to Ashes: The History of Smoking and Health*. Edited by Stephen Lock, Lois Reynolds, and E. M. Tansey. Amsterdam: Rodopi, 1998.

Brownson, Ross C. et al. "Environmental Tobacco Smoke: Health Effects and Policies to Reduce Exposure." *Annual Review of Public Health* 18 (1997): 163–185.

Chapman, Simon. "Banning Smoking Outdoors is Seldom Ethically Justifiable." *Tobacco Control* 9 (2000): 95–97.

Feldman, Eric A., and Ronald Bayer, eds. *Unfiltered: Conflicts Over Tobacco Policy and Public Health*. Cambridge, Mass.: Harvard University Press, 2004.

Goodin, Robert. *No Smoking: The Ethical Issues*. Chicago: University of Chicago Press, 1989.

Kluger, Richard. *Ashes to Ashes: America's Hundred-Year Cigarette War, the Public Health, and the Unabashed Triumph of Philip Morris*. New York: Knopf, 1996.

Nathanson, Constance. "Social Movements as Catalysts for Policy Change: The Case of Smoking and Guns." *Journal of Health Politics, Policy and Law* 24, no. 3 (1999): 421–488.

Rabin, Robert L., and Stephen Sugarman, eds. *Regulating Tobacco*. Oxford: Oxford University Press, 2001.

Repace, James. "Banning Outdoor Smoking Is Scientifically Justifiable." *Tobacco Control* 9 (2000): 98.

Sullum, Jacob. *For Your Own Good: The Anti-Smoking Crusade and the Tyranny of Public Health*. New York: Free Press, 1998.

Smuggling and Contraband

contraband trade traffic in a banned or outlawed commodity. Smuggling.

There is no comprehensive study of tobacco smuggling and contraband in the sixteenth, seventeenth, and eighteenth centuries but the work that has been done for particular countries mentioned here suggests that smuggling and contraband can be broken down into three main spheres of activity. In the New World *rescate*, or **contraband trade,** between colonial producers and foreign interlopers challenged the efforts of mercantilist European states to monopolize the trade of their colonies in the interest of the metropolitan economy. In Europe, with the exception of Holland and the Spanish Netherlands, most governments classified tobacco as both a pernicious and luxury commodity and drew considerable revenues from it by levying heavy taxes and assigning trade, manufacture, and distribution to either private or state monopolies. Domestic tobacco cultivation was either prohibited or subjected to severe restrictions. European smuggling might be directed either to evade prevailing duties and regulations on imported tobacco or to the illegal cultivation and distribution of the home-grown product. Since the whole objective of tobacco smugglers was to avoid notice, historians have to assume that many of them were successful in doing so and that the official complaints and accounts of their activities only represent the tip of a much larger iceberg.

Contraband Trade at New World Plantations

CARIBBEAN REGION. French and English vessels were present in the Caribbean from the 1560s engaged either in privateering or in contraband trade at Spanish or indigenous settlements on the islands and the Spanish Main. Although it is likely that tobacco—featured with hides, sugar, and other commodities—acquired by illegal barter during these early years, it was not until the 1590s that Spanish colonial officials reported that the *rescate* in tobacco was reaching crisis proportions. By then the traffic was particularly noticeable at Caracas, Cumaná, and Cumanagoto in eastern Venezuela, at Port-of-Spain on Trinidad, and at San Tomé de la Guayana on the lower Orinoco River. French, English, Dutch, and occasional Irish traders were seen there. Their interest in the trade reflected growing consumer demand for tobacco in northwestern Europe and the increasing risks of acquiring it from Spain or Portugal.

After the outbreak of the Anglo-Spanish war in 1585 periodic embargos on foreign shipping in Iberian ports seriously disrupted trade with the peninsula. Faced by mounting costs of maintaining and protecting its annual *flota* (Atlantic convoy) to the Caribbean, the Spanish monarchy abandoned any attempt to maintain direct trade with its more marginal colonies in the region. The struggling settlements in eastern Venezuela, Trinidad, and the Orinoco were left to pay high prices for sparse supplies of European goods received either from coastal traders dispatched from the official ports-of-call of the *flota* or from other colonies privileged to receive special trade vessels with Spain. It is in these circumstances that the marginal colonies chose to trade with interlopers. Censorious Spanish officials dismissed the settlers as "riffraff who had no other source of income than the tobacco crop that was so esteemed in Flanders and England" (Andrews, p. 227). In fact the trade was vital to the impoverished communities and all, including local magistrates, participated in it. The planters at Cumanagoto were reported to have sold some 30,000 pounds of tobacco in 1603.

By 1606 the volume of the *rescate* had grown to such proportions that the Spanish Crown prohibited all tobacco growing in eastern Venezuela and the Windward Islands for ten years. Facing summary execution if they were captured in the latter colonies, foreign interlopers turned their attention instead to the tiny Trinidad and Orinoco settlements that continued to provide them with approximately 200,000 pounds of tobacco each year between 1605 and 1612. Don Fernando de Berrio, governor of the tiny settlements on Trinidad and the Orinoco, it was reported, "conducts the business of bargaining, divides the goods among his companions and pays for them." At Trinidad "in Lent they say four ships arrived, the crews whereof lodged in the town as they might in their own country. Some of them stayed in the monastery of St. Francis, where they say a good friar provided them with a chicken on Friday, and others stayed in the governor's house, allegedly to treat him when he was sick" (Andrews 1978). In 1612 the onset of unwelcome investigations by a specially commissioned Spanish judge immediately diminished and, within four or five years, shut down the illegal trade.

Illicit barter at undersupplied Spanish colonies resurrected in the second quarter of the seventeenth century, operating from the Dutch colony on Curaçao and the network of Dutch, English, and French settlements on Barbados, St. Christopher, Guadeloupe, Martinique, Montserrat, Nevis, and Tobago. Cut off from trade with Spain by the effective blockade enforced by Dutch privateering fleets, Spanish colonists relied particularly on the Dutch to supply them with slaves and vital European goods in exchange for tobacco, hides, cacao, and other commodities. In the 1640s and 1650s the catastrophic drop in tobacco prices and disruptions of civil war at home made the French and English tobacco colonies in the Lesser Antilles almost exclusively dependent on Dutch traders. The unwillingness of the English West Indian planters to comply with acts of parliament of 1650 and 1651 that prohibited foreign vessels from trading at English colonies prompted Oliver Cromwell to send fleets to the Caribbean to deal with Dutch interlopers in 1651, 1654, and 1655. It was Dutch traders who kept French and English planters going when their own national merchants showed no interest and it was the slaves, equipment, and practical knowledge supplied by the Dutch that allowed these colonies to make the transition from tobacco to sugar production after the mid-seventeenth century.

BRAZIL. Commercial production of tobacco in Brazil seems to have begun around 1600, concentrated thereafter mainly around Bahia. The first monopoly contracts for export appear to have been granted to private entrepreneurs in the early 1630s. In 1644, however, prompted by the serious shortage of slaves for the Brazilian sugar **plantations** after the loss of Angola to the Dutch, the restored Portuguese monarchy authorized the Bahian planters to trade their product directly to Mina. This allowed Brazilian producers to develop a thriving black market in the waters off the West Africa coast, selling their tobacco to foreign merchants rather than exchanging it for slaves.

plantation historically, a large agricultural estate dedicated to producing a cash crop worked by laborers living on the property. Before 1865, plantations in the American South were usually worked by slaves.

SMUGGLING AND CONTRABAND

Woodcut showing United States revenue officers attacking smugglers at Masonborough, North Carolina, 1867.
© BETTMANN/CORBIS

BERMUDA, VIRGINIA, MARYLAND. English settlers on Bermuda seem largely to have complied with the shipping monopoly of the Bermuda Company but, knowing their own tobacco to be inferior, made their profits by trading for tobacco grown at Spanish West Indian plantations and shipping it to England as their own less heavily taxed product. Settlers in Virginia appear to have been interested in trade with the Dutch from the very onset of commercial production. From 1621 to 1776 all tobacco exported from the English Chesapeake colonies of Virginia and Maryland was required to be shipped to England first, no matter what its ultimate destination. The Dutch settlement of New Amsterdam, established on the Hudson River in 1624, offered a means of avoiding English customs duties which were higher than those prevailing in Holland. The Dutch were also willing to buy in bulk and give long-term credit. A brisk trade developed between Virginia and New Amsterdam in the 1620s and 1630s, supplemented by a growing number of vessels from Holland in the 1640s when the connections with England were disrupted by the Civil War.

Chesapeake planters vigorously opposed the exclusion of the Dutch and continued to ship with them in spite of the restrictions imposed by the Cromwellian legislation of 1650 and 1651. They were further encouraged by fact that New Amsterdam had abolished duties on tobacco in 1653. In 1660 the governors of Virginia and New Amsterdam concluded a free trade treaty. Neither stern warnings from London nor the passage of the Navigation Act of 1661 dissuaded the English Chesapeake colonists, as the Council for Foreign Plantations complained in 1662, from conveying "both by land and water . . . great quantities of tobacco to the Dutch whose plantacons [plantations] are contiguous" (Van der Zee 1978). The **acquisition** of New Amsterdam (New York) in 1667 put an end to this intercolonial trade; however, by this time Virginia and Maryland producers were engaged in illicit direct trade to Scotland, Ireland, and Europe.

acquisition the purchase—sometimes called a merger—of a smaller company by a larger one. During the late twentieth century, major tobacco companies diversified their holdings through acquisition of nontobacco products.

Examples of Smuggling in Western Europe

GREAT BRITAIN. Tobacco imported into England from 1606 on was subjected to significant customs and impost dues that were differentiated to favor the product of English over foreign colonies. Royal proclamations throughout the reigns of James I and Charles I indicate that smuggling was already a serious problem. During the seventeenth century, although tobacco was quietly run ashore in obscure harbors and creeks, most smuggling took place on the customs quays in London and the outports by collusion between underpaid and overworked customs officers and merchants. Foreign tobacco was misidentified as originating from English plantations. Until 1713 tobacco declared unfit for consumption was free from **duty** and merchants commonly bribed customs agents to declare good tobacco as damaged and therefore duty free. Under-weighing of imported and exported tobacco **hogsheads** was rife. As London retailers complained in 1625, "Lewd persons under pretence of selling tobacco keep unlicensed alehouses and others barter with mariners for stolen and uncustomed tobacco" (*Calendar of State Papers Domestic, 1625–6* 1897).

duty a tax, usually a tax on certain products by type or origin. A tariff.

hogshead a large wooden barrel formerly used to store and transport cured leaf tobacco. A hogshead typically held approximately 800 to 1000 pounds (350 to 450 kg) of tobacco.

In the early eighteenth century systematic efforts to crack down on customs fraud shifted the focus of smuggling to the re-landing trade. Customs dues were remitted for tobacco re-exported for sale abroad. Merchants reclaimed their duty by re-exporting their tobacco and subsequently conveyed it to Dunkirk, Ostend, the Channel Islands and the Isle of Man, from where it was clandestinely re-landed in Devon, Cornwall, Dorset, the eastern and western Midlands, Ireland, and Scotland. Most smugglers were peaceable, honest merchants and seamen, although the profits of the proximity to London led to the emergence of criminal armed gangs in Kent and Sussex by the early eighteenth century. By 1750 some one-third of the total 8.6 million pounds of tobacco consumed in England and Scotland had not been subjected to full duties. Excise duties were imposed in 1789 ensuring that illegal trafficking would continue.

FRANCE. In France tobacco use developed very slowly and by the 1670s consumption levels were only one-tenth of those prevailing in England. The import, manufacture, distribution, and retail of tobacco came under a state monopoly in 1674. The jurisdiction of the monopoly extended to all but the eastern frontier provinces and domestic cultivation was prohibited within it except for certain parishes in Normandy and the region around Bordeaux and Montauban in the southwest. External trade in tobacco was free but its import was restricted to specific ports where it could be sold only to the agents of the monopoly. For the latter maritime smuggling proved to be the least problematic in the period before the French Revolution. Although legitimate merchants did occasionally make clandestine landings, serious maritime smuggling was

SMUGGLING AND CONTRABAND

not a sideline of legitimate commerce but the work of specialists using small boats to convey cargo to organized networks ashore. In the Mediterranean domestic grown tobacco from the southwest was routinely exported and then re-landed on the Provencal coast. Smuggling was particularly rife in Brittany at the turn of the eighteenth century, where lesser landed gentleman, supported by complacent local magistrates and clergy, maintained armed bands to run in cargoes from the Channel Islands.

Yet, the massive land smuggling of tobacco grown in the exempt eastern border provinces was much more difficult to control. That traffic tended to be carried by gangs formed from the extended families of the rural poor. Soldiers in frontier garrisons also sold contraband tobacco to civilians to supplement their pay. Officers of the monopoly found it impossible to control the thousands of small retailers who would mix smuggled and legal tobacco. In order to protect its revenues the French Crown increasingly encouraged the import of foreign tobacco and strictly curtailed domestic cultivation.

SPAIN AND PORTUGAL. Tobacco imported into Spain moved from free exchange subject to excise taxes in the sixteenth century, to an exclusive monopoly of licensed private contractors in the seventeenth century to a state-administered monopoly in 1701. Government records indicate that the Crown was especially preoccupied by the need to eradicate tobacco contraband in Spain during the period from 1654 to 1786. In the second half of the eighteenth century lax authorities in French Catalonia harbored roving bands of smugglers who regularly ran tobacco down to Spain. Portugal offered another source of contraband tobacco. Tobacco was commercially grown in Portugal from 1570s. The first monopoly contract for the sale of tobacco there was probably granted in the early 1630s and within ten years it was clear that early efforts at regulation were already being undermined by expanding contraband. Desperate for revenue and anxious to protect the Brazilian plantations, the newly restored Portuguese monarchy continued to sell regional monopoly contracts for Portugal and the Atlantic islands, overseen after 1674 by the Junta da Adminstração do Tabaco.

In spite of these efforts, large quantities of tobacco legally shipped from Brazil were surreptitiously unloaded from the Brazil fleets before they were inspected, or filched from state or private warehouses. Sailors on the fleets also smuggled tobacco ashore to sell on the black market. Others on the India fleets furtively carried it out to Goa where its sale undercut the monopoly that had been in existence since 1624. Foreigners who had purchased Brazilian tobacco off West Africa quietly slipped it into Lisbon and other ports. Although no tobacco could be grown domestically without licence, both male and female religious houses became notorious centers of production and sale, dispatching large quantities of it over the land border into Spain. Smuggling was particularly active between the Algarve and Andalucia and efforts to control it led to riots in which local officials as well as ordinary citizens participated.

See Also State Tobacco Monopolies; Taxation; Trade.

▌JOYCE LORIMER

BIBLIOGRAPHY

Andrews, Kenneth R. *The Spanish Caribbean: Trade and Plunder 1530–1630.* New Haven, Conn.: Yale University Press, 1978.

Calendar of State Papers, Domestic Series, of the Reign of Charles I, 162–6. London: HMSO, 1897.

Goslinga, Cornelis Christiaan. *The Dutch in the Caribbean and on the Wild Coast, 1580–1680.* Gainesville: University of Florida Press, 1971.

Hanson, Carl. A. "Monopoly and Contraband in the Portuguese Tobacco Trade, 1624–1702." *Luso-Brazilian Review* XIX 2 (1982): 149–168.

Klooster, Wim. *Illicit Riches: Dutch Trade in the Caribbean, 1648–1795.* Leiden, the Netherlands: KITLV Press, 1998.

Kupp, Jan "Dutch Notarial Acts Relating to the Tobacco Trade of Virginia, 1608–1653." *William and Mary Quarterly* 30, no. 4 (1973): 653–655.

Lorimer, Joyce. "The English Contraband Tobacco Trade from Trinidad and Guiana, 1590–1617." In *The Westward Enterprise: English Activities in Ireland, the Atlantic, and America 1480–1650.* Edited by Kenneth R. Andrews, Nicholas P. Canny, and P. E. H. Hair. Liverpool, England: Liverpool University Press, 1978.

MacInnes, Charles M. *The Early English Tobacco Trade.* London: Kegan Paul, Trench, Trubner, 1926.

Menard, Russell R. "The Tobacco Industry in the Chesapeake Colonies, 1617–1730." *Research in Economic History* 5 (1980): 109–177.

Nash, Robert C. "The English and Scottish Tobacco Trades in the Seventeenth and Eighteenth Centuries: Legal and Illegal Trade." *Economic History Review* 35, no. 3 (1982): 354–372.

Price, Jacob M. *France and the Chesapeake: A History of the French Tobacco Monopoly, 1674–1791, and of its Relationship to the British and American Tobacco Trades.* Ann Arbor: University of Michigan Press, 1973.

Rive, Alfred. "A Short History of Tobacco Smuggling." *Economic Journal* 4 (1929): 554–569.

Sluiter, Engel. "Dutch-Spanish Rivalry in the Caribbean Area, 1594–1609." *Hispanic American Historical Review* 28 (1948): 165–196.

Van der Zee, Henri. *A Sweet and Alien Land: The Story of Dutch New York.* New York: Viking, 1978.

Williams, Neville. "England's Tobacco Trade in the Reign of Charles I." *Virginia Magazine of History and Biography* 65, no. 4 (1957): 403–449.

Snuff

Rather than constituting a short-lived historical anomaly, as it is often portrayed, snuff-taking was the most popular mode of consuming tobacco in European societies of the eighteenth and early-to-mid-nineteenth centuries, and in some countries it maintained this position well into the twentieth century. In fact, nasal snuff—a dry, powdered form of tobacco—was present from the very earliest introduction of the plant into Europe (Rogozinski 1990). Snuff was also widely used by Native American peoples for millennia in the pre-Columbian era, and the practice of snuffing was mentioned in some of the earliest accounts of Amerindian tobacco use by European explorers of the New World.

SNUFF

A native wagon driver takes snuff by rubbing it on his teeth. In the 1840s and 1850s, many Dutch farmers immigrated to the Cape of Good Hope, Africa, and hired native laborers to help with the journey. © CORBIS

Somewhat paradoxically however, at its zenith in eighteenth-century, Europe snuff was understood to be a distinctively aristocratic, refined mode of consuming tobacco, one which had its origins within the French court. While the spread of the practice did gain impetus from its esteem among the French aristocracy, it was in fact Spain that was the first Old World nation to make widespread use of tobacco in this form (Rogozinski).

Nonetheless, following from its purported French courtly origins, the practice of snuffing by the European elite developed into a civilized art involving highly elaborate rituals and codes of etiquette. Even the seminal manners text of the time, Antoine de Courtin's *Rules of Civility*, contains an entry on its use: "If you see Tobacco, either in Snuff or cut, you must not run presently to his Box, and either chew or thrust it up into your Nose; you must rather expect till he offers it, and in that case 'tis civil to pretend to take it, though of your self you have no inclination" (Courtin 1703).

Offering snuff involved a courtly dandy adopting the correct stance, holding the snuffbox appropriately, and presenting it to others such that his wrist cuffs and jewelry were displayed to their best advantage. If accepted, the recipient would take a pinch in a manner which again allowed for a similar display of refinement. Snuff would be administered into the nostrils, sometimes with a specially designed ladle, and snorted so as to induce a sneeze or a series of sneezes. The manner in which one took a pinch announced one's pedigree through displaying awareness of the etiquette surrounding taking snuff and, to a degree, one's individuality, through the addition of discretionary personal touches to the rituals involved. There were even snuff schools in the early eighteenth century established for the sole purpose of teaching the fashionable the socially correct ways to use snuff.

The Pinch of Snuff. Lithography by John James Chalon. © STAPLETON COLLECTION/CORBIS

Snuff Accoutrements

The ritualized ostentation of aristocratic snuffing was by no means simply confined to the practices involved. Snuffing at this time also involved equipment which itself was ornate and elaborately adorned. Before snuff was widely manufactured it was created by hand-grating tobacco (freshly grated snuff was known as rappee) from rolled blocks known as *carrots* by rasps made of ivory, wood, or metal which were highly ornamented, painted, or engraved. By the middle of the eighteenth century, snuff was more commonly bought in a pulverized form, and accordingly such equipment became less necessary. Snuffboxes, nonetheless, maintained their position as the snuffer's essential fashion accessory and were so lavishly produced—often made from gold and silver and inset with precious stones—that they were considered items of jewelry and exchanged as gifts among the aristocracy.

Snuff concoctions were made from a broad array of substances in addition to different varieties of tobacco and tailored to individual tastes and constitutions. Snuff was regarded in a similar manner to fine wines today, with connoisseurs well-versed in the multitude of types available and their relative costs and origins. It was often perfumed and flavored by substances such as orange oil, rose leaves, musk, ginger, and even

opium an addictive narcotic drug produced from poppies. Derivatives include heroin, morphine, and codeine.

gallant a well-dressed, well-spoken gentleman, attentive to the needs and concerns of ladies, but in a proper way. Rhett Butler is a gallant.

Further Reading

For an excellent all-round discussion of snuff and other forms of smokeless tobacco, see Jan Rogozinski's *Smokeless Tobacco in the Western World, 1550–1950* (1990). See also Jordan Goodman's *Tobacco in History: The Cultures of Dependence* (1993), particularly pages 69–89; Jason Hughes's *Learning to Smoke: Tobacco Use in the West* (2003), particularly pages 66–75; James Walton's *The Faber Book of Smoking* (2000), particularly pages 49–57; and Iain Gately's *La Diva Nicotina: The Story of How Tobacco Seduced the World* (2001), particularly chapter 6.

pepper and mustard. Snuff was also spiked with a range of substances to enhance its properties. Some historical sources suggest that the adulteration of snuff extended to highly toxic and psychotropic substances. Wlodzimierz Koskowski, in his *The Habit of Tobacco Smoking* (1955), for example, provides a list that includes lead, arsenic, hydrogen cyanide, cocaine, hashish, and **opium.**

Societal Aspects

The excesses of the snuffing **gallant** invited lampoon from antitobacco writers of the day. Many seized upon the potentially slovenly aspects of snuff-taking. In his 1720 publication *Lust of the Longing Nose* Johann Heinrich Cohausen, for example, remarked that snuffers had "dust heaps" for noses. Other satirists referred to pretty young ladies wearing moustaches of scented powder (Brooks 1952). Indeed, such writings highlight the inherent contradiction of snuff-taking: on the one hand, it was considered to mark the height of refinement, a mode of tobacco use a world away from plebeian pipe smoking, yet on the other, snuffing involved the transgression of conventional manners and mores through the public expulsion of mucus and saliva, the insertion of fingers into orifices, and so forth.

As suggested above, however, the practice of snuffing was not confined to the European aristocracy. Historical evidence suggests that it was used by all levels of society and both genders by the middle of the eighteenth century. Snuffing by members of the working classes was not necessarily undertaken simply in imitation of courtly and aristocratic figures, but also because it was an economical means of consuming ground tobacco stalks and low grade leaves; though, even at the height of snuff-taking, smoking remained an important mode of consumption by this social group. Snuffing was also practiced in the Indian subcontinent, Tibet, Africa, and Japan. In China, where tobacco smoking had been forbidden soon after the beginning of the Qing Dynasty (in 1644), the use of Snuff gained rapid popularity—snuff was deemed acceptable as it was considered to be an effective medicinal remedy for a range of ailments. Partly because of Climatic conditions, the Chinese stored snuff in sealable bottles which were often crafted from precious materials and were intricately decorated—the technique for painting the inside of snuff bottles became an art form in itself. Like European snuff boxes, these snuff bottles came to be highly prized, such that by the nineteenth century they were used as currency within Chinese society for the purchase of favors, and as a source of leverage to social positioning.

In Sweden, Denmark, and Norway wet oral snuff—a moister, coarse- or ribbon-cut form of tobacco placed between the lip and the gums—became one of the most popular forms of tobacco from the mid-nineteenth to around the mid-twentieth century. Oral snuff still constitutes a significant proportion of tobacco sales in these countries, particularly in Sweden.

North Americans have shown a historical preference for chewing tobacco, but they too came to use wet snuff—which was applied to the gums, sometimes by dipping with a stick, or held in the cheek and sucked, and also chewed—by the end of the nineteenth century. Today, remnants of the golden age of snuff can still be seen. For example, there are filled snuffboxes in both the House of Commons in the United Kingdom and in the U.S. Senate Chamber.

See Also China; Native Americans; Social and Cultural Uses.

■ JASON HUGHES

BIBLIOGRAPHY

Arents, George, Jr. *Tobacco, Its History Illustrated by the Books, Manuscripts, and Engravings in the Library of George Arents, Jr.* 5 vols. New York: Rosenbach Company, 1937–1952.

Brooks, Jerome E. *The Library Relating to Tobacco, Collected by George Arents, by Jerome E. Brooks.* New York: The New York Public Library, 1944.

———. *The Mighty Leaf: Tobacco through the Centuries.* Boston: Little, Brown, 1952.

Cohausen, Johann Heinrich. *Lust of the Longing Nose.* Leipzig, Germany, 1720.

Courtin, Antoine de. *The Rules of Civility; or, The Maxims of Genteel Behaviour, As They Are Practis'd and Observ'd by Persons of Quality, upon Several Occasions*, 12th ed. London: Printed for Robert Clavell and Jonathon Robinson in St. Paul's Church-Yard, and Awnsham and John Churchill, in Pater Noster Row, 1703.

Curtis, Mattoon M. *The Story of Snuff and Snuff Boxes.* New York: Liveright Publishing Corporation, 1935.

Fairholt, Frederick William. *Tobacco: Its History and Associations.* London: Chapman & Hall, 1859.

Gately, Iain. *La Diva Nicotina: The Story of How Tobacco Seduced the World.* London: Simon & Schuster, 2001.

Goodman, Jordan. *Tobacco in History: The Cultures of Dependence.* London: Routledge, 1993.

Hughes, Jason. *Learning to Smoke: Tobacco Use in the West.* Chicago: University of Chicago Press, 2003.

Koskowski, Wlodzimierz. *The Habit of Tobacco Smoking.* London: Staples Press, 1955.

Mack, Peter Hughes. *The Golden Weed: A History of Tobacco and of the House of Andrew Chalmers, 1865–1965.* London: Newman Neame, 1965.

Rogozinski, Jan. *Smokeless Tobacco in the Western World, 1550–1950.* New York: Praeger, 1990.

Walton, James, ed. *The Faber Book of Smoking.* London: Faber, 2000.

Social and Cultural Uses

It is helpful for explanatory purposes to categorize the socio-cultural development of tobacco use according to three main stages: pre-Columbian, describing tobacco use among the indigenous peoples of the Americas prior to contact with European explorers; modern, referring broadly to tobacco use in European societies—and its spread beyond to other parts of the globe—during the sixteenth to nineteenth centuries;

and contemporary, concerning tobacco use from the late-nineteenth century to the present day. However, notwithstanding this separation into historical stages, it is also useful to understand these developments as a progressive whole. While it is impossible to provide here anything but a very general account of long-term global changes in the social and cultural uses and associations of tobacco, such a broad-brush approach serves to highlight an overall pattern to such changes, which is in itself significant.

At the most general level, this pattern can be understood as being marked by a shift from the use of tobacco to lose control—as an intoxicant—as a characteristic of pre-Columbian use, and a move toward the use of tobacco as a means of self-control, particularly within the contemporary era. That is to say, the overall direction of change can be seen to involve a move away from the use of tobacco to escape so-called normality and toward its use as an instrument to return to normality. This transformation is in turn historically premised on the consumption of progressively less potent strains, species, and varieties of tobacco, and modes of consuming these.

An associated long-term trend has been, generally speaking, a move away from the idea of tobacco use as a mark of general sociability and as a plant of great symbolic and spiritual significance, and toward the idea of tobacco as a commodity of great material significance—a drug to be consumed increasingly in an individual manner, the practice of which might only be considered sociable within the context of highly specific peer groups.

This brief and greatly simplified synopsis of the long-term development of socio-cultural uses and understandings of tobacco is intended to serve here as a framework to help introduce and contextualize the distinctive characteristics of each stage as discussed below.

Pre-Columbian Understandings and Uses

The pre-Columbian era of tobacco use was one characterized by the consumption of highly potent strains of the plant in accordance with rituals and beliefs that stressed its importance in mediating the bonds between humans and between humans and the spiritual world, as well as beliefs concerning the fundamental importance of tobacco to health and healing within Amerindian cosmology.

AMERINDIAN ORIGINS. While a few scholars suggest that tobacco may have also been known in ancient Egypt (Balabanova et al. 1993), it is widely agreed that tobacco was unknown outside of the Americas until the late-fifteenth century. Yet the plant had been in use by indigenous peoples for millennia prior to contact with Columbus and other explorers from Europe (Wilbert 1991). Prevailing present-day understandings of tobacco might lead us to believe that **snuff** and cigars, for instance, have their origins in courtly and aristocratic circles in Europe. However, these and almost every other conceivable mode of tobacco consumption—chewing; drinking (the juice of the tobacco plant); licking (rubbing tobacco resin against gums and teeth); topical application (to wounds, bites, and stings); ocular absorption; anal injection; and, of course, smoking—had been developed by Native American peoples long before Columbus first encountered tobacco on his voyage to San Salvador.

snuff a form of powdered tobacco, usually flavored, either sniffed into the nose or "dipped," packed between cheek and gum. Snuff was popular in the eighteenth century but had faded to obscurity by the twentieth century.

SOCIAL AND CULTURAL USES

An engraving of a tattooed Iroquois Indian holding a snake and smoking a peace pipe, c. 1701. © CORBIS

SPIRITUAL ALLIANCES. The use of tobacco by different Native American peoples varied enormously. What follows is a generalized account, one which focuses as far as possible on relatively widespread practices and understandings.

Tobacco had enormous spiritual significance within Amerindian belief systems: it was understood as a plant that had supernatural origins as a substance that could facilitate transportation into and within the spiritual world, as the locus of certain spiritual beings, and as an offering to appease the spirits. Tobacco was characteristically offered as part of what was understood to be a reciprocal exchange: Spirits themselves were seen to have an insatiable hunger for tobacco, which, since they could not grow the plant for themselves, made them dependent on humans; in return for tobacco, spirits could bestow good favor and fortune upon their providers. Both material tobacco and tobacco smoke were used to this end.

In ritual offerings, tobacco leaves would be tossed into the air or blown from the hand and accompanied by appropriate words of bequest and request. The Huron, for example, would throw tobacco into fires to implore health; over rocks to ensure the safety of their villages; or into lakes to request safe passage (Tooker 1964). Tobacco was smoked on almost every formal occasion; the rising tobacco smoke itself held great symbolic value—as akin to an ascending petition or prayer to those spirits believed to have resided in the sky.

Tobacco was of equal importance in cementing alliances between humans and was often used as a peace offering. Among many Amerindian peoples, any bond, treaty, or agreement was not considered binding unless it was undertaken in conjunction with tobacco consumption, particularly through the passing of a pipe. Indeed, the ceremonial sharing of the **calumet pipe** on important socio-political occasions is perhaps one of the most celebrated (in film and literature) and widely known aspects of Native American tobacco use. Less widely known, however, is the broader cosmology of which this distinctive ritual formed part. While tobacco played a significant role in maintaining the bonds between men (in general, it was much less common for women to smoke), its social importance and value went far beyond this.

calumet pipe a highly ornamented ceremonial pipe used by American Indians.

TOBACCO AND NATIVE AMERICAN COSMOLOGY. Even peoples who grew no other crops cultivated the tobacco plant. Indeed, among some Amerindians, those who planted and grew tobacco were held in the highest regard. It was understood by the Crow peoples, for example, that the plant was crucial to their nation's survival. Thus, the role of Tobacco Planter was not to be undertaken lightly; in fact, an extensive ordeal was involved in attaining the position. Applicants had to endure an arduous ceremony in which they would be extensively and severely burned and cut, forced to go without food and water for a number of days, and then, if they had survived the ordeal, would exchange their every worldly possession for some tobacco seeds. Those who continued the Crow's tradition of tobacco planting were understood to be endowed with a broad array of supernatural powers, including the capacity to control the weather (especially the wind and rain); to bring buffalo and other game near their settlements; and, most importantly, to prevent and to heal a range of diseases (Denig 1953). Indeed, it is in this last respect—in relation to the link between tobacco and healing—that the Crow's beliefs are typical of Native American understandings of tobacco use.

shamanism an ancient religion based on commune with animal spirits and characterized by magic, healing, and out of body experiences. Shamanism was widely practice American Indians.

Within Native American cosmology, the spiritual and social importance of tobacco were intimately bound up with the plant's role in relation to medicine and healing within **shamanistic** ritual. According to the understandings and practices of shamanism, illness and disease could be caused by spiritual forces. For example, a malevolent spirit might cause illness by introjecting itself or a magical object into the body, or by drawing a sick person's soul into the spirit world. In order to cure such illness and disease, therefore, the shaman's task was to enter the spirit world to either remove the offending object or retrieve the ill person's soul (Goodman 1993).

SOCIAL AND CULTURAL USES

One of ten sixteenth-century engravings by Theodor de Bry that provided the first visual impressions of life in the immense wilderness across the sea. Dried "tapaco" leaves will be smoked from a pipe by the men on the benches to cure their infections. © BETTMANN/CORBIS

The use of hallucinogenic plants was central to the shaman's task, since only these, it was understood, would facilitate altered states of consciousness, allowing access to the world of the spirits. Perhaps surprisingly, the one plant used more than any other for this end was tobacco. That this fact might be surprising reveals a great deal about the long-term development of tobacco use. If we consider, for example, the present-day western smoker's experience of a cigarette, it seems impossible that tobacco could produce a strong enough effect to significantly alter consciousness.

However, there is a wealth of evidence to suggest that the tobacco used by Native Americans, particularly the *Nicotiana rustica* strains, were fully capable of producing hallucinations. In order to produce the consciousness-altering effects required, shamanistic practice demanded the consumption of tobacco on a scale considerably larger than that of recreational use. According to Johannes Wilbert in *Tobacco and Shamanism in South America* (1987)], a central tenet of shamanistic belief is that it is only by overcoming death that one is capable of curing others, and thus a common theme throughout ethnographic accounts of Native American tobacco use is that of shamans taking themselves to the verge of death by acute nicotine poisoning.

RECREATIONAL USE. Recreational use of tobacco by Native American peoples, while considerably less dramatic than that of shamanistic use,

SOCIAL AND CULTURAL USES

nonetheless frequently involved considerably stronger, more pronounced effects than those which are commonly associated with the drug in present-day western societies. Even within recreational use, tobacco use was on the whole a highly ritualized activity, one which required an extensive process of habituation and one which followed very different patterns of consumption from those of the present day. John P. Harrington's excellent 1932 study of tobacco use among the Karuk provides one of the few direct translations of a Native American account of tobacco smoking. While the account was documented in the twentieth century, it describes a tradition of smoking that stretches back into the antiquity of the Karuk people:

> He sucks in . . . then quickly he shuts his mouth. For a moment he holds the smoke inside his mouth. He wants it to go in. For a moment he remains motionless holding his pipe. He shakes, he feels like he is going to faint, holding his mouth shut. It is as if he could not get enough. . . . He shuts his eyes, he looks kind of sleepy-like. His hand trembles, as he puts the pipe to his mouth again. Then again he [inhales]. . . . He just fills up the pipe once, that is enough, one pipeful. He rests every once in a while when smoking. . . . He feels good over all his meat when he takes it into his lungs. Sometimes he rolls up his eyes. And sometimes he falls over backward.
>
> HARRINGTON 1932

As can be observed from this account, even a seasoned habituated Karuk user would frequently faint after smoking just one pipeful of tobacco. Indeed, the Karuk actively sought out intoxication, both through the technique of smoking and through the continued cultivation of the highly potent strains of tobacco. The extract also serves to illustrate that tobacco use among the Karuk involved, on a day-to-day basis, the relatively infrequent consumption of highly potent tobacco (smoked over a long period), a pattern which contrasts starkly with that of the present-day western cigarette smoker, who smokes much milder tobacco, generally more frequently and more quickly.

POSTCONTACT CHANGES. It is significant to note that, in accordance with the overall direction summarized above, during the post-contact era Native American tobacco use developed in a number of respects: there is evidence to suggest that the socio-cultural associations of tobacco increasingly shifted from the spiritual to the profane; it became far more common for Amerindian women to smoke; and finally, tobacco use became more hedonistic, involving more frequent consumption outside of ritual and ceremony of generally less potent tobaccos.

Modern Understandings and Uses

The modern era in the development of tobacco use can be understood as characterized by both an extension of Native American social and cultural uses of tobacco—in, for example, the initial adoption of smoking as a symbol of sociability and the use of the plant as a medicinal remedy—and an increasing move away from these in the ever-developing quest to find a sophisticated and refined means of tobacco consumption that could help

Tobacco In History and Culture
AN ENCYCLOPEDIA

An elderly Hakka Chinese woman smokes a pipe in Hong Kong. China is the largest market for cigarettes and the world's leading tobacco producer. However, the majority of tobacco smokers in China are men; fewer than ten percent of smokers are women. © DAVE BARTRUFF/CORBIS

Because of South East Asia's ideal climate and soil, tobacco cultivation and consumption were common there even before European colonists introduced tobacco as a cash crop. Today cheroots are part of the socioeconomic fabric. Young girls are expected to learn cheroot making, and mothers have been known to hold cheroots to their babies' lips. It is therefore not surprising to see young girls like the one from Myanmar pictured to the right, her face painted with thanaka bark powder, selling cheroots from a basket, or old women like the one from Burma pictured below smoking cheroots.

© NEVADA WIER/CORBIS

© CHRISTOPHE LOVINY/CORBIS

Tobacco In History and Culture
AN ENCYCLOPEDIA

Tobacco must be dried and cured before manufacturing can begin. On this modern-day tobacco plantation in the Philippines, rows of tobacco leaves hang from poles to dry in the open air. © PAUL ALMASY/CORBIS

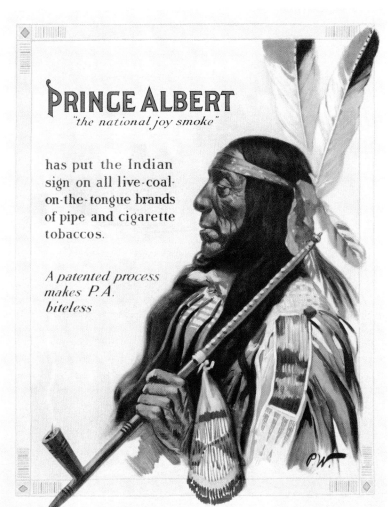

Native Americans were commonly depicted in tobacco advertisements, like this one of a chief holding a calumet (ceremonial pipe), from Prince Albert. Commercial artists often adapted images of people, including Native Americans, from files of prints or from photographs. They seldom used live models, because copying and adapting an existing image was easier than working from life. © SWIM INK/CORBIS

Tobacco In History and Culture
AN ENCYCLOPEDIA

This eighteenth-century illustration "Faiseur de tabac" by Engelbrecht is an imaginative depiction of what appears to be a tobacconist, or tobacco shopkeeper. The man has a tobacco plant on his head and a pipe in his hand, and is wearing a tobacco-leaf kilt. In the background, two men are processing tobacco. © GIANNI DAGLI ORTI/CORBIS

Advertising in the last decade of the twentieth century increasingly used people and vehicles as mobile advertisements. This baby in Senegal, wearing a Marlboro sweatshirt, is typical of women and children in third-world nations who are frequent targets of multinational tobacco companies. PHOTO BY ANNA WHITE

Tobacco In History and Culture
AN ENCYCLOPEDIA

By the early 1500s, European colonists and African slaves in the Caribbean adopted the indigenous custom of smoking cigars. Before the nineteenth century, there was no particular association between masculinity and cigars. This woman in Havana carries on the cigar-smoking tradition. © DAVE G. HOUSER/CORBIS

The brilliant colors of this cigarette paper advertisement poster by Galicello (c. 1930) are characteristic of French commercial art of the period, which tended to use humor and playful style. France was a style setter in the visual arts, whether fine arts or commercial, during the nineteenth century and part of the twentieth. Another major center for lithography, commercial art, and printing techniques was Germany, whose style was known especially for its bold design and simplicity. © SWIM INK/CORBIS

Tobacco In History and Culture
AN ENCYCLOPEDIA

Harvesting and drying tobacco is done much the same way today as it was centuries ago. Here a boy in Maryland places tobacco leaves on a chestnut spear. They will later be hung to dry, probably in a barn or a shelter. © LOWELL GEORGIA/CORBIS

Many ethnic groups throughout Asia have long traditions of smoking tobacco, either hand rolled, as in a cheroot, or in a simple pipe, as this Akha tribe woman from northern Thailand demonstrates. © CHRISTINE KOLISCH/CORBIS

Tobacco In History and Culture
AN ENCYCLOPEDIA

Tobacco merchants sell tobacco products tailored for specific uses, including in pipes, in hand-rolled cigarettes, and in flavored bidis. Here tobacco is being sold loose in a market in Meghalaya, India. © EARL AND NAZIMA KOWALL/CORBIS

European and European-style pipes of the eighteenth and nineteenth centuries were made from materials such as wood, clay, and meerschaum, a fine, white clayey material. This Massachusetts man smokes a pipe whose bowl appears to be a meerschaum bust of a New England fisherman. © DAVE G. HOUSER/CORBIS

Tobacco In History and Culture
AN ENCYCLOPEDIA

Tobacco was introduced by the Spanish, Portuguese, and, principally, Dutch to various parts of the Pacific Islands beginning in the early seventeenth century. Today, cigarette smoking has largely replaced more traditional forms of tobacco use. This Huli tribeman in traditional costume takes a cigarette break during a 1990s performance in Papua, New Guinea. © BOB KRIST/CORBIS

 Tobacco In History and Culture
AN ENCYCLOPEDIA

SOCIAL AND CULTURAL USES

Satirical print, *A Smoking Club*, late eighteenth century. © HULTON-DEUTSCH COLLECTION/CORBIS

elite groups distinguish their smoking from its plebeian commonality and Amerindian origins. This stage was also marked by a move toward an increasing preoccupation with the effects of tobacco on the senses and the brain—snuff, administered through the nose, was considered the most direct route to the seat of consciousness—and a move away from, but by no means an absence of, understandings and uses of tobacco as a medicinal remedy for the body more generally. Finally, the modern period can be seen as characterized by a gradual move away from the understanding and use of tobacco as an intoxicating agent. Where, for example, in the seventeenth century both tobacco and alcohol were said to cause drunkenness, toward the end of the eighteenth century it was understood that the chief danger of intoxication in relation to tobacco use (through both smoking and snuffing) was that the practice would make one "too dry" and thus more likely to consume alcohol.

THE TRANSFER OF TOBACCO. When one considers the great spiritual and medicinal significance of tobacco to Amerindian peoples, it is hardly surprising that the plant was the focus of much attention from the first New World explorers, particularly given that, at the time of first contact, there was a widespread belief in the existence of a universal panacea—a substance that could cure any condition or ailment—one which merely awaited discovery. Indeed, by 1571, the leading Spanish physician Nicolas Monardes had published a detailed study of all the plants brought back from the Americas to that date titled *Joyfull Newes Out of the Newe Founde Worlde* in which he prescribed tobacco for almost every common ailment of the time, including toothache, carbuncles, flesh wounds, chilblains, "evill" breath, headaches, and even "cancers" (Monardes 1925), and so provided all the medical justification needed to suggest that the search for the miracle cure-all was over. Thus, in the early stages of the modern era, tobacco was widely understood to be a "divine sent" medicinal remedy (Pego et al. 1995).

It was not until the end of the sixteenth century—some fifty or more years after tobacco's first appearance in European courts—that its more recreational use became widespread. However, while the usage of tobacco may have become more diverse and hedonistic, during this period there was no clear cut dividing line, as there is today, between the consumption of substances for leisure and for medicinal purposes.

In considering the global spread of, and regional variations in, tobacco, a focus on European usage is revealing since it was Europeans who were central to the diffusion of tobacco; to establishing the plant as an important medicinal remedy (an understanding that spread alongside material tobacco); and in establishing practices and modes of consumption that were copied and ultimately adapted elsewhere (Goodman). European nations adopted their modes of consuming tobacco from those of the Amerindians they encountered in the New World. For example, the English, like the majority of European nations, generally smoked pipes after encountering this mode among the indigenous peoples of North America, whereas the Spanish more commonly smoked cigars, chewed, and snuffed tobacco after the Amerindians of Central and South America.

EARLY EUROPEAN USES AND UNDERSTANDINGS. Explorers of the New World had returned to Europe with only the mildest, and to their tastes (and constitutions), most palatable species and varieties of tobacco. Yet even these were still considerably more potent and more capable of producing intoxication than those widely consumed within the present-day west. Smokers of the time were known as tobacco "drinkers" and "dry drunks," not just because no other model was available to make sense of the practice, but possibly because the effect of the tobacco consumed at this time more closely approximated that of alcohol in terms of its narcotic capacity.

To some degree, Europeans also adopted from Amerindian cultures the idea that tobacco was a mark of sociability. Historical accounts documenting the early stages of the development of tobacco use in Europe describe an era in which smokers would meet and share pipes in tobacco sellers' shops, taverns, apothecaries, and (later in the seventeenth century) in coffee houses. Other sources point toward smokers across the social spectrum smoking almost constantly with only factors such as price and availability containing the practice, though there remains some controversy as to whether certain social proscriptions over, for example, age and gender existed.

But this widespread popularity of smoking presented a problem for those smokers belonging to elite groups within society. Tobacco use had become common, a fact that only served to compound the more general problem that smoking had its origins amongst Native American peoples who were seen to be uncivilized heathens—to whom King James I in his famous *Counterblaste to Tobacco* referred as "godlesse Indians" (James I 1954). There emerged within this period a growing need for members of such elite groups to distinguish their use of tobacco from that of, whom they considered to be, their inferiors' within plebeian society, or worse, that of the peoples of the New World.

SMOKING GALLANTS. In this historical context, the practice of smoking by some social groups became increasingly elaborate and sophisticated. In his *Social History of Smoking* (1914) G. L. Apperson, the social historian of tobacco, describes at length the ornate apparatus required by the smoking **gallant** or dandy of the seventeenth century, which included, along with a range of clay pipes, precious tobacco boxes made of silver and gold; a pipe case; a pipe pick (for cleaning); a knife to shred tobacco; and tongs to lift coals from a fire (needed to light a pipe).

gallant a well-dressed, well-spoken gentleman, attentive to the needs and concerns of ladies, but in a proper way. Rhett Butler is a gallant.

Equally elaborate were the practices that came to be adopted by these smokers: all manner of techniques involving the inhalation and exhalation of tobacco smoke, complex sets of practices concerning filling and lighting the pipe, and so forth. Moreover, in parallel there emerged professional tutors, professors, and even schools devoted to teaching such techniques and practices, along with an extensive body of knowledge concerning species and varieties of tobacco, their botanical features, their prices, their regional sources, and more.

SNUFFING AS REFINEMENT. By the beginning of the eighteenth century, however, snuffing had began to replace pipe smoking as an important means of tobacco consumption across Europe. The practice of snuffing, which had its European origins in Spain, spread to other parts of Europe, most notably France, where it had come into vogue within courtly society comparatively early. During the seventeenth and eighteenth centuries, the French court became established as a model-setting center for the European upper classes as a whole. Elite groups looked to France for standards of behavior and etiquette, and soon adopted snuffing in favor of smoking. Snuffing, in turn, involved even more elaborate practices and codes of etiquette surrounding its use than those of the smoking gallants.

The practice of snuff taking also marked a move toward more private and individualized use. Individuals would prepare their own particular and distinctive snuff concoctions, which they would consume both alone and within the context of highly specific social groups. A bewildering array of snuffs became available, which, even at the point of source, were often highly doctored and adulterated (sometimes with decidedly toxic substances), often as part of what were understood to be attempts to refine them.

With snuffing came a growing shift in socio-cultural uses and understandings of tobacco. From being considered a mark of general sociability, snuff increasingly marked individuality (both in terms of how one "took the pinch" and in the preparation of highly individualized snuff mixtures) and refinement (in terms of showing reflexive awareness of socially distinctive codes of etiquette). While there were

SOCIAL AND CULTURAL USES

This undated image from the *Fondation Napoleon* and released by the New Orleans Museum Art shows a gold and ivory tobacco box with a portrait of French Emperor Napoleon Bonaparte. © AFP/CORBIS

many treatises for and against the therapeutic value of snuff (particularly concerning its effects on sight and on the brain), the increasing use of snuff can be viewed, generally speaking, as a move away from the understanding of tobacco solely as a self-administered medicinal remedy—which itself was becoming increasingly contested in relation to smoking—and a move toward the use of tobacco as at once a marker to others and a marker of oneself.

Following a pattern that is repeated throughout the history of tobacco use, the popularity of snuff began to precipitate through the social strata such that by the end of the eighteenth century, and well into the nineteenth century, it had become the most popular form of tobacco in Europe, and no longer the exclusive preserve of a fashionable elite. The rise of snuff, however, definitely did not lead to the total demise of the pipe. Amongst the common people, and within specific occupational sectors, such as academia and the military, the pipe remained relatively popular throughout the zenith of snuff.

Indeed, the pipe began rapidly to gain broader popularity once again during the nineteenth century: This happened alongside a more general resurgence of smoking that included the rise of cigars as a popular mode of consumption, and ultimately cigarettes toward the very end of the century, so marking the beginning of the contemporary era of smoking, as discussed below. In some countries, Sweden in particular, snuffing remained popular well into the twentieth century;

in other countries, other modes of consumption prevailed before being replaced by smoking. For example, in the United States, chewing tobacco was the most popular mode of consumption from the early nineteenth century until World War I.

Contemporary Understandings and Uses

The contemporary era of tobacco use has been characterized by a shift toward the increasing, but by no means exclusive, use of tobacco as an instrument of self-control and an associated move toward progressively milder forms of tobacco. The development of cigarettes themselves continued in this direction with the twentieth-century emergence of filter cigarettes, "light" (or lower **tar** and lower nicotine) cigarettes, and more recently the advent of "super low" and "ultra low" alternatives. It is important to note, however, that the move toward "milder" and "lighter" cigarettes has involved changing the public perception of cigarettes as apparently more healthy, less dangerous products as much as—some would say more than—changing their pharmacological yield.

tar a residue of tobacco smoke, composed of many chemical substances that are collectively known by this term.

Such developments, while intimately related to changing medical understandings of tobacco and increasing regulation of the tobacco commodity, can also be understood to relate to shifts in the sociocultural uses of tobacco, specifically toward the increasing use of tobacco to return to, rather than escape from, normality—requiring tobacco which was in some ways significantly different from the highly potent, narcosis-inducing varieties consumed by, for instance, Amerindian tobacco shamans. Smoking also became an increasingly symbolic and discursive act within the contemporary era—a means of individual self-expression and a source of personal identity and a practice that has retained its associations with sociability but now more exclusively within the context of a defiant community of smokers.

THE RETURN TO SMOKING. The resurgence of smoking in the nineteenth and twentieth centuries related to a broad range of social processes. In Britain the Revolutionary and Napoleonic wars had made the ways of the French (including their penchant for snuff taking) considerably less appealing. European soldiers fighting in the Peninsula War (1808–1814) encountered cigars from the Spanish and brought these home, and cigars ultimately replaced snuff as the choice of the aristocracy. Changes in the apparatus of tobacco consumption also facilitated the return to smoking—including the development of more practical **briar** and meerschaum pipes; the invention of safety matches; and, ultimately, the growing availability of the cigarette. While by no means an invention of the nineteenth century west, the cigarette was now being produced in a form bearing many of the hallmarks (the use of milder tobaccos and a paper wrapper) of a novel mode of consumption—one which would eventually come to be central to the rise of smoking in the contemporary era.

briar a hardwood tree native to southern Europe. The bowls of fine pipes are carved from the burl, or roots, of briar trees.

SMOKING AS A SUPPLEMENTARY TOOL. Writers of this period, and many subsequent authors, partly explain the rise of new modes of consumption—the briar pipe, cigars, and early cigarettes—in terms of their convenience, their practical advantages. For example, the briar pipe was more durable than the clay pipe; cigars required hardly any of the paraphernalia of the pipe, and the invention of matches meant that one could now smoke on the move; and the cigarette had the additional advantage of only requiring a fraction of the time to consume than that needed for a cigar or pipe. However, that these characteristics were seen as advantageous can also be seen as fundamentally interrelated to a broader transformation in uses and understandings of tobacco during this period: a move away from the understanding of smoking solely as an activity in and of itself and a move toward the understanding of smoking as having a supplementary and enhancing function.

The cigarette in particular marked not just a move toward greater mobility and convenience for the smoker but arguably toward a milder and less potent mode of consumption that generated effects that were potentially less dramatic and more ambiguous than those of previous eras. Such a change was a prerequisite for the practice to become undertaken in conjunction with a range of other tasks and activities as a means of augmenting these. Smoking was increasingly being understood and used as an aid to concentration while working. Its associations with the practice of writing, for example, appear frequently in the historical literature. The Russian writer Leo Tolstoy, writing in 1890, attests to the widespread understanding that "smoking facilitates mental work" and to the common belief that "If I do not smoke I cannot write. I cannot get on; I begin and cannot continue" (quoted in Walton 2000). Smoking was also increasingly coming to be understood as a source of stimulation to counter monotonous activities, a boredom breaker, and as a **nervine** to counter stressful situations. The latter use was particularly important to soldiers fighting in World War I.

nervine a plant remedy that has a beneficial effect upon the nervous system in some way. Some herbal compounds have been used successfully in smoking cessation programs.

dysphoria a feeling of unhappiness and discomfort; being ill at ease. Cigarette smokers can experience dysphoria when deprived of cigarettes.

In combination with this broader change, and following from increasing medical and popular interest in the effect of tobacco on the brain, tobacco became widely understood as a psychological tool, one which could counter the ills of civilization—a drug which could return one to normal from a range of **dysphoric** states relating to both emotional arousal and under-arousal. Thus a defining characteristic of the contemporary era, in relative contrast to the pre-Columbian and, to a lesser degree, modern stages, was the increasing move toward understandings and uses of tobacco as a means of self-control and the further move away from the use of tobacco to lose control. This move can also be understood as marking a further development in the individualization of tobacco use. Whereas snuffers of the eighteenth and early-nineteenth centuries individualized their snuff concoctions, cigarette smokers of the twentieth century (and those of the present-day) increasingly came to individually tailor the effects and functions of tobacco.

The rise of the cigarette as a popular mode of consumption, in fact as a mass commodity, is a complex phenomenon involving a combination of social, economic, and technological changes. Significant geographical variations in the rate at which, and the degree to which, cigarette smoking replaced other modes of consumption make generalization concerning the practice difficult. Nonetheless, the cigarette was central to major transformations in the socio-cultural uses and associations of tobacco within the contemporary era.

CIGARETTES AND FEMININITY. Understandings of the cigarette itself changed dramatically from the time of its first introduction. In England,

SOCIAL AND CULTURAL USES

World War I-era advertisement for Helmar Turkish cigarettes featuring servicemen in the design. © BETTMANN/CORBIS

as an archetypal example, it was initially almost exclusively men who smoked cigarettes. However, by the end of the nineteenth century, the cigarette had become known as a feminine mode of smoking—the "female cigar" and a "weaker vessel" of tobacco (Odhner 1894). This development, in part, related to a broader transition in the gender associations of tobacco and a related rise in women smoking. Tobacco, at one stage viewed as a female commodity to be bought by men—Rudyard Kipling, for example, was known to refer to his cigars as a "harem of dusky beauties tied fifty to a string" (quoted in Mitchell 1992)—was increasingly coming to be understood as means to become like a man. Indeed, by the 1920s, cigarettes began to emerge as an emblem of women's emancipation—as a symbol of women's equality to men (Greaves 1996).

Tobacco companies of the time gradually came to exploit and harness such associations as they realized the potential of what was then a relatively untapped market. For example, in 1929 one such company organized a publicity march in New York City in which a group of women smoked "torches of freedom" as a protest against their inequality (Greaves).

TOBACCO AS A COMMODITY: MASS CONSUMERISM AND BRANDING. The eventual mass production of cigarettes brought with it advertising and mass marketing campaigns that both utilized and informed contemporary understandings and uses of tobacco. For example, the manufacturers of Lucky Strike targeted women smokers with the slogan "Reach for a Lucky Instead of a Sweet" (Goodman), thus drawing upon and reinforcing the associations between smoking and weight control.

The use of branding, especially in the early stages of the contemporary era, also became a widely popular practice. Brands presented not just the type of cigarette, but also its image and broader associations: tobacco users were more and more able just to choose a form of tobacco which suited their taste but one which expressed who they were—their individuality, their identity. Indeed, the image of any given brand might transform over time as manufacturers both engineered and responded to consumer associations with their product: Marlboro, for example, was initially a brand which signaled luxury, then femininity, and later masculinity under the guise of the Marlboro Man.

SMOKING AND GLAMOUR. As the example of Marlboro also serves to demonstrate, particularly after World War II, such brands increasingly became gendered: names such as Virginia Slims, and later, Eve Lights, were distinctively female and again evoked the notional relationship between smoking and slimness or lightness. Such brands are also indicative of a further shift in the gender associations of tobacco—a shift toward the understanding of smoking as something women do to attract men, not just through becoming thinner, but through the use of the cigarette as an expressive erotic prop (Greaves). This link between smoking, glamour, and self-expression was widely reinforced in Hollywood. Stars such as Marlene Dietrich and Bette Davis evoke a golden age of smoking when many onscreen actors and actresses appeared to breathe little else. In the era of silent movies in particular, the cigarette was widely used as an allegorical subtext to films, often to depict forbidden sexual activity (Gately 2001).

epidemiological pertaining to epidemiology, that is, to seeking the causes of disease.

TOBACCO, DEATH, AND DEFIANCE. Associations between smoking and glamour, however, were to become eclipsed by developments in medical understandings of the practice. The publication of findings from a series of **epidemiological** studies in the 1950s and 1960s that linked tobacco smoking to fatal diseases provided the impetus behind a series of dramatic shifts in popular perceptions of tobacco. From at one stage being understood as a panacea, tobacco increasingly came to be understood as in itself a pandemic—an addictive, fatal, disease that spanned the globe.

Epidemiological studies marked not just new knowledge about the long-term effects of smoking but also the predominance of a new

SOCIAL AND CULTURAL USES

medical orthodoxy in which statistical associations rather than biological mechanisms were taken as evidence of cause and effect, and with it a new way of understanding tobacco use. Partly in relation to the rise of this new medical paradigm, prevailing associations of tobacco as a source of personal control became almost inverted, and smokers came to be understood as helpless addicts enslaved to a set of biological processes over which they had little choice. That is to say, a central feature of the contemporary era is the understanding that people smoke because they are addicted. Following the more recent era of debates about passive smokers, they are, not for the first time in the history of tobacco, also prohibited from conducting the practice in an increasing number of public spaces. Thus where smoking was once a highly social activity, it is now, particularly within North America, increasingly a **marginalized** practice, one which would only be sociable within the context of specific peer groups.

However, not all present-day smokers have internalized the notion that they are powerless addicts and social outcasts. As a matter of fact, within the context of increasing opposition (on both medical and moral grounds) to the practice, smoking has increasingly come to signal an act of defiance, an idea more than complimentary to contemporary associations between smoking, freedom, and individual self-expression. Such associations form central themes within the advertising campaigns of present-day tobacco corporations. For example, the slogan "come to Marlboro country," invariably accompanied by images of an open landscape, invokes the visual metaphors of freedom and rugged **individualism.** Similarly, while the association between glamour and smoking has become considerably more equivocal and complex in Hollywood film, its

Cartoon showing the influence of smoking on British society, c. 1827. The cartoon is entitled "Puff. Puff. It is an Age of Puffing, Puff, Puff, Puff." © HISTORICAL PICTURE ARCHIVE/CORBIS

marginalization the act of shunning or ignoring certain ideas or behavior that results in it being pushed outside the mainstream of the peer group.

individualism an independence of spirit; the belief that self-interest is (or should be) the goal of all human actions.

use as a symbol of defiant personal power can be seen, for instance, in films such as *Basic Instinct*, where the leading female character coolly smokes a cigarette in the face of police interrogation.

Somewhat paradoxically, associations between cigarettes, risk, and defiance may well hold particular appeal to younger smokers. Crucially, the risk from smoking is a long-term invisible one: for a teenager the risk is apparently safe as it stands at a significant temporal distance from youth, with most tobacco-related diseases taking effect only much later in life.

See Also Addiction; Calumets; Chewing Tobacco; Cigarettes; Cigars; Class; Connoisseurship; Film; Hallucinogens; "Light" and Filtered Cigarettes; Literature; Medical Evidence (Cause and Effect); Native Americans; Pipes; "Safer" Cigarettes; Shamanism; Smoking Restrictions; Snuff; Soldiers; Therapeutic Uses; Visual Arts; Women; Youth Tobacco Use.

■ JASON HUGHES

Further Reading

For a more general discussion of the socio-cultural uses of tobacco and a sociological exploration of how these changed over time, see Hughes (2003). For an authoritative social, cultural, and economic history of tobacco, see Goodman (1993). Wilbert (1987) provides an extensive ethnopharmacological discussion of tobacco use in the pre-Columbian era, particularly its relationship to shamanism. Finally, Walton (2000) provides an interesting and lively collection of historical writings on tobacco from the first European encounters to the present day.

BIBLIOGRAPHY

Adams, K. R. "Prehistoric Reedgrass (Phragmites) 'Cigarettes' with Tobacco (Nicotiana) Contents: A Case Study from Red Bow Cliff Dwelling, Arizona." *Journal of Ethnobiology* 10, no. 2 (Winter 1990): 123–139.

Apperson, G. L. *The Social History of Smoking*. London: Ballantyne Press, 1914.

Balabanova, S., et al. "Was Nicotine Known in Ancient Egypt?" *Homo: The Journal of Comparative Human Biology* 44, no. 1 (June 1993): 92–94.

Bean, Lowell John, and Sylvia Brakke Vane. "Cults and Their Transformations." In *Handbook of North American Indians*. Vol. 8, *California*. Edited by Robert F. Heizer. Washington, D.C.: Smithsonian Institution, 1978. Pp. 662–672.

Brandt, Allan. "The Cigarette, Risk, and American Culture." *Daedalus* 119, no. 4 (fall 1990): 155–176.

Brooks, Jerome E, ed. *Tobacco: Its History Illustrated by the Books, Manuscripts and Engravings in the Collection of George Arents, Jr.* 5 vols. New York: The Rosenbach Company, 1937–1952.

Denig, Edwin Thompson. *Of the Crow Nation*. Edited by John C. Ewers. 1953. Reprint, New York: AMS Press, 1980.

Elferink, Jan G. R. "The Narcotic and Hallucinogenic Use of Tobacco in Pre-Columbian Central America." *Journal of Ethnopharmacology* 7 (1983): 111–122.

Fairholt, F. W. *Tobacco: Its History and Associations*. London: Chapman and Hall, 1859.

Gately, Iain. *La Diva Nicotina: The Story of How Tobacco Seduced the World*. London: Simon & Schuster, 2001.

Goodin, R. E. "The Ethics of Smoking." *Ethics* 99 (1989): 574–624.

Goodman, Jordan. *Tobacco in History: The Cultures of Dependence*. London: Routledge, 1993.

Greaves, Lorraine. *Smoke Screen: Women's Smoking and Social Control*. London: Scarlet Press, 1996.

Gusfield, Joseph R. "The Social Symbolism of Smoking and Health." In *Smoking Policy: Law, Politics, and Culture*. Edited by Robert L. Rabin and Stephen D. Sugarman. New York: Oxford University Press, 1993. Pp. 49–68.

Haberman, Thomas W. "Evidence for Aboriginal Tobaccos in Eastern North America." *American Antiquity* 49, no. 2 (1984): 269–287.

SOCIAL AND CULTURAL USES

Hackwood, Frederick William. *Inns, Ales, and Drinking Customs of Old England.* London: T. Fisher Unwin, 1909.

Harrington, John P. *Tobacco among the Karuk Indians of California.* Washington, D.C.: U.S. Government Printing Office, 1932.

Hughes, Jason. *Learning to Smoke: Tobacco Use in the West.* Chicago: University of Chicago Press, 2003.

James I. *A Counterblaste to Tobacco.* 1604. Reprint, London: Rodale Press, 1954.

Janiger, Oscar, and Mariene Dobkin de Rios. "Suggestive Hallucinogenic Properties of Tobacco." *Medical Anthropology Newsletter* 4 (1973): 6–11.

———. "Nicotiana an Hallucinogen?" *Economic Botany* 30 (April–June 1976): 149–151.

Kiernan, V. G. *Tobacco: A History.* London: Hutchinson Radius, 1991.

Klein, Richard. *Cigarettes Are Sublime.* Durham, N.C.: Duke University Press, 1993.

Koskowski, Wlodzimierz. *The Habit of Tobacco Smoking.* London: Staples Press, 1955.

Kroeber, Alfred L. "Culture Element Distributions: XV. Salt, Dogs, Tobacco." *Anthropological Records* 6, no. 1 (1941): 1–20.

Krogh, David. *Smoking: The Artificial Passion.* New York: W. H. Freeman, 1991.

Lohof, Bruce A. "The Higher Meaning of Marlboro Cigarettes." *Journal of Popular Culture* 3 (1969): 441–450.

Mackenzie, Compton. *Sublime Tobacco.* London: Chatto and Windus, 1957.

McCracken, Grant. *Culture and Consumption: New Approaches to the Symbolic Character of Consumer Goods and Activities.* Bloomington: Indiana University Press, 1988.

Mitchell, Dolores. "Images of Exotic Women in Turn-of-the-Century Tobacco Art." *Feminist Studies* 18, no. 2 (summer 1992): 327–350.

Monardes, Nicolás. *Joyfull Newes Out of the Newe Founde Worlde.* Translated by John Frampton. London: Constable, 1925.

Morgan, Lewis H. *League of the Ho-dé-no-sau-nee or Iroquois.* New York: Dodd, Mead and Company, 1901.

Odhner, Carl Theophilus. *Tobacco Talk by an Old Smoker, Giving the Science of Tobacco.* Philadelphia: The Nicot Publishing Company, 1894.

Pego, Christina M, et al. "Tobacco, Culture, and Health among American Indians: A Historical Review." *American Indian Culture and Research Journal* 19, no. 2 (1995): 143–164.

Penn, W. A. *The Soverane Herbe: A History of Tobacco.* London: Grant Richards, 1901.

Porter, Cecil. *Not Without a Chaperone: Modes and Manners from 1897 to 1914.* London: New English Library, 1972.

Robert, Joseph C. *The Story of Tobacco in America.* New York: Alfred A. Knopf, 1949.

Rogozinski, Jan. *Smokeless Tobacco in the Western World, 1550–1950.* New York: Praeger, 1990.

Schivelbusch, Wolfgang. *Tastes of Paradise: A Social History of Spices, Stimulants, and Intoxicants.* Translated by David Jacobson. New York: Pantheon Books, 1992.

Schleiffer, Hedwig. *Narcotic Plants of the Old World Used in Rituals and Everyday Life: An Anthology of Texts from Ancient Times to the Present.* Monticello, New York: Lubrecht & Cramer, 1979.

Springer, J. W. "An Ethnohistory Study of the Smoking Complex in Eastern North America." *Ethnohistory* 28 (1981): 217–235.

Tate, Cassandra. *Cigarette Wars: The Triumph of "the Little White Slaver."* New York: Oxford University Press, 1999.

Tooker, Elisabeth. *An Ethnography of the Huron Indians, 1615–1649*. Washington, D.C.: U.S. Government Printing Office, 1964.

von Gernet, A. "Hallucinogens and the Origins of the Iroquoian Pipe/Tobacco/Smoking Complex." In *Proceedings of the 1989 Smoking Pipe Conference*. Research Records No. 22. Edited by Charles F. Hayes III. Rochester, New York: Research Division of the Rochester Museum and Science Service, 1992. Pp. 171–185.

Walton, James, ed. *The Faber Book of Smoking*. London: Faber and Faber, 2000.

Wilbert, Johannes. "Tobacco and Shamanistic Ecstasy among the Warao Indians of Venezuela." In *Flesh of the Gods: The Ritual Use of Hallucinogens*. Edited by Peter T. Furst. New York: Praegar, 1972. Pp. 55–83.

———. *Tobacco and Shamanism in South America*. New Haven, Conn.: Yale University Press, 1987.

———. "Does Pharmacology Corroborate the Nicotine Therapy and Practices of South American Shamanism?" *Journal of Ethnopharmacology* 32, no. 1–3 (April 1991): 179–186.

Soldiers

physiology the study of the functions and processes of the body.

Soldiers have been a major force in the diffusion of tobacco use globally. This is partly because the **physiological** properties of tobacco lend it to use in wartime and partly because travel to different countries offers soldiers the opportunity to trade goods. Scholars have attributed the introduction of the cigar into Britain to a mixing of soldiers from different countries—primarily Spain, Portugal, and England—during the Peninsular War (1808–1814). The Crimean War (1853–1856) served a similar purpose for cigarettes. In the United States, soldiers fighting in Mexico in 1848 brought back cigars, while soldiers brought back cigarettes from lands gained through the Spanish-American War of 1898. During the Boer War (1899–1902), superstitions about smoking practice developed among the soldiers. It was considered unlucky to light three cigarettes from the same match, as this gave the enemy time to spot, target, and shoot the third smoker.

However, it was World War I (1914–1918) that was pivotal to the expansion of cigarette smoking, as cigarettes came to be seen as an essential rather than a luxury item. For the troops, smoking was an escape from the reality of war, and a way of establishing a rapport with fellow soldiers, regardless of rank. Smoking was also thought to calm nerves and cigarettes were more convenient to smoke than pipes or cigars. As an article in the *Tobacco Trade Review* noted in 1915, "[the cigarette] requires no pipe, there is nothing to lose except the match." The packets and tins in which cigarettes came also played a role at the front—backs of cigarette packets were stuck to torn banknotes to hold them together. Player's, a British firm, received a number of letters from soldiers claiming to owe their lives to cigarette and tobacco tins in their breast pockets. The tins stopped shrapnel from hitting their flesh, and they enclosed the bullet-pierced tins with their letters to prove it.

For those back home, sending cigarettes became a way of showing support for the troops and the war effort. In December 1914 in Britain, for example, the people of Glasgow held a Tobacco Day with a fancy dress

This woman is packing cigarettes at a London tobacco factory in 1939. Parcels of duty-free cigarettes were sent to the British field forces free through the Customs daily and were admitted into France free of French duty. All the best-known cigarette manufacturers were accepting orders on this basis. © HULTON-DEUTSCH COLLECTION/CORBIS

parade and torchlight procession, collecting money to buy cigarettes for the troops. Cigarette funds were established across the country, and women bought and sent cigarettes to loved ones. Advertising drew on military imagery and the cigarette became a symbol of patriotism and unity, linking the civilian population and the armed forces. The cigarette manufacturer British American Tobacco started up a weekly bulletin to keep troops in touch with families and colleagues back home. In Britain by 1915, the cigarette trade had doubled, mostly through exports to the front. In 1916, the British government took on the task of supplying **duty**-free tobaccos and cigarettes directly to the troops. By the end of World War I the cigarette had overtaken pipe tobaccos and cigars in popularity.

duty a tax, usually a tax on certain products by type or origin. A tariff.

World War I had a similar effect on smoking in the United States. When the United States entered the war in 1917, cigarettes were included in soldiers' rations and subsidized in post exchange stores at home. Cigarette manufacturing was deemed an essential industry, and civilians and charitable organizations collected cigarettes to augment government supplies to soldiers abroad. Recruiting posters carried images of men with cigarettes. Through these associations with patriotism, cigarette smoking gained respectability and became the most popular way of consuming tobacco.

Cigarette consumption among soldiers was also criticized, however. In Britain, there was concern that valuable shipping space was being taken up by tobacco rather than food, while in the United States concerns were raised over the effects of tobacco on the health of servicemen. Such concerns were not new: In 1845, the Duke of Wellington tried to dissuade troops from smoking, and in the aftermath of the British defeat in the Boer War juvenile smoking was one reason suggested for the poor physical health of recruits.

Smoking rates continued to rise in the decades following World War I, and for many men service in the armed forces provided the introduction. During World War II (1939–1945), steps were taken to secure the supply of cigarettes to the front immediately. In Britain, the industry was brought under government control, a tobacco controller was appointed, and home supplies were cut in order to meet demand for the troops. Even in Germany, where the Nazi regime was antismoking, soldiers were provided with tobacco rations during the war. But it was the American soldier, with his generous supply of cigarettes, who was most strongly associated with cigarettes in Europe. Girls dating GIs received gifts of nylons, cigarettes, and lipstick, while in Germany after the war American brands fetched the highest value on the black market and nurtured the taste for American-blended cigarettes that prevails in the 2000s.

opium an addictive narcotic drug produced from poppies. Derivatives include heroin, morphine, and codeine.

Cigarettes remained an essential part of soldiers' kits during later wars of the twentieth century, although during the Vietnam War marijuana, **opium,** and heroin often replaced tobacco as the drugs of choice. However, the close association of cigarettes and soldiers has left a long-term mark on the health of both the veterans of conflict and the countries they saw service in. Successive mortality studies by the Australian government's Department of Veteran's Affairs have found elevated mortality from smoking-related cancers among Korean and Vietnam War veterans. In the 2000s, the reunified German government is implementing public health programs to tackle the problem of smoking among the population.

See Also Intellectuals; Missionaries; Sailors.

▮ ROSEMARY ELLIOT

BIBLIOGRAPHY

Elliot, Rosemary. "Destructive, but Sweet: Smoking among Women in Britian, 1890–1990." Ph.D. diss., University of Glasgow, 2001.

Tate, Cassandra. *Cigarette Wars: The Triumph of "the Little White Slaver."* New York: Oxford University Press, 1999.

Times (London), 4 October 1939, pp. 3, 5.

Tobacco Trade Review (London), 1 January 1915, p. 7.

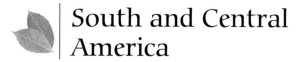

South and Central America

Tobacco was cultivated throughout Latin America prior to colonization, and during the colonial (1500–1800) and postcolonial periods (from 1800 onward). Small-scale farming of the crop in rural areas occurred in countries as diverse as Honduras, Ecuador, Argentina, and Colombia. Gradually, as demand for tobacco abroad increased, commercial cultivation of tobacco took place.

Colonial Beginnings

Commercial tobacco farming started in the late eighteenth century when tobacco cultivation became an important component of the economy in countries such as Mexico, Colombia, and Cuba. In an effort to

SOUTH AND CENTRAL AMERICA

maintain control over commercial tobacco production in the New World, the Spanish Crown designated specific zones in which tobacco farming was permitted. It established tobacco monopolies in the larger countries by which tobacco trade was the exclusive monopoly of the Spanish government. This often implied the building of large warehouses in which the tobacco was processed before being exported to Europe.

In Brazil tobacco production was strongly controlled by the Portuguese Crown. At the same time, tobacco was an important component in the illegal trade that started to flourish in the more remote parts of the Spanish Empire in the eighteenth century. Everywhere in the Spanish territories around the Caribbean small farmers had started to cultivate tobacco from the late sixteenth century onward. British and Dutch buccaneers avidly bought this tobacco. The Spanish Crown depopulated whole tobacco producing regions in present-day Venezuela and the island Hispaniola in a fruitless effort to counter these incursions on its territory. Illegal tobacco trade was in this way instrumental in the undermining of Spanish colonialism and eventually played an important role in its end.

After Latin American independence at the beginning of the nineteenth century northwestern Europe—particularly London, Amsterdam, and the independent Hanseatic towns in northern Germany, Hamburg, and Bremen—became the most important consumers of Latin American tobacco. The urban markets for cigars and later cigarettes were instrumental in the rapid expansion of the tobacco trade in the nineteenth

Guatemalan Mayan gives a cigar to pagan saint Maximon during a ceremony in Santiago Atitlan, 2000. Once a year Mayans celebrate this pagan saint, who represents a mixture of Catholic and Mayan beliefs. During the celebration, the people who visit Maximon must make offerings of alcoholic beverages, tobacco, and money. © REUTERS NEWMEDIA INC./CORBIS

SOUTH AND CENTRAL AMERICA

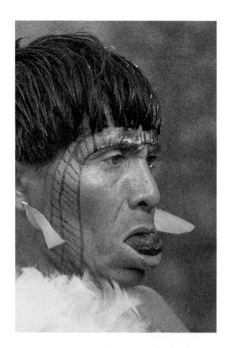

Latin American countries like Chile and Brazil have implemented antismoking legislation since the late 1990s. However, tobacco consumption in Latin America is still growing. Here, a Brazilian warrior chews tobacco, 1990. © CORBIS SYGMA

sharecropping a form of agricultural labor that gained popularity after the Civil War. Laborers, usually families, lived and worked on land belonging to a proprietor. They grew staple crops like tobacco and cotton. Rather than regular cash wages, they were paid with shares of the crop at harvest time.

and twentieth centuries, making tobacco an important international commodity. The European traders had offered moral and material support to the Latin American cause of independence and helped them to break the Spanish commercial monopoly. Now they benefited greatly when the new Latin American republics opened their commercial borders.

Black Tobacco Cultivation

Prior to the turn of the twentieth century, tobacco cultivation in Latin America was limited to a great variety of local types of dark (or black) tobacco, which was used for cigars and, later, for the dark cigarettes smoked by the proletarian masses of the industrialized world. Farming dark tobacco was an undertaking well suited to the subsistence-based peasant agriculture prevalent in most Latin American countries at the time, for the rural families were easily able to integrate the cultivation of tobacco into their subsistence agriculture. Planters in Cuba and in parts of Brazil and Argentina produced better quality tobacco, but tobacco cultivation in these regions also remained a small-scale operation.

Until well into the twentieth century, dark tobacco was usually produced on small family farms, which managed the cultivation and most of the processing. In regions with open land tobacco was often produced through slash-and-burn agriculture, but in regions where access to land was limited, tobacco farmers often engaged in various kinds of **sharecropping** called *aparcería*. According to this system, farmers had to give part of their harvest to the landowner.

Tobacco leaves were dried in primitive sheds, sometimes in and around the rural dwellings, before being piled up, fermented, and sold to local middlemen. These *corredores*, as they were called in the Dominican Republic, were usually small traders with good relationships among the rural population. They offered the peasant families access to goods and small loans in exchange for their tobacco harvest. They later tried to sell the tobacco they accumulated in this way to large export houses called *especuladores*. On all levels of this commercial chain, tobacco was exchanged for money in an intricate and often complex set of negotiations in which personal trust and loyalty, profit, and expectations were inextricably intertwined.

Rural Industries

A portion of the peasant-grown tobacco was always intended for local consumption. This often led to artisanal ways of processing the tobacco. Primitive cigars, crumbed tobacco (*polvo*), or tobacco leaves pressed and processed in tight roles for smoking in rural pipes were some of the ways by which tobacco was consumed locally. Most of the tobacco was destined for the international market, however. Apart from Cuba, which acquired fame for its fine tobacco and well-manufactured cigars, the primary Latin American exporters of tobacco leaves (Brazil, Colombia, and the Dominican Republic) generally produced low-quality, relatively cheap bulk tobacco leaves (*tabaco en rama*) in the late nineteenth and twentieth centuries. The tobacco was avidly sought for in the late nineteenth and twentieth centuries because it was well suited to mix with more expensive tobacco from the Dutch East Indies.

SOUTH AND CENTRAL AMERICA

The European cigar industry required different types of tobacco, which were mixed to give a specific taste to each brand of cigars. After the 1860s, the Netherlands Indies (now Indonesia) became the most important producer of cigar tobacco, but European cigar manufacturers continued to seek out other varieties to combine with the tobacco from Sumatra or Java and soon found what they required in the Latin American countries of Colombia, the Dominican Republic, and Brazil. The Dominican Republic exported some 10 million kilograms in 1920. Exportation from Bahia, a region in northeastern Brazil, normally oscillated between 20 and 30 million kilograms in the same period. The income these exports generated had significant economic effects for the regional and national economy of the tobacco producing countries. It provided work for men and women in the region and brought in much-needed cash. In this way, entire tobacco producing region breathed on the rhythm of the tobacco trade.

Because of the segmented nature of the tobacco market, in which each variety of tobacco had its own niche, Latin American tobacco was normally easily sold on the international market in spite of increasing competition from other parts of the world. The market for dark tobacco changed in the twentieth century when the consumption of cigarettes overtook that of cigars. The cigar industry became more sophisticated in this period, and there was increased demand for better quality leaves. In an effort to improve the quality of the tobacco, governments and tobacco exporters attempted to curtail the autonomy of the peasant producers and to implement changes in the cultivation and processing of the tobacco leaf. Everywhere state institutions for tobacco were established. Moreover, European and U.S. importers began moving into the producing regions.

Tobacco Industry

Beginning in the nineteenth century, tobacco producing Latin American countries began manufacturing tobacco for local consumption in small cigar factories that often employed only a handful of cigar makers. In the late nineteenth and twentieth centuries some of these small enterprises expanded their operations and evolved into large businesses that produced both cigars and cigarettes. Just as the Cuban author Fernando Ortiz tried to point out in his famous study *Cuban Counterpoint*, (first published in 1940) many Latin Americans remember the early tobacco industry as a sane, human, and truly national economic sector.

Cigar makers at this time were often relatively well-educated individuals with good knowledge of the world and high professional standards. The quiet environment of the traditional workplaces often allowed for a reader (*lector*), who was paid by the workers, to read aloud newspapers or popular novels. This romantic image of the early tobacco industry acquired force in the second half of the twentieth century when cigarettes began replacing cigars, and the cultivation and consumption of tobacco radically changed. In countries like Brazil, Argentina, Mexico, Colombia, and the Dominican Republic large industrial enterprises came into existence, which often under the license of big companies such as Phillip Morris or the British-American Tobacco Company (BAT) started to produce cigarettes on a large scale.

La Información, a newspaper published in the Dominican Republic, described the local tobacco scene in 1917 this way: "Daily teams of mules (*recuas*) enter the city loaded with the aromatic leaf. The storehouses are being filled and the large halls of the storehouses are busy and animated by the presence of the women who select, store and classify the tobacco. The female workers arrive in the early morning to take their places and sweeten the monotonous atmosphere of the trade with their happy songs. Similarly, the men who classify the tobacco and the sweaty and content porters shout the *to le lá* which, rhythmical and melodious, incites them to the work which they do happily because they know it guarantees them their daily bread."

Modern Developments

In the latter half of the twentieth century, increasing cigarette consumption led to greater demand for blond tobaccos such as Burley or Kentucky, locally designated as *tabaco rubio*. Initially, Latin American nations did not produce blond tobacco and had to import it from the United States. In an effort to end this dependence, Latin American governments began to stimulate the cultivation of blond tobacco by introducing new seeds and giving technical advice to the farmers. By the 1940s Latin American production of Burley and **flue-cured tobaccos** was more than 45 million kilograms, which was mostly consumed in Latin America.

flue-cured tobacco also called Bright Leaf, a variety of leaf tobacco dried (or cured) in air-tight barns using artificial heat. Heat is distributed through a network of pipes, or flues, near the barn floor.

Blond tobacco cultivation required closed drying sheds and more sophisticated technology that was beyond the means of the traditional small producers, and this sector therefore became dominated by large-scale capitalist producers, usually companies that were financed partly by foreign capital. Over time, cigarette production in Latin America fell under the control of a small group of large multinational firms such as Philip Morris and British American Tobacco. Often these firms bought existing local companies, maintaining their names. In the 2000s, these two companies control a large part of the cigarette industry in Latin America.

Efforts to ban or restrict smoking have surfaced in Latin America since the late 1990s and countries like Chile and Brazil have implemented antismoking legislation. However, they have not been as effective as similar campaigns in the United States or Europe. Indeed, tobacco consumption in Latin America is still growing and therefore offers interesting opportunities for these multinational companies. At the start of the twenty-first century, Brazil and Argentina were the largest producers of tobacco in Latin America, and Guatemala, despite the fact that its largely indigenous population has a low smoking rate, was the largest producer of tobacco in Central America.

In the shadow of Latin America's multinational cigarette industry, peasant cultivation of dark tobacco continues to be important in many countries. In some areas, local entrepreneurs have succeeded in taking advantage of the restructuring of the global cigar market after the 1959 Cuban Revolution. Because the United States no longer imports Cuba's high-quality cigars, a new, thriving cigar industry has developed in the Dominican Republic. Brazilian companies also successfully captured a larger share of the internal cigar market in the latter part of the twentieth century.

See Also Brazil; British American Tobacco; Cigars; Cuba; Mexico; Philip Morris; Portuguese Empire; Spanish Empire.

■ MICHIEL BAUD

BIBLIOGRAPHY

Baud, Michiel. *Peasants and Tobacco in the Dominican Republic, 1870–1930*. Knoxville: University of Tennessee Press, 1995.

———. "A Colonial Counter Economy: Tobacco Production On Española, 1500–1870." *Nieuwe West-Indische Gids* (New West Indian Guide) 65, no. 1–2 (1991): 27–49.

Deans-Smith, Susan. *Bureaucrats, Planters, and Workers: The Making of the Tobacco Monopoly in Bourbon Mexico*. Austin: University of Texas Press, 1992.

Harrison, John Parker. "The Colombian Tobacco Industry, from Government Monopoly to Free Trade: 1778–1876." Ph.D. diss., University of California at Berkeley, 1951.

Ortiz, Fernando. *Cuban Counterpoint: Tobacco and Sugar.* Translated by Harriet de Onis. Durham, N.C.: Duke University Press, 1995.

Stubbs, Jean. *Tobacco on the Periphery: A Case Study in Cuban Labour History, 1860–1958.* New York: Cambridge University Press, 1985.

South Asia

Tobacco was first brought to the South Asian region in the early 1600s by the Portuguese, who had a colony in the port of Goa in southern India. The Mughals, powerful Central Asian Turks who ruled India at this time, also used tobacco. When one of the great Mughal leaders of South Asia, Akbar, received a gift of tobacco and a pipe, his physician forbade him to inhale the smoke of tobacco, which was an unknown, potentially harmful substance. He suggested it would be safer if Akbar passed the tobacco through water, and thus the *hookah*, or water pipe, was created. Smoking tobacco through a *hookah* became popular at social and religious functions under the Mughal courts and among the aristocratic and elite classes in the region. The sharing of a *hookah* was viewed as a form of friendship and as a measure of social acceptance.

Tobacco became a popular commodity in South Asia. Persian and South Asian traders carried tobacco and other goods along the Silk Road for distribution among the courts of the Persian, Mughal, and Chinese rulers. Initially, only the Mughal rulers and the wealthy upper classes, predominantly men, smoked tobacco. However, within a century, tobacco was incorporated into the existing habit of chewing areca nut (*Areca catechu*) and was enjoyed by both men and women across socioeconomic classes. Tobacco was generally used to stave off hunger during travel, to suppress pain from toothaches, and to sustain long hours of difficult work.

The Portuguese introduced tobacco cultivation into India on a small scale in the seventeenth century, but it was not until the British colonial period that the crop began to be cultivated extensively. As tobacco growing in North America became disrupted by the American Revolution (1776), British colonies around the globe began to take over the production and growing of tobacco. By the 1800s, South Asia had its own tobacco **plantations** growing Virginia tobacco and the product was cheaply available. Between 1890 and 1920, the area under tobacco cultivation in British India tripled, and it continued to grow into the next decades. By 1930, India had begun to flue-cure tobacco and had established itself as a major global producer of tobacco. In fact, in 1938, the area known as British India (which in the twenty-first century comprises Pakistan, Bangladesh, Sri Lanka, and Burma) ranked third in world tobacco production.

During the Indian Independence movement, Mahatma Gandhi was a staunch critic of British rule. He preached that the values and agricultural practices of the British had to be done away with and replaced by a simpler, more spiritual lifestyle. In his writings and speeches, Gandhi

plantation historically, a large agricultural estate dedicated to producing a cash crop worked by laborers living on the property. Before 1865, plantations in the American South were usually worked by slaves.

SOUTH ASIA

Bidis, small hand-rolled cigarettes, are the most popular tobacco product on the Indian market. They are the cigarette of choice among agricultural laborers because they are inexpensive. Men often smoke them as a form of relaxation and to pass time. This photograph shows packs of bidis displayed at a New Delhi sidewalk stall, 2002. © AFP/CORBIS

criticized the growing of cash crops like tobacco, which had been introduced by the British and did not provide food for India's people. He urged that tobacco be replaced by food crops and cotton, which would directly feed and clothe the people of India.

Modern Cultivation and Production of Tobacco

After India gained independence from Britain in 1947, tobacco continued to be an important crop. In the 2000s, India is the third largest producer of tobacco in the world, following China and the United States, and the eighth largest exporter of tobacco and tobacco products. India produces 600 million kilograms of tobacco annually, which fall under two botanical species, *Nicotiana tabacum* and *Nicotiana rustica*. Two-thirds of the tobacco grown in India is the variety used for making bidis (a hand-rolled cigarette), for making chew products, and for use in a *hookah*. One-third of the tobacco grown in India is **flue-cured** Virginia tobacco, used in the making of commercial cigarettes.

flue-cured tobacco also called Bright Leaf, a variety of leaf tobacco dried (or cured) in air-tight barns using artificial heat. Heat is distributed through a network of pipes, or flues, near the barn floor.

Approximately 3 percent of agricultural land in India is devoted to tobacco growing. The major tobacco growing regions of India are the states of Andhra Pradesh, Gujarat, and Karnataka. Over 3.5 million people are estimated to be engaged full time in tobacco manufacturing, which accounts for almost 12 percent of all manufacturing work in the country. Almost 1 million people work in growing and curing tobacco.

Tobacco Consumption

India and more generally, South Asia, is a region in which tobacco is consumed in a multitude of ways.

BIDIS. Bidis are small cigarettes that consist of indigenously grown tobacco wrapped in a dried leaf (*Diospyros melanoxylon*) and tied with a thread. They are hand rolled by women at home as a cottage industry and sold in packets of twenty to thirty bidis. Bidis account for about 50 percent of tobacco consumption in the region.

Although smaller than cigarettes, bidis yield more than three times as much carbon monoxide and more than five times as much nicotine and **tar** as cigarettes. Since the leaf is not porous, the bidi smoker has to inhale often and deeply to keep it lit. A bidi smoker must take three to four times as many puffs as one does with a cigarette.

Bidis are the most popular tobacco product in India, Pakistan, Nepal, and Bangladesh, particularly among agricultural laborers. They are also widely smoked in Sri Lanka. Bidis are exported all over the world. In Western countries, flavored and filtered bidis have become popular among youth.

tar a residue of tobacco smoke, composed of many chemical substances that are collectively known by this term.

CIGARETTES. In India, about 20 percent of the total tobacco consumed is in the form of cigarettes and more than 65 percent of cigarette sales in India are for single sticks. The cost of a single cigarette ranges from 4 to 6 cents, depending on the brand. Although cigarettes are relatively inexpensive, they are eight to ten times more expensive than bidis. Cigarettes have come to be associated with higher socioeconomic status and modern lifestyles. It is usually men with higher education who smoke cigarettes, while bidis are consumed by the uneducated. Cigarettes are distributed by a highly sophisticated marketing network which reaches even the most remote village shops.

Cigarette production has increased steadily in India. In 1970, 62,900 million sticks were produced. In 2000 and 2001, 91,400 million sticks were produced and this number exceeded 100,000 million sticks in 2001 and 2002. The consumption of cigarettes is high in India with approximately 110 billion cigarettes sold each year (the equivalent of $2 billion in sales). Between 1990 and 1995, per capita consumption of cigarettes in India increased, a distinction it shares with only two other countries: China and Indonesia.

SMOKELESS TOBACCO. Oral use of smokeless tobacco is very common in India and other countries in South Asia, and is both prepared by the user as well as available prepackaged. *Pan*, also known as "betel quid," is a product hand rolled at the time of consumption. It consists of one of several varieties of betel leaf (*Piper betle*), in which areca nut (*Areca catechu*) and slaked lime are added, often along with tobacco. Slaked lime is added

SOUTH ASIA

Antitobacco activists in New Delhi display a skeleton and sachets containing locally produced chewing tobacco, areca nut, lime, spices, and additives known as *gutkha*. *Gutkha* has been a source of public health concern in India because regular chewers experience a rapid progression to precancerous oral lesions. In India, more than 2,200 people are estimated to die every day from tobacco-related causes. © AFP/CORBIS

because it increases the absorption of the betel nut and tobacco for the consumer. Condiments and sweetening agents may also be added.

Pan chewing is a widespread cultural practice engaged in at important events such as marriages, funerals, and ritual performances. In rural areas of South Asia, it is common for guests to be offered a plate containing betel leaves, areca nut, and tobacco shortly after entering a home. It is also chewed commonly as a pastime and by agricultural laborers who enjoy its mildly stimulant effects.

Gutkha, a prepackaged mixture of chewing tobacco, areca nut, lime, and aromatic spices, is sold in small packets. It is widely available in small roadside shops and costs between 4 and 11 cents per packet, depending on the size of the packet. Rapidly distributed through a network of agents, *gutkha* is popular among youth and manual laborers, who find this product fast and convenient to use, and people in urban areas who wish to chew but do not want to stain their mouth red, which occurs as a result of chewing betel quid.

Khaini, a packaged chewing preparation containing tobacco flakes, slaked lime, and aromatic spices is cheaper than *gutkha*, costing only only 4 cents per packet. *Khaini* is stronger and harsher than *gutkha* and one packet contains sufficient tobacco for three chews. Consumption patterns of the many tobacco products available in India vary widely by age, geographic region, and by economic class.

OTHER TOBACCO FORMS. Other popular forms of tobacco in India are *cheroots* and *chutta,* which are small cigars made by rolling a tobacco leaf. These are popular in particular regions. Another popular product is **snuff,** which is powdered tobacco that is inhaled through the nose. There are also several smokeless tobacco preparations (*mishri,* creamy snuff, *bajjar*) which are intended primarily for cleaning one's teeth but are potentially addictive.

snuff a form of powdered tobacco, usually flavored, either sniffed into the nose or "dipped," packed between cheek and gum. Snuff was popular in the eighteenth century but had faded to obscurity by the twentieth century.

DEMOGRAPHICS. Prevalence data on tobacco consumption in India is minimal. The World Health Organization estimates that between 52 and 70 percent of males and between 3 and 38 percent of females currently use tobacco in some form in different areas of India and South Asia. In general, South Asian men smoke as well as chew tobacco, whereas the vast majority of women who use tobacco are chewers (usually hand-rolled *pan* with tobacco). Cigarette smoking among women is not widely acceptable in South Asia, although it is gaining some popularity among the elite in urban centers of India such as Mumbai, Delhi, Bangalore, and Chennai.

▌MIMI NICHTER

BIBLIOGRAPHY

Bhonsle, R. B., P. R. Murti, and Prakash C. Gupta. "Tobacco Habits in India." In *Control of Tobacco-Related Cancer and Other Diseases: International Symposium.* Edited by Prakash C. Gupta, James E. Hamner, III and P. R. Murti. Bombay: Oxford University Press, 1992.

Centers for Disease Control and Prevention. *Global Youth Tobacco Survey,* 2000.

Kaufman, Nancy J., and Mimi Nichter. "The Marketing of Tobacco to Women: Global Perspectives." In *Women and the Tobacco Epidemic.* Edited by Jon M. Samet and Soon-Young Yoon. Washington, D.C.: World Health Organization, 2001.

Kumar, S. "India Steps Up Anti-Tobacco Measures." *Lancet* 356 (2000): 1089.

Nichter, Mimi, Mark Nichter, and David Van Sickle. "Popular Perceptions of Tobacco Products and Patterns of Use Among Male College Students in India." *Social Science and Medicine* (2004).

Ray, Cecily Stewart, with Prakash Gupta and Joy de Beyer. *Research on Tobacco in India (Including Betel Quid and Areca Nut): An Annotated Bibliography of Research on Use, Health Effects, Economics and Control Efforts.* Washington, D.C.: World Bank Publications, 2003.

World Bank. *Curbing the Epidemic: Governments and the Economics of Tobacco Control.* Washington, D.C.: World Bank, 1999.

South East Asia

While the per capita consumption of tobacco remains high throughout South East Asia, the industry faces many challenges in the twenty-first century, including rising excise taxes, tough antismoking legislation, increasingly competitive markets, and shrinking demand.

SOUTH EAST ASIA

Two young women making cheroots. These cheroots are about eight inches in length and are composed of small pieces of pith combined with tobacco leaf, which is enclosed in a covering made of a thick, fibrous, white leaf. Traditionally, Burmese cheroots were the choice of most Burmese smokers, but conventional cigarettes have overtaken cheroots in popularity.
© BETTMANN/CORBIS

The climate and soil in many areas of South East Asia lend themselves to the cultivation of tobacco, and since its introduction as a cash crop by European colonists tobacco has played a major role in shaping the economies of the region. However, tobacco consumption, especially the smoking of cheroots (cigars cut square at both ends), predates the formal introduction and cultivation of tobacco in many Asian countries, including Sumatra (now Indonesia), the Philippines, Burma (now Myanmar), and Sarawak (Eastern Malaysia).

Sumatra

Sumatra was one of the first areas where formal tobacco cultivation was introduced. Sumatrans had known and enjoyed tobacco prior to the arrival of colonizers and were smoking hand-rolled cheroots long before the Europeans started developing a tobacco industry in the country. In 1863, the Dutch businessman Jacobus Nienhuys traveled to Sumatra to buy tobacco, but was disappointed with the cultivation and production techniques in use there. His efforts to improve Sumatran tobacco production are credited with establishing the Indonesian tobacco industry, which by the turn of the twentieth century ranked as one of the world's top producers.

Indonesia is among the top five tobacco-consuming nations in the world, and it exports various grades and types of tobacco to many countries. Indonesian cigar wrappers are widely used in many international cigar brands and are recognized as having superior qualities. Several different types of cigar filler tobaccos are also cultivated in Indonesia for export. The local consumption of cigars in Indonesia, as in the rest of South East Asia, is very small, and cigarettes dominate the region's markets.

SOUTH EAST ASIA

Burmese cheroot makers at work, c. 1886.
© HULTON-DEUTSCH COLLECTION/CORBIS

The introduction of clove-flavored cigarettes, called kretek in the local tongue, toward the end of the eighteenth century in Indonesia marked a turning point in Indonesian smoking demographics, but it was not until the 1970s that *kretek* consumption became widely popular in urban areas. Previously, *kretek* use was mainly confined to poorer rural areas and white cigarettes dominated the market. Over the following decades, *kretek* sales slashed into the **market share** of conventional white cigarettes and by the end of the 1990s *kretek* sales accounted for 90 percent of all tobacco products sold throughout Indonesia. Despite the efforts of manufacturers, *kretek* still remains largely unknown in other countries.

As with all South East Asian economies, Indonesia relies heavily on its tobacco industry for both government revenues from excise taxes and employment opportunities for its people. An estimated 500,000 Indonesians are employed in the production of *kretek* cigarettes, most of them working in hand-rolling factories situated in central Java. The government mandates that any *kretek* manufacturer must produce a range of hand-rolled *kretek*, assigning lower tax brackets to these products in an effort to ensure stable employment opportunities for hand-rollers. A side effect of this policy is that tobacco regulations mandating lower **tar** and nicotine levels in cigarettes has been postponed indefinitely as manufacturers have yet to find ways to lower these values in the low-tech hand-rolling industry.

market share the fraction, usually expressed as a percentage, of total commerce for a given product controlled by a single brand; the consumer patronage for a given brand or style of product.

tar a residue of tobacco smoke, composed of many chemical substances that are collectively known by this term.

Tobacco in History and Culture
AN ENCYCLOPEDIA

Relatively high taxation and regular increases in excise taxes on cigarettes, coupled with the low spending power of the population, has seen cigarette sales decline across the board throughout Indonesia. However, the country exports manufactured cigarettes as well as increasingly large amounts of tobacco leaf. There are several corporate programs working to develop and improve leaf production in Indonesia, but little government involvement in the agricultural side of the tobacco industry.

Vietnam, Thailand, and Myanmar

Some countries, such as Vietnam, Thailand, and Myanmar (formerly Burma), have followed the Chinese example of creating state monopolies to control the production and sale of tobacco. However, with increasing international pressure to ease trade barriers, in some cases mandated by entry into the World Trade Organization, and an increasingly sophisticated and affluent consumer base, most of these traditionally closed markets are being forced to open up and allow the entry of foreign brands and technologies to remain competitive.

While Myanmar remains effectively closed to most foreign products, and the quality of domestically produced Myanmar tobacco is widely regarded as inferior on the world market, there are a few international companies, such as Rothmans of Pall Mall (Myanmar), producing cigarettes for local consumption.

By contrast, Vietnam's state-owned Vinataba has made great improvements in its domestic leaf production, and has allied itself with international companies in continuing efforts to improve its crop and production technologies. Because almost half of the Vietnamese male population smokes, most tobacco companies view Vietnam as a highly desirable potential market. However, the Vietnamese government, having allowed a dozen or so international companies to gain a foothold in the economy, is, as of 2004, not allowing anymore companies to enter the market until further notice, and is instead concentrating on boosting the export potential of its leaf and cigarettes.

Thailand's leaf production remains relatively small and of relatively low quality, despite efforts to improve the crop base throughout the last three decades of the twentieth century. The Thailand Tobacco Monopoly (TTM) saw its fortunes ebb significantly during the 1990s as imported cigarettes, which were not legally permitted to be sold in Thailand until the late 1990s, found their way onto the market. Consumers started demanding higher quality products, even though locally produced TTM cigarettes sell for approximately 30 percent less than imported brands. Despite the former ban on imported cigarettes, most popular international brands were sold openly throughout Thailand, highlighting two issues that afflict the region: smuggling and counterfeiting. Porous borders and the huge amounts of money to be made in the illegal trade of cigarettes have led to an increase in illegal cross-border trade of legitimate cigarettes and the production of counterfeit brands, many of which are thought by industry insiders and government officials to emanate from China. Revenue loss to the affected governments in the region is enormous, with Malaysia announcing in 2003 that smuggled cigarettes accounted for approximately 21 percent of the total domestic market, costing the country approximately $1.3 billion annually in lost revenues.

Compañie General de Tobaccos de Filipinas made cigars from local tobacco as well as leaf imported from other Spanish colonies. CGDTF was founded in 1848 and remained a Spanish monopoly until the late 1870s, when the cigar division was bought out by La Flor de Isabella, which still makes cigars in its Manila plant today. © BETTMANN/CORBIS

Malaysia, the Philippines, and South Korea

Malaysia is another country with a history of tobacco usage predating colonization, and in the 2000s it is still possible to see the occasional hookah, a middle eastern water pipe, being smoked at coffee shops throughout the country, a reminder of its Middle Eastern links. In the 2000s, Malaysians, together with their neighbors in Singapore, Vietnam, and South Korea, prefer English-style Virginia tobacco cigarettes, unlike Thailand and the Philippines, where U.S. tastes predominate.

The Philippines is one of the largest consumers of tobacco in the region, with a long history of tobacco cultivation. Filipinos, especially those in the northern provinces of the main Island Luzon, have been smoking tobacco for centuries, and Spanish colonists established tobacco cultivation in several areas of the country, including Luzon, the Visayas, and Mindanao. During the Japanese occupation of World War II, many rural women took to smoking black *negritas*, a thin, cigarillo type of cigarette. They would frequently smoke these with the burning end in their mouths, anecdotally to dissuade Japanese soldiers from forcing their attentions on them. One may still occasionally find older women in the provinces smoking *negritas* this way.

SOUTH EAST ASIA

flue-cured tobacco also called Bright Leaf, a variety of leaf tobacco dried (or cured) in air-tight barns using artificial heat. Heat is distributed through a network of pipes, or flues, near the barn floor.

menthol a form of alcohol imparting a mint flavor to some cigarettes.

The Philippines produces some fine export quality **flue-cured** Virginia and Burley tobacco, as well as lower quality local variants such as *saplak*, a burley and native tobacco crossbreed accounting for up to 70 percent of the crop grown in some provinces such as Pangasinan. The region is one of several low-lying tobacco-producing provinces where high salinity in the soil can result in low-quality tobacco that does not burn well and has poor taste characteristics. Coupled with poor farming techniques and haphazard grading, the quality of the Philippines tobacco crop in general is regarded as inconsistent by many domestic and international buyers.

However, there are some fine tobaccos produced in the Philippines and several success stories, many of which have been brought about by the infusion of capital and technological knowledge by international companies, such as Dimon, one of the leading buyers of tobacco leaf globally; Philip Morris; and others. Filipinos are one of South East Asia's largest **menthol** cigarette smokers; several brands, local and foreign, are produced locally. The domestic trade and production of cigars in the Philippines has declined dramatically over the last few decades of the twentieth century, but Philippine cigars have carved a niche for themselves with aficionados across the world, and compare favorably with Cuban and Dominican Republic brands, although with a milder and more flavorful taste.

South Korea was until the end of the twentieth century another country that imposed a state-run monopoly on its tobacco industry, but by the beginning of the twenty-first century it had started easing restrictions on foreign companies. There is a longstanding cigarette culture in South Korea; conscripted soldiers were for many years paid partly in cigarettes, and over 50 percent of the male population of the country are smokers. In the 2000s, foreign brands are capturing increasing shares of the domestic market, but improvements in South Korean leaves and an aggressive marketing strategy have resulted in increased exports of South Korean cigarettes, especially in the Middle East.

Antitobacco Legislation

Antitobacco legislation is playing an increasingly important role in the development of the tobacco industry throughout South East Asia. While governments earn significant percentages of their annual revenues from tobacco taxes, health concerns and pressure from antismoking groups have contributed to the formulation of policies restricting the sale of cigarettes to minors, mandating lower tar and nicotine levels, requiring manufacturers to cover ever larger portions of their packaging with health warnings, and, most importantly, because of the difficulty of establishing new brands in these markets, banning tobacco companies from advertising or sponsoring sporting and cultural events. As a result, the price of cigarettes has risen throughout the region as additional excise revenues are applied, and there is an overall decrease in the number of cigarettes sold. Due to rising prices and health concerns, reportedly fewer new smokers are taking up the habit, and as international prices for tobacco continue to remain low, many farmers throughout the region are looking to other crops to replace tobacco.

■ HENEAGE MITCHELL

BIBLIOGRAPHY

Lockwood Publications Ltd. of New York. Published the world's first tobacco journal, *Tobacco*, in 1886. Produces subscription-based tobacco trade magazines, including *Tobacco International*, *Smokeshop*, *Smoke*, and *Distribution International*. Online. Available: <http://www.lockwoodpublications.com>.

Tobacco Asia magazine. Subscription-based magazine published by Lockwood Trade Journal Co., Inc., Bangkok, Thailand. Online. Available: <http://www.tobaccoasia.com>.

Tobacco Journal International. Subscription-based magazine published by Verlagsgruppe Rhein Main, Germany. Online. Available: <http://www.tobaccojournal.com>.

Tobacco Merchants Association (TMA). Online subscription-based resource for current news and economic data on tobacco worldwide. Online. Available: <http://www.tma.org>.

Tobacco Reporter Magazine. Subscription-based magazine published by Tobacco Reporter, Raleigh, N.C. Online. Available: <http://www.tobaccoreporter.com>.

Spanish Empire

Tobacco had been used by the pre-Columbian peoples since time immemorial, but it was unknown in Europe until its discovery by the Spaniards in 1492. Christopher Columbus made specific reference to this event in his journal entry describing the exploration carried out by Luis de Torres and Rodrigo de Jerez on 5-7 November in the area adjoining the Bay of Givara, in Cuba: "They saw women and men, the men always with a lighted stick in hand, and certain dry herbs wrapped in a certain leaf, also dried, so as to form something like a paper musket . . . and lighting it on one end, they suck or swallow from the other, taking in that smoke with their breath, which numbs their flesh and nearly makes them drunken, and they say that thus they feel no fatigue" (Fernández de Navarrete, p. 202). From this date on for the next hundred years, almost all that scholars know about tobacco are suppositions.

The Spaniards soon saw that tobacco use was common and widespread among all the indigenous communities they encountered throughout the Americas. Before long, the more daring among the Spaniards began to imitate the practice. It also became evident that Amerindians used the substance in pagan rites and ceremonies, which required a wholesale rejection on the Spaniards' part. Thus the first polemic over tobacco emerged.

Between the initial discovery of tobacco and its later "official" recognition by way of the establishment of the royal monopoly in Castile in 1636, three successive periods may be distinguished. Each period reflects a distinct change in the intensity and diversity of tobacco use. The first period, enduring some sixty or seventy years, lasted until the middle of the sixteenth century. It was a phase of first contacts and of sporadic experiences with tobacco. Previously unknown, tobacco was not a product either sought or desired in its own right. Therefore it required time for acquaintanceship and acceptance. After a few decades,

SPANISH EMPIRE

chroniclers' writings testify to the prevalence of tobacco consumption among the settlers in the new colonies, sometimes to the point of abuse. "I knew Spaniards on the island of Hispaniola," wrote Bartolomé de las Casas "who had the habit of using [tobacco], and who, on being reprimanded and told this was a vice, replied that they could not stop." (Casas). This attitude was completely different from that generally observed among the native peoples, who held it as sacred as well as enjoyable. For the Europeans, tobacco use was for pleasure, without any more transcendent goal. Thus the modern smoker emerged. Records of tobacco's appearance in the metropolis during this time are scant. Some writers mention certain shipments of leaves arriving in Seville, and even allude to significant quantities, but these are mere conjectures. What are known are attempts to adapt the plant to the Iberian peninsula, and the initial study of its properties by doctors and botanists there.

The second period of tobacco's rise took place between the middle and the end of the sixteenth century. This period was shorter—barely fifty years—but much more intense. It took place not only in Spain but also in Portugal, France, the Netherlands, and England, for now tobacco use began to take root. Although there exists no proof that tobacco was as yet used for pleasure in Spain, its growing medicinal use is evident. In the last decades of the sixteenth century, texts attest to tobacco's rise in the American colonies and the first clear indication of its acceptance in the metropolis. Nicholas Monardes, the Sevillan doctor, trader, and writer, alludes to tobacco having been "brought in the past few years to Spain" (Monardes). The regulation of tobacco cultivation in Cuba dated from 1580, and in these same years the first mills for grinding tobacco leaves (*polvomonte de Indias*) were constructed on the outskirts of Havana. All these developments seem to coincide.

The third period was briefer still—just the first third of the seventeenth century—and may be said to have ended with the establishment of the royal tobacco monopoly at the end of 1636, for such a far-reaching measure implied a spectacular growth of the tobacco habit on both sides of the Atlantic. This period saw colonists expand tobacco cultivation in the American colonies and the development of a prosperous trans-Atlantic trade. There are many testimonies to this expansion in almost all the American territory under Spanish control, including Barinas (Venezuela), Puerto Rico, Cuba, and Hispaniola. It should be noted that the first tobacco harvests of Virginia date from the second decade of the seventeenth century as well.

Spanish explorer Christopher Columbus (1451–1506). His journal entries in 1492 were among the first by Europeans to describe tobacco smoking by Native Americans, a practice long established in the New World but previously unknown in Europe. LIBRARY OF CONGRESS

Some results of this initial expansion were the royal decree in 1606 prohibiting the planting of tobacco in the Caribbean islands and the continental Caribbean coast (in Puerto Rico, Hispaniola, Cuba, the Margaritas, and the provinces of Venezuela, Cumaná, and Nueva Andalucía). Angry planters and merchants lobbied successfully for the repeal of this decree some years later (1614), and even secured tax exemptions to promote cultivation in some zones (Trinidad and Guyana, 1625). During this period, there was also a failed attempt to impose a monopoly in Venezuela (1620), so as to require planters to sell their harvests to royal agents at set prices, and to stop the illegal trade with foreign merchants. A short-lived monopoly was also imposed in Puerto Rico in 1632, though with a much more limited character than that which would develop in the eighteenth century. The early seventeenth century also saw the establishment of the tobacco processing factory of

San Pedro in Seville (1620) and the appearance of the first customs duties on American tobacco imported into the metropolis. Finally, this period ended with the creation of the Castilian monopoly, which marked a watershed between a "before" and an "after" in the Spanish tobacco realm.

At the beginning of the seventeenth century, tobacco had become an American product of great importance. As a result of tobacco's rapidly accelerating popularity, on 28 December 1636 the Spanish Crown established a royal monopoly in Castile, the largest of Spain's kingdoms, although in the remaining Spanish territories tobacco continued to be bought and sold freely. Spain's tobacco monopoly was later imitated by states throughout Europe and beyond.

The monopoly gave the Spanish Crown the exclusive right to process, distribute, and sell tobacco in Castile. As such, the monopoly provided the Crown a new source of revenue, and allowed the chronically penurious royal treasury to borrow more money from international creditors guaranteed by future monopoly proceeds. Yet, the state did not directly operate the tobacco monopoly. Instead, it administered the monopoly by leasing it to private contractors, known as *arrendadores*, which was a common practice in the Spanish fiscal system of the time. Accordingly, the Crown publicly auctioned the new monopoly and gave it and its management to the highest bidder. The first lease lasted only fifteen months, rather than the ten years stipulated. Such developments were characteristic throughout the seventeenth century as competing lease holders successively outbid the previous lessees, a consequence of the extraordinary growth in demand for tobacco in Castille. The exponential increases in the value of tobacco monopoly leases attest to the spectacular expansion of demand. The first lease was valued at 23 million *maravedis*; twenty years later, it was 57 million; in 1675, it was 285.3 million, and by 1698 it was 304.5 million. Although the Spanish monetary crisis lessens the real value of these figures, the expansion and the high volume of business cannot be disputed. Nonetheless, the inflated tobacco prices set by the monopoly ensured that rampant contraband consumption co-existed with legal sales.

In practice, the monopolist, or lessee, would subcontract with other entities who would operate the monopoly in different territories and provinces. In turn, these would further subdivide their regions into smaller districts. At the lowest level were the *puestos estancos* or *estanquillos*, in other words, the retail tobacconists. These vendors would contract with the monopolist (or his subfactors) to sell a minimum, predetermined amount of tobacco at a set price in a set time period. This was known as the "tabaco de obligación" (tobacco of obligation), though the tobacconist could also sell tobacco in excess of that amount, known as tobacco "fuera de obligación" (tobacco outside of obligation).

The majority of monopoly lease holders, or *arrendadores*, were Portuguese *conversos* (descended from Jews), a mercantile community who played an important role in helping the government avoid bankruptcy. Their presence in Spanish banking was largely due to the protection of the Philip IV's Spanish prime minister, the Count-Duke of Olivares, who helped them avoid entanglements with the Inquisition. After Olivares' fall in 1643, the Inquisition launched a persecuting campaign against the prominent monopolists, the target of popular anti-Semitism, and succeeded in prosecuting some of the most notables, such as Antonio

snuff a form of powdered tobacco, usually flavored, either sniffed into the nose or "dipped," packed between cheek and gum. Snuff was popular in the eighteenth century but had faded to obscurity by the twentieth century.

de Soria (in 1654) and Francisco López Pereira (in 1658), bringing their monopoly contracts to an end.

Late in the seventeenth century, between 1683 and 1687, the Crown experimented with taking direct control of the tobacco monopoly, seeking to eliminate the middlemen. The state created the *Junta de Tabaco* to oversee the monopoly, named a director to manage the Seville factory (where **snuff**, and later cigars, were manufactured), adopted measures against fraud and contraband, and prepared guidelines for procuring, manufacturing, and distributing the product. The new system did not last long, and soon the monopoly returned to the method of leases, although that trial run became the precursor to the system that would prevail in the eighteenth century.

Throughout its existence, fraud and contraband beset the monopoly, from within and without. The high retail prices set by the monopoly—exponentially above the "market" price—encouraged contraband sales. Critics also blamed tobacco producers in the Spanish colonies who illegally avoided selling their tobacco to monopoly agents in favor of selling tobacco in the international market where it fetched higher prices. Moreover, they complained of the monopolists who adulterated tobacco with cheaper and illegally imported varieties, and accused them of faking bankruptcies as a way to avoid fulfilling their obligations to the Crown. Ecclesiastics were also notorious participants in contraband, cultivating tobacco clandestinely in convent fields and selling it illegally to consumers below the monopoly prices.

When Charles II produced no heir, the Habsburg dynasty ended in Spain with his death in 1700 and a Bourbon, Philip V (grandson of Louis XIV), came to the throne. Beginning with decrees in April of 1701, the royal tobacco monopoly was transformed under the Bourbon regime: The "leasing" system gave way to a system of central administration, at least partly modeled after the measures implemented between 1683 and 1687. However the War of Succession to the Spanish Crown (1702–1713) delayed the shift to direct state control to a more favorable moment.

hegemony control or superior influence over; dominion.

It took several decades before the monopoly's transition to a system of central administration was completed. Gradually, more and more provinces and districts came under the direct control of the state treasury, and eventually the remaining Spanish territories were integrated into the monopoly: Valencia, Aragon, Catalonia, Balearic Islands, Canaries, and later Navarre. The monopoly instituted stronger safeguards against fraud; in 1725, it asserted the **hegemony** of the Sevillan tobacco manufacturing with the New Factories Project, and it strengthened the provisioning system to ensure it received the best tobacco from the Spanish colonies.

Over the course of the eighteenth century the monopoly flourished, bringing in substantial revenue for the treasury. Official consumption grew from 1.1 million pounds of tobacco to almost 4 million in 1730–1731. Then a drastic price hike resulted in a drop in legal consumption to 2.5 million pounds in 1742; by 1779 legal demand had recovered and topped 4 million pounds once again. In that same year another drastic price increase led annual consumption to drop to 3 million pounds. From 1780 until the end of the century, annual sales ranged between 2.5 and 2.7 million.

Despite the drop and stagnation in monopoly sales, monopoly profits grew throughout the century. At the beginning of the century, the state received 7 million *reales* in profit, which rose to 46.5 million by 1730–1731. Even after sales dampened, the price increases ensured that profits continued to rise, averaging 74 million *reales* in the decade between 1740 and 1749, 90 between 1750 and 1759, 99 between 1760 and 1769, 115 between 1770 and 1779, 125 between 1780 and 1789, and 123 million between 1790 and 1798.

Contraband and fraud was an even greater problem in the eighteenth century, precipitated by the great expansion in tobacco consumption and the disproportionate price hikes set by the monopoly. In response, the monopoly developed a militia, known as the *resguardo*, to diminish the illegal traffic, but did not did have much success.

During the nineteenth century, the tobacco monopoly suffered the consequences of the political crises that erupted in Spain. Though the wars, coups d'etat, and revolutions that devastated the entire territory made it impossible to obtain accurate output figures, it is clear that the output of the tobacco factories was deeply affected by all these vicissitudes. After recovering from the Napoleonic Invasion and the subsequent War of Independence (1808–1814), monopoly sales rebounded, only to fall again with the first Carlist War (1833–1839). After this war there was another moderate recovery that lasted until the crisis period that began with the Revolution of 1868. After 1874 the monopoly flourished again until its transformation in 1887.

Despite the recurring crises, the institutional structure of the monopoly saw few changes until the end of the century, although there were several efforts to transform it. At various times, Progressive governments tried to end the monopoly, viewing it as an anachronism, inconsistent with their liberal ideology. Progressive governments tried to eradicate the monopoly, such as when the Cortes of Cadiz repealed it (3 December 1813) and later during the Constitutional Triennial (9 November 1820), but they did not achieve the desired success and returned again to the same monopoly system (16 February 1824). Similarly, in 1855, as in 1869, plans to repeal the monopoly never were put in place. On the contrary, an 1876 law expanded the monopoly to include the three Basque provinces.

Finally, in 1887, the monopoly was reorganized with the creation of the *Compañia Arrendataria de Tabacos* (State Leasing Tobacco Company), or CAT. Formed of state and private capital, the company took over control of the monopoly's activities, including procurement, production, distribution, and sales. CAT undertook a modernization program by improving the industrial process, drastically reducing labor requirements, and creating a better system of fraud control. Because of these changes the industry started a new phase of recovery and achieved substantial growth at the beginning of the new century.

The Civil War (1936–1939) brought an end to this program, after which Tabacalera replaced CAT in 1945. Tabacalera was charged with repairing the monopoly during the critical years after the war. During the early years of the Franco regime, the monopoly struggled for survival, but the 1960s the monopoly returned to a long period of expansion. With its entrance into the European Economic Community (EEC) in 1986, Spain had to bring an end to the state tobacco monopoly.

Etching of Philip V (1683–1746), founder of the Bourbon dynasty in Spain. In 1701, a year after he came to the throne, the royal tobacco monopoly was transformed when the "leasing" system was changed to a system of central administration. However the War of Succession to the Spanish Crown (1702–1713) delayed the shift to direct state control for several decades.
© BETTMANN/CORBIS

Tabacalera privatized and, later merged with Seita (the old French Monopoly), together constituting Altadis.

The Tobacco Monopoly in the Spanish Colonies

The establishment of the royal tobacco monopoly is one of the most noteworthy economic aspects of colonial policies governing the Spanish Empire. Despite the dramatic success of the Castilian royal tobacco monopoly and the growth of consumption in the colonies, the Crown did not extend this peninsular model to its American possessions during the seventeenth century. Only in the period between 1684 and 1687, during which the Royal Treasury administered the monopoly directly, were more rigid measures on colonial tobacco activities adopted; these included greater control over the supply sustaining the Royal Factory in Seville and a requirement that the Indies consume products manufactured in the metropolis. To these moves may be added the failed attempt to establish a tobacco monopoly in Mexico (1642) and South America (1647), as well as some other measures of more minor interest.

The policy of expanding the monopoly to Spanish America arrived in force with the Bourbons after 1700. Its implementation, however, suffered delays of almost two decades and, in most cases, more than half a century. When implementation began in Cuba after the end of the War of the Spanish Succession, it turned out to be fraught with peril. The monopoly was imposed on 11 April 1717, in the form of an exclusive royal agency (the Factoría) to whom planters were required to sell their tobacco at prices set by the state. However, planters mounted three successive rebellions—in 1717, 1720, and 1723—which required the Crown to moderate its decrees. After this difficult beginning, the monopoly took the form of the Intendencia General (1727–1734), whose function was to guarantee the supply; followed by a system of contracts with particular privileged merchants; and finally, after 1740, an exclusive arrangement with the Royal Havana Company. In 1760 this regulation too was rescinded, and the Factoría was restored (1760–1817), although in a less extreme form than during the first attempt.

By then, the extension of the monopoly to other colonial territories was in full swing. The changed attitude toward American commerce after 1740 (the War of Jenkins' Ear) and the need for a growing source of revenue to sustain the empire after the Treaty of Paris (1763) precipitated the monopoly's expansion throughout the Spanish colonies. The monopoly began by incorporating the viceroyalty of Peru: first in Lima and a year later in Santiago (Chile), then in 1754 in Lower Peru (Cuzco, Trujillo, Arequipa), and in 1755 in Upper Peru (La Paz, Charcas). At the end of that year, the regime was extended to Buenos Aires, although under the jurisdiction of the Junta of Chile.

In New Spain, where the rise of tobacco consumption had continued uninterrupted, the monopoly was decreed in 1764, and the resulting revenues were always spectacular: some 417,000 pesos in 1767, 1.2 million pesos in 1775, and nearly 3.5 million pesos in 1790. The provinces of Guatemala were incorporated into the monopoly in 1766. The Intendencia of Venezuela first became subject in mid-1777, with the monopoly's establishment in its subdivisions, including that of Caracas, Maracaibo, Cumaná, and Guyana; the definitive stage came in May 1779.

In that same year, the monopoly also spread to Paraguay, and in 1785 the Factoría of Puerto Rico was created.

In the Philippines the monopoly began in 1782 with the same goal as in other areas: to promote self-financing. At first only the capital and a group of surrounding provinces were affected (the productive zones of the provinces of Gapán, Nueva Écija, Marinduque, and Cagayán); six years later monopoly control reached Ilocos Norte, Olocos Sur, La Unión, Abra, and La Isabela.

The blueprint that was followed to extend the reach of the monopoly throughout the empire was similar, in all cases, to the metropolitan one: creation of a Supreme Council of the Royal Tobacco Monopoly (Junta Superior de la Real Renta del Tabaco) and the General Administration of the Monopoly (Dirección General de la Renta) for each province; and establishment of various *factorías* and structures for their control (based on the long experience accumulated in the Peninsula), such as general and **subsidiary** administrative bodies, exclusive tobacco warehouses, retail stores, and police forces. These were run according to successive sets of regulations and ordinances based on the *Instrucciones y Reglas Universales* imposed in Spain in 1740. Significant manufacturing developed in some areas, especially in New Spain, with factories in Mexico City, Orizaba, Puebla, Oaxaca, Guadalajara, and Querétano, but also in Cuba from quite early on, and in Peru. Only in Cuba were the resultant products shipped to the Old World; otherwise, all the local production was distributed within the colonies themselves.

The independence movements that sprang up within the Spanish Empire in the early nineteenth century put an end to the royal monopoly in most of the areas where it had held sway. Only Cuba and Puerto Rico in America and the Philippines in Asia maintained their ties with the metropolis, and only in the latter did the monopoly last until its repeal in 1881. The Factoría de Puerto Rico succumbed during the Independence War, while in Cuba the monopoly was abolished in 1817.

subsidiary in commerce, a branch or affiliate of a larger unit that provides components or support services.

Tobacco and Mercantilism

Within this overview of the tobacco monopoly in the Spanish Empire, certain general themes emerge. One surprising fact is that Spain never achieved complete control over tobacco activity in its immense colonial realm. Tobacco was certainly a vital economic sector, transforming the agrarian economy and commercial circles in many parts, becoming a source of population flows and colonial expansion wherever its cultivation took root, and providing colossal sums to sustain the hegemony of the Crown. Nonetheless, both before and after the spread of the monopoly throughout these territories, the tobacco trade to a large extent escaped the policies promoted in the metropolis, in part because of the reality of colonial exploitation itself. The interests of the planters—most often medium-sized proprietors, if not simple peasant families—always tended to favor fraud or even open contraband trading of their products, because they made greater profits if the products went elsewhere other than Spain (to other European countries or other Spanish colonies). Though tobacco planters did not acquire much economic power as individuals, as a group they achieved undoubted importance in many of these regions. Thanks to this power and, in the majority of cases, to the collaboration of colonial authorities, the tobacco planters were able to

SPANISH EMPIRE

skirt restrictive regulations issued from the metropolis, or fight them directly, even by force of arms.

Despite the high demand and profitability, tobacco cultivation was restricted for a number of reasons. Oftentimes it competed with other cash crops, such as sugar, cacao, ginger, and coffee, in many of these territories. And in many areas cattle ranchers and tobacco cultivators fought over control of the land. In addition, an important part of colonial tobacco production (in all of New Spain and Peru, but also in parts of Santo Domingo, Puerto Rico, and Cuba) remained for internal consumption, sometimes generating strong commercial interests there.

For these and other political and strategic reasons, imperial Spain never managed to acquire all its tobacco imports from its own colonies, even though it controlled many of the most important tobacco-producing regions of the time. From the beginnings of the Spanish tobacco industry at the end of the sixteenth century—during which time there existed the temporary union of Spain and Portugal—evidence may be found of the presence of rolled tobacco from Brazil and soon of Virginia tobacco as well, and later came to include supplies from Louisiana and Kentucky. The proportion of Spain's tobacco imports originating outside the empire reached very high levels at many points; at the same time, a significant proportion of the Spanish colonies' crops were diverted to other European markets. This led Spain to lose specie as it bought tobacco from its imperial rivals—and the state sought to change the situation by continually issuing new regulations and decrees, which attests to how little effect these measures actually had.

Another factor that must be taken into account is the high proportion of the colonial tobacco shipments to Spain made up by already manufactured goods. This reality completely contradicts the mercantile theories that, it has been thought, guided imperial policy. During the eighteenth century at least 25 percent of the annual supplies of snuff tobacco had already been processed in Havana (Cuba), and this figure reached more than 90 percent in some years. All told, the average was between 40 and 45 percent per year. Is it possible, in view of these figures, to keep speaking of a "colonial pact"—the mercantilist notion in which the metropolis would exclusively benefit from the import of raw materials from its colonies, and, in turn, process and export them as manufactured goods to protected American markets? On the contrary, Spain failed to monopolize economic activities in the New World, nor could it thwart the empowerment of economic groups, such as colonial tobacco producers, whose interests were in direct conflict with those of the metropolis. Certainly, the monopoly did become established throughout the empire, and it was an outstanding revenue-raising success for the state. But it could not manage to eliminate certain problems or to cure many of the ills inherent in the system of management imposed on the colonies by Spain. Its example serves as a model for a new interpretation of colonial policy.

See Also Brazil; British Empire; Caribbean; Cuba; Dutch Empire; French Empire; Portuguese Empire; Smuggling and Contraband; South and Central America; Taxation; Trade.

▮ JOSE MANUEL RODRÍGUEZ GORDILLO

BIBLIOGRAPHY

Casas, B. de las, *Historia de las Indias*. Madrid: Biblioteca de Autores Españoles, 1957.

Comín Comín, F., and Martín Aceña, P. *Tabacalera y el estanco del tabaco en España, 1636–1998*. Madrid: Fundación Tabacalera, 1999.

Fernández de Navarrete, M. *Colección de los viajes y descubrimientos que hicieron por mar los españoles desde fines del siglo XV.* Madrid, 1858.

García de Torres, J. *El Tabaco: consideraciones sobre el pasado, presente y porvenir de esta Renta*. Madrid: Imprenta de Juan Noguera, 1875.

Monardes, N. *Segunda parte del libro de las cosas que se traen de nuestras Indias Occidentales, que sirven al uso de medicina*. Sevilla, 1571.

Ortiz, Fernando. *Cuban Counterpoint: Tobacco and Sugar*. Trans. Harriet de Onís. Durham: Duke University Press, 1995.

Pérez Vidal, J. *España en la historia del tabaco*. Madrid: CSIC, 1959.

Rivero Muñíz, José, *Tabaco. Su historia en Cuba*, 2 vols. La Habana: Instituto de Historia, 1965.

Rodríguez Gordillo, J. M. *Diccionario histórico del Tabaco*. Madrid: Cetarsa, 1993.

———. *La creación del estanco del tabaco en España*. Madrid: Fundación Altadis, 2002.

———. *La difusión del tabaco en España*. Sevilla: Universidad de Sevilla/Fundación Altadis, 2002.

Spanish Tobacco Monopoly *See* Brazil; British Empire; Caribbean; Colonialism; Cuba; Dutch Empire; French Empire; Portuguese Empire; Smuggling and Contraband; Spanish Empire, State Tobacco Monopolies; Taxation; Trade.

Sponsorship

Tobacco industry sponsorship largely evolved once other forms of cigarette promotion were restricted or no longer permissible. Researchers Stephen Townley and Edward Grayson, in *Sponsorship of Sport, Arts and Leisure* (1984), define sponsorship as "a mutually acceptable commercial relationship between two or more parties in which one party (called the sponsor) acting in the course of a business, trade, profession or calling seeks to promote or enhance an image, product or service in association with an individual, event, happening, property or object (called the sponsee)" (Townley and Grayson 1984). Following an analysis of various industry efforts to define sponsorship, the Global Media Commission of the International Advertising Association defined commercial sponsorship as "an investment, in cash or in kind, in an activity, in return for access to the exploitable commercial potential associated with that activity" (Larson and Park 1993).

Sponsorship should not be confused with patronage. A patron makes a donation, whereas a sponsor makes an investment. Companies usually

SPONSORSHIP

Camel and Marlboro ads at the Macau Grand Prix gain widespread television exposure. It is estimated that 300 million television viewers watch each Formula One race. © EARL & NAZIMA KOWALL/CORBIS

make donations to support and improve their public image and to demonstrate that they are caring and good corporate citizens. Patronage implies funding and support on a noncommercial basis and such generosity is usually based on personal satisfaction and a belief in the worthiness of the cause. The sponsor's motives for making its investment, however, are commercially based rather than altruistic. The sponsor is involved in a business activity to gain benefits and meet particular objectives.

A Shift from Broadcast Advertising to Sponsorship

The early 1970s represents the defining period when sports and cultural sponsorship became an increasingly important form of promotion for American tobacco manufacturers. Virginia Slims, for example, began sponsoring women's professional tennis in 1970, while Winston Cup auto racing and Marlboro Cup horse racing started in 1971 and 1973, respectively. The shift toward sponsorship largely reflected the fact that cigarette firms were no longer permitted to advertise on radio and television. In the United States, federal legislation, the Public Health Cigarette Smoking Act, stipulated that broadcast advertising for cigarettes was banned, commencing 2 January 1971.

Tobacco manufacturers were aware that cigarette brand exposure could persist on radio and television if broadcast sporting and cultural events were sponsored. Studies have illustrated that by sponsoring sports such as auto racing, U.S. tobacco companies continue to receive millions of dollars worth of national television exposure. A videotape recording of the Marlboro Grand Prix on 16 July 1989, for example, revealed that Marlboro was seen or mentioned 5,933 times. Cigarette billboards remained present in several sports stadiums until the mid-1990s and gained considerable television exposure. During the 1992

Sugar Bowl football game, Marlboro signs located near the scoreboard gained forty-eight seconds of camera exposure, which was valued at $144,000 in advertising time. The billboards were strategically placed to maximize the extent of their exposure. Tobacco manufacturers have used sponsorship as a means of circumventing television advertising regulations or restrictions.

Key Sponsorship Objectives

Internal tobacco industry documents, which are publicly accessible through various court proceedings, reveal that primary objectives for sponsoring sports and cultural events include increasing brand awareness (through continued brand visibility) and enhancing or reinforcing brand image. In an attempt to enhance or reinforce brand imagery, tobacco firms identify events and brands that possess complementary symbolic properties, seeking a transfer of the imagery associated with the event, participants, or sponsorship partners to the sponsoring cigarette brand. Marlboro, the most popular cigarette brand in the world, is promoted as a symbol of masculinity, ruggedness, independence, and self-reliance. Thus, the brand typically sponsors individual rather than team sports (making a link with independence), as well as activities that convey masculine overtones. According to Ellen Merlo, vice president of marketing services at Philip Morris, quoted in a 1989 Marlboro advertisement, "We perceive Formula One and Indy car racing as adding, if you will, a modern-day dimension to the Marlboro Man. The image of Marlboro is very rugged, individualistic, heroic. And so is this style of auto racing. From an image standpoint, the fit is good."

Sponsoring cultural and sports events may serve a number of additional marketing objectives for tobacco companies and their products. Through highlighting the community benefits that result from sponsorship, yet meanwhile placing a de-emphasis on the commercially based motives of the investment, sponsoring events is viewed by tobacco companies as an opportunity to be regarded as good corporate citizens. Sponsorships, it is argued, allow cultural and sports events to be staged which might otherwise be denied. Charity associations are also commonly accentuated. For example, Rick Sanders, president of Sports Marketing Enterprises at R.J. Reynolds Tobacco Company, emphasized in a 2000 letter to racing fans that more than $1.8 million was raised for charities over a ten-year span from the NASCAR Winston Cup Preview. Furthermore, tobacco companies may gain allies for lobbying efforts by forming sponsorship-affiliated front groups and asking those sponsorship organizations that they financially support to vehemently oppose any proposals to ban sponsorship.

There are several other potential corporate motives of sponsorship. First, event sponsorship can enhance employee relations and morale. Productive employees may be rewarded with tickets to various events and feel that their employers are good corporate citizens by supporting community events. Second, political contact and support can be strengthened because politicians are frequently invited as guests to sponsored events. Third, sponsorship provides hospitality opportunities, with the events serving as a forum for customer or employee entertainment. Fourth, demographic data may be collected from target consumers that are attending sponsored events. Name databases can be established for direct marketing purposes. Fifth, event sponsorship provides sampling opportunities so that consumers are provided with convenient opportunities to try products at no financial risk. Finally, sponsorship can provide cross-promotional opportunities, as well as options for revenue deductions or tax write-offs.

The Current Legislative Environment

Tobacco brand sponsorship has been banned in several countries, including Algeria, Bulgaria, Canada, Finland, France, Italy, Jordan, Norway, Saudi Arabia, and Sweden. Domestic tobacco sponsorships are prohibited in Australia, New Zealand, and the United Kingdom, but exemptions have been made for events of international significance. The Framework Convention on Tobacco Control (FCTC), which was negotiated through the World Health Organization, calls for countries to undertake a comprehensive ban of all tobacco promotion directed toward consumers, including brand sponsorships, in accordance with each country's respective constitution. A minimum of forty nations must ratify the treaty before its provisions take effect (the FCTC would be legally binding for only those countries that ratify the treaty).

See Also Advertising; Marketing; Public Relations; Sports.

■ TIMOTHY DEWHIRST

Distinguishing Between Brand and Corporate Sponsorships

Compared to tobacco brand sponsorship, corporate sponsorships (for example, Philip Morris is identified as the sponsor rather than Marlboro) are more often integrated with public relations strategies, with objectives such as generating positive publicity and enhancing the image of the firm. Corporate names are more commonly used than brand names for identifying the support of "issue" sponsorships, which often relate to social programs, education, health care, and environmental concerns.

BIBLIOGRAPHY

Blum, Alan. "The Marlboro Grand Prix: Circumvention of the Television Ban on Tobacco Advertising." *The New England Journal of Medicine* 324 (1991): 913–917.

Cornwell, T. Bettina. "The Use of Sponsorship-linked Marketing by Tobacco Firms: International Public Policy Issues." *Journal of Consumer Affairs* 31 (1997): 238–254.

Crompton, John L. "Sponsorship of Sport by Tobacco and Alcohol Companies: A Review of the Issues." *Journal of Sport and Social Issues* 17 (1993): 148–167.

Dewhirst, Timothy. "Smoke and Ashes: Tobacco Sponsorship of Sports and Regulatory Issues in Canada." In *Sports Marketing and the Psychology of Marketing Communication*. Edited by Lynn R. Kahle and Chris Riley. Mahwah, N.J.: Lawrence Erlbaum, 2004.

Hoek, Janet, and Robert Sparks. "Tobacco Promotion Restrictions: An International Regulatory Impasse?" *International Marketing Review* 17 (2000): 216–230.

Hwang, Suein L. "Some Stadiums Snuff Out Cigarette Ads." *Wall Street Journal*, 17 July 1992, p. B6.

Larson, James F., and Heung-Soo Park. *Global Television and the Politics of the Seoul Olympics*. Boulder, Colo.: Westview Press, 1993.

Lavack, Anne M. "An Inside View of Tobacco Sports Sponsorship: An Historical Perspective." *International Journal of Sports Marketing and Sponsorship* 5 (2003): 33–56.

Shafey, Omar, Suzanne Dolwick, and G. Emmanuel Guindon, eds. *Tobacco Control Country Profiles 2003*. Atlanta, Ga.: American Cancer Society, 2003.

Townley, Stephen, and Edward Grayson. *Sponsorship of Sport, Arts and Leisure: Law, Tax and Business Relationships*. London: Sweet & Maxwell, 1984.

Sports

Smoking has always had an important link with sports. In the early 1900s, and also in Nazi Germany in the 1930s, social reformers argued that children who smoked cigarettes would be unable to participate in sports. However, in the 2000s the link between smoking and sport has been increasingly fostered and exploited by advertisers around the world.

Smoking and Social Reform

In the early 1900s, part of the argument used against smoking among young people was that it would impair their ability to participate in sports. In this respect, smoking was bound up with wider debates about physical deterioration and citizenship. Writing of football in *Scouting for Boys* (first published in 1908), Robert Baden-Powell enjoyed watching the players "but my heart sickens at the reverse of the medal—thousands of boys and young men, pale, narrow-chested, hunched-up, miserable specimens, smoking endless cigarettes, numbers of them betting" (Baden-Powell). Similarly, in Nazi Germany, smoking among young people was criticized because it was seen to be sapping the strength of the German people—at work, at school, in the bedroom, in the field of battle, and on the sports field. Experiments conducted in Nazi Germany in the 1930s showed that smoking impaired the ability of a soldier to march long distances.

Smoking and Advertising

However, sports and tobacco advertising have always been closely connected. Sports is one of those common denominators in society, much like patriotism, the nation's past, health and beauty, and optimism about the future, that has been exploited by advertisers. Sports was used as a vehicle for advertising more because of its general associations with youth, fitness, and citizenship than as part of a more deliberate attempt to combat the moral and health claims made about cigarettes.

This was certainly the case in the United States, where baseball was one of the most popular themes featured on **cigarette cards.** Some cards featuring baseball were issued as early as 1886, but it was over 1887–1888 that a series was released by the Goodwin Company of New York City. More than 2,000 cards were featured in packets of Old Judge and Gypsy Queen cigarettes. Uniformed athletes assumed a variety of poses simulating action, along with gallery poses, and these were used on the covers of giveaway scorecards for National League games. Cigar companies sponsored local baseball teams; cigar tins included such brands as Home Run; and a billboard showed a swimmer clambering out of the pool with the caption, "Now for a Chesterfield" (Petrone). Streetcar advertising for Chesterfield cigarettes associated them with tennis, while for Lucky Strike the link was with baseball. Other baseball stars testified to the superiority of particular brands: "Old Gold hits a homer for Babe Ruth in blindfold cigarette test"(Petrone).

Sports and tobacco advertising have remained closely linked. In a survey of tobacco advertising in Australian newspapers and periodicals,

"Luckies do not affect my wind," says baseball player Paul Waner, voted National League's most valuable player in 1927. "When I first started to smoke I was anxious to find a cigarette that would give me pleasure without taxing my wind or irritating my throat. I soon discovered Lucky Strikes. I am very fond of the excellent flavour of these cigarettes and they keep my throat clear and do not affect my wind in the least" (Petrone).

"For some men things always go smoothly. Such as getting first night tickets. Getting par for the course in golf. Getting a rare piece of art. And getting the experience of a truly great smoke. Wills Filter Kings. Rich, mellow—and very, very smooth. For some men, it's most satisfying" (Chapman 1986).

cigarette cards paper trading cards sometimes featuring sports personalities or movie stars packaged with cigarettes and offered as an incentive for purchase.

SPORTS

Sports and tobacco advertising have always been closely linked as shown by this billboard advertisement from 1961 of a football player exhaling smoke from a Camel cigarette. © ROGER WOOD/CORBIS

researchers found that sports made up 60 advertisements or 5.8 percent of the total. Since roughly 1970, tobacco advertising has been controlled more strictly in western Europe and America, although this has been more about restricting advertising in general than any explicit attempt to target images of sport and fitness. Moreover, in eastern Europe and developing countries, the association between tobacco and sports has continued. A brand of cigarettes called Popularne, manufactured by the Polish Tobacco Industry Combine, were formerly named Sport, and in the 1980s cigarette marketing in developing countries was shown to be directly associating its products with sports. In Indonesia, the Kreek brand Djarum lent its name to the country's badminton team, and it also sponsored boxing matches. In India, the VazirSultan Tobacco Company spent a significant part of its budget on cultural and sporting events. Rothmans Royals (Fiji) were promoted through an illustration of a cricketer; Wills Filter Kings (India) were marketed with references to golf; and Bastos Blonde (Cameroon) used tennis as its sport of association. Meanwhile, Virginia Slims sponsored tennis, and John Player Special, Marlboro, and Rothmans were involved in Grand Prix motor racing worldwide. Since the 2000s, there have been efforts internationally to restrict tobacco industry sponsorship of sporting events as a means of reducing smoking initiation among youth.

Throughout the twentieth century, tobacco has had a complex and contradictory association with sports. Throughout the century, even before smoking was linked to cancer in the 1950s, smoking has been discouraged among young people on the grounds that its effects would render them unable to participate in sports activities. But at the same time, advertisers have exploited the popularity of sport as a vehicle for tobacco advertising through using popular sports figures in advertising or sponsoring sporting events.

See Also Advertising; Marketing; Sponsorship.

■ JOHN WELSHMAN

BIBLIOGRAPHY

Baden-Powell, Sir Robert. *Scouting for Boys: A Handbook for Instruction in Good Citizenship*, 2nd ed. London: Arthur Pearson, 1909.

Chapman, Simon. *Great Expectorations: Advertising and the Tobacco Industry*. London: Comedia, 1986.

Hilton, Matthew. *Smoking in British Popular Culture 1800–2000: Perfect Pleasures*. Manchester, England: Manchester University Press, 2000.

Petrone, Gerard S. *Tobacco Advertising: The Great Seduction*. Atglen, Pa.: Schiffer Publishing Ltd., 1996.

Proctor, Robert N. *The Nazi War on Cancer*. Princeton, N.J.: Princeton University Press, 1999.

Welshman, John. "Images of Youth: The Issue of Juvenile Smoking 1880–1914." *Addiction* 91, no. 9 (1996): 1379–1386.

State Support of Agriculture See Chesapeake Region, Kentucky.

State Tobacco Monopolies

Until the end of the twentieth century the tobacco monopoly was a fiscal one. That is, it was not an administrative monopoly, as the Post Office might be, because its exclusive purpose was to supply revenue to the state that controlled it. The tobacco monopoly did not create new needs among consumers, but rather built upon the increasing production and sale of a product that already met widespread demand among the population; individuals were consuming tobacco products and would have gone on consuming them with or without the monopoly.

The tobacco monopoly appeared in Castile in 1636 and later in its colonies, at which time certain taxes on the product's sale (the *alcabala*) and customs duties on its importation (the *almojarifazgo*) were already in existence. The government set prices for the products offered by the tobacco monopoly. For the state monopoly to be effective, aside from prohibiting anyone else from carrying out this economic activity, state treasuries had to establish revenue-collecting procedures—either farming out or direct administration by the government—as well as specific repressive regulations to combat smuggling and other infractions,

STATE TOBACCO MONOPOLIES

which were already widespread even before the implementation of the monopoly. As a result, enforcement necessitated police squads employed by customs agencies, or even monopoly police in particular.

In the nineteenth century, economic and social ideas changed. As public opinion no longer found revenue generation a sufficient argument to justify maintenance of the tobacco monopoly, liberal governments had to offer more specific justifications, such as to penalize the consumption of articles considered dangerous to health. Nonetheless, like all state monopolies, the tobacco monopoly found its main justification in its usefulness to the public treasury, which had imposed both the monopoly and the tobacco taxes, and which continued to collect its share by means of the eminent power of the state.

Both historically and geographically, the tobacco monopoly has been the most prevalent and most profitable of the state monopolies, as well as the most studied. But such monopolies have been established also for other products and for other purposes in addition to revenue collection. There are state monopolies whose objective is to supply public services, like telecommunications, transportation, water, gas, and electricity. Fiscal monopolies have been established over many products, such as salt, alcohol, gunpowder, lotteries, codfish, and playing cards.

The historical success of state monopolies was based on their simplicity and efficiency (low revenue-collection costs and no need for a tax-collecting apparatus). Furthermore, the revenue obtained through state monopolies was less evident to the payers than was the burden imposed by taxes.

Fiscal monopolies are as old as the existence of the state, but as a modern fiscal instrument they were born in 1636, when the tobacco monopoly was set up in Spain. From there they spread through Europe and the American colonies during the seventeenth century. Even when the liberal revolutions of the eighteenth and nineteenth centuries overthrew the ancient regimes, liberal politicians in some countries continued to generate revenue through these monopolies even though they ran counter to the politicians' principles. Early tobacco monopolies were in Spain, France, Austria-Hungary, Italy, Portugal, the Papal States, Turkey, and the eastern European countries. Sweden created a tobacco monopoly in 1919. Outside the European continent, the outstanding tobacco monopolies of the twentieth century were in Japan, the Soviet Union, and communist China.

Generally, state monopolies covered all activity related to the monopolized product. Thus, the tobacco monopoly covered all activity from production or importation of the leaf to the manufacturing, distribution, and marketing of the finished tobacco goods. Fiscal monopolies could be exploited directly by the state, or their management could be farmed out to companies that paid for the privilege. Farming out the management has predominated from the nineteenth century into the twenty-first century. In countries such as Spain, Turkey, Portugal, and Sweden, private companies leased the tobacco monopoly from the state. In eighteenth-century Europe, direct management by the state was more the rule. In the twentieth century, direct exploitation of the tobacco monopoly was practiced in France, Italy, and the communist countries. On the other hand, in the majority of non-European countries, led by large producers like the United States, India, and Australia, the tobacco industry has historically been characterized by free trade. Since the nineteenth century there has been no tobacco monopoly in Great Britain and Ireland, Germany, the Benelux countries, Switzerland, Greece, and some Scandinavian countries.

Fiscal monopolies have tended to be imposed on products whose consumption was strongly rooted in the habits of the population, as in the case of tobacco and its smokers. Demand has thus been relatively inelastic with respect to price, which has rendered the collection of revenue easier for the treasury. Although consumption was voluntary, the consumer's choice was limited to the products offered by the monopoly, and price levels were set unilaterally by the state. State monopolies always generated **contraband trade** in the given products and, in response, governments often established police bodies charged with combating illegal tobacco traffic. The higher the monopoly prices and the lower the quality of the goods, the greater the incentives for contraband trade, which was a crime created by the existence of the monopoly itself. In addition, high taxes and customs duties also created an incentive for contraband. Although revenue collection by state monopolies was carried out through commercial contracts, still the product or service user was "an authentic contributor" by way of both the tax passed on by the monopoly and the elevated monopoly price. Besides the costs of administration, production, and marketing, the sales price included the indirect tax on its tobacco consumption as well as the entrepreneurial profits of the company that operated the monopoly. In effect, the tobacco monopoly illustrated the four types of income that a state monopoly could include:

contraband trade traffic in a banned or outlawed commodity; smuggling.

- The monopoly profit on tobacco, the state's net income from industrial or commercial exploitation after deducting any percentage or commission paid to leasing companies;
- The state's share (as a shareholder of the leasing corporation) on the profit of the managing company when the monopoly was farmed out;
- The income received by the state from consumption taxes imposed specifically on tobacco products;
- The sums collected by the state through the profit tax paid by the leasing company to the treasury.

Fiscal monopolies have appeared to be an efficient revenue-gathering mechanism for treasuries with limited administrative systems, systems whose resources are few and whose structures are in need of modernization. The collection of revenue from a state monopoly was simpler than collection of indirect taxes and customs duties, because the monopolist was a single passive subject. Once in the power of the state, the monopoly could be regulated either to reduce its social cost (that is, reduce the monopoly profits by selling the product in conditions more like those of perfect competition of the free marketplace), or to keep the profits high, as state profits, giving rise to a state monopoly with an exclusively income-generating goal. Logically, this last solution is the one generally adopted in modern Europe in the seventeenth and eighteenth centuries.

At the turn of the twenty-first century, state monopolies were abolished or were in retreat, given the fall of the communist regimes and the abolition of monopolies in the countries that make up the European Union. Since the 1957 Treaty of Rome, which established the Common Market, state monopolies and monopolies on public services have been viewed as running counter to the rules of competition and free trade among the member states. With the implementation of the Common Agricultural Policy in 1970, the Council of the European Economic Community (EEC) decided to regulate the disappearance of state monopolies, including that of tobacco, in their commercial aspect. The incorporation of Spain and Portugal in 1986 and Sweden in 1995 into the EEC accelerated the retreat of state monopolies within the European Union.

See Also Brazil; British Empire; Caribbean; Cuba; Dutch Empire; French Empire; Portuguese Empire; Smuggling and Contraband; Spanish Empire; Taxation; Trade.

▌FRANCISCO COMÍN

BIBLIOGRAPHY

Aharoni, Yair. *The Evolution and Management of State-Owned Enterprises.* Cambridge, Mass.: Ballinger, 1986.

Comín, Francisco. "Los monopolios fiscales." In *Historia de la empresa pública en España.* Edited by Francisco Comín and P. Martín Aceña. Madrid: Espasa Calpe, 1991, pp. 139–175.

Comín, Francisco, and Pablo Martín Aceña. *Historia de la empresa pública en España.* Madrid: Espasa Calpe, 1991.

———. *Tabacalera y el estanco de tabacos (1636–1998).* Madrid: Fundación Tabacalera, 1999.

Gálvez-Muñoz, Lina, and Francisco Comín. "Business and Government: Labour Management in the Spanish Tobacco Monopoly (1887–1935)." In *Business and Society.* Edited by A. M. Kuijlaars et al. Rotterdam: Center of Business History, 2000, pp. 159–170.

Toninelli, Pier Angelo, ed. *The Rise and Fall of the State-Owned Enterprise in the Western World.* New York: Cambridge University Press, 2000.

Taxation

Tobacco taxes are centuries old, yet they remain a crucial element in the debate surrounding the tobacco industry, smoking, and public health. Taxation began as soon as European governments began importing and trading tobacco in the sixteenth century. These taxes (primarily in the form of import duties) were levied to generate revenue for the government. This is an important role that tobacco taxes play to this day. However, the goals of tobacco taxes have changed since they were first used more than 400 years ago, and their use is not as straightforward as it might seem. In the 2000s, in addition to raising general revenue, tobacco taxes may also be used as "user fees" paid by smokers to offset the social costs of smoking or as regulation to directly reduce tobacco consumption. Since they can be successful in each role, governments around the world continue to aggressively use them, tempered mainly by issues of fairness and intergovernmental tax avoidance and smuggling.

A Brief History

Tobacco, once introduced to Europe in the sixteenth century, quickly became popular. Soon after, governments recognized that it could become a lucrative basis of tax revenues. However, at least one European sovereign sought to reduce its use. In 1604 the king of England, James I, published anonymously a booklet titled *Counterblaste to Tobacco*, which condemned smoking as a dirty and unhealthy custom (and even pointed out the problems of secondhand smoke). Coupled with this publication was a huge increase on the import **duty** on tobacco, which raised the rate more than 40 times its original level, from 2 pence per pound of tobacco to 82 pence per pound. This represents the first use of taxation on tobacco to reduce its consumption.

duty a tax, usually on certain products by type or origin. A tariff.

The actions of King James had important consequences for the issues of tobacco taxes, tobacco use, and government revenue. To avoid paying the increased tax rate, importers simply smuggled tobacco into England. Tobacco grew in popularity, but the royal coffers were not filling with tax revenue. Seeing the potential value of tobacco imports,

TAXATION

James changed his mind, and reduced the tobacco tax to 12 pence. This paid off in the form of increased tax collections, and tobacco has been recognized as a good source of government revenue ever since. Thus, by using tobacco taxes to generate revenue, King James set much of the groundwork for the politics of tobacco taxes that still exist in the twenty-first century.

However, it would have to wait several centuries before the political debate over tobacco would encompass public health. The delay was due to tobacco's popularity and its economic and fiscal importance, as well as the fact that there was no authoritative consensus that the health effects of smoking were negative. Instead, tobacco was seen as a vital source of revenue for many governments. Countries including Spain, Venice, France, Austria, Russia, Poland, Morocco, and the Papal States taxed tobacco beginning in the seventeenth century. Great Britain used tobacco taxes to help finance its role in the Napoleonic Wars, and within a few years of ratifying the Constitution, the United States in 1794 adopted tobacco taxes that subsequently rose and fell with revenue needs.

Tobacco Taxes as Revenue Policy in the Past Century

Historically, the federal government in the United States has used tobacco taxes as revenue policy, rather than for regulation or as user fees. The evidence of this is outlined in the following table (adapted from U.S. Department of Health and Human Services), which shows that tax increases on cigarettes coincide with increases in revenue needs due primarily to war and budget deficits. (Note: When two rates are reported, the lower rate is for inexpensive cigarettes.) Excise taxes on tobacco have been popular because historically it has been considered a luxury item. Although the federal government collected only a small portion (.4%) of its total revenue with tobacco taxes in 2002, tobacco tax increases are still proposed, for example by the Congressional Budget Office, as a way to increase revenue by meaningful amounts.

Changes in U.S. Cigarette Tax Rates

Date	Tax Per Pack (cents)	Notes About Adjustment
30 June 1864	8, 2.4	Revenue for Civil War
1 April 1865	2.4, 4.0	Revenue for Civil War
1 August 1866	4.0, 8.0	Alleviate Civil War debts
2 March 1867	10.0	Alleviate Civil War debts
20 July 1868	3.0	Revenue needs due to war eased
3 March 1875	3.5	
3 March 1883	1.0	
15 August 1897	2.0	

Date	Rate	Reason
14 June 1989	3.0	Revenue for Spanish-American War
1 July 1901	1.08, 2.16	Revenue needs due to war eased
1 July 1910	2.5	
4 October 1917	4.1	Revenue for World War I
25 February 1919	6.0	Alleviate debts from World War I
1 July 1940	6.5	Military expansion
1 November 1942	7.0	Revenue for World War II
1 November 1951	8.0	Revenue for Korean War
1 January 1983	16.0	Reduce federal budget deficit
1 January 1991	20.0	Reduce federal budget deficit
1 January 1993	24.0	Reduce federal budget deficit
1 January 2000	34.0	Maintain balanced budget
1 January 2002	39.0	Reduce federal budget deficit

State governments in the United States have also used tobacco taxes as revenue policy to great success. Although most states now tax a wide range of tobacco products, including chewing tobacco, cigars, and **snuff,** cigarette taxes get the most attention because cigarette smoking accounts for the vast majority of all tobacco consumed in the United States, according to the U.S. Department of Agriculture. In 1921, Iowa was the first state to adopt a tax on cigarettes and several other states quickly followed. As a result of declining income tax revenues during the Great Depression, nineteen more states adopted a cigarette tax between 1927 and the late 1930s. By 1969, all fifty states had a cigarette tax. In 2000, state cigarette tax collections generally represented between .5 percent and 1 percent of state revenue, not including federal government transfers, although Michigan collected 1.5 percent of its revenues with cigarette taxes, according to 2002 figures from the Centers for Disease Control and Prevention. Michigan collects a greater portion of its revenue from cigarette taxes than other states because in the mid-1990s it used a large increase in the cigarette tax rate as a key component of funding of property tax relief. Finally, studies since the 1990s, including those conducted by the World Bank, consistently show that governments around the world can raise tobacco taxes substantially in order to generate revenue. Thus, taxes on tobacco remain an important source of government revenue to this day. For example, Greece generates 9 percent of its total tax receipts from cigarette taxes.

snuff a form of powdered tobacco, usually flavored, either sniffed into the nose or "dipped," packed between cheek and gum. Snuff was popular in the eighteenth century but had faded to obscurity by the twentieth century.

TAXATION

There are several reasons for the usefulness of tobacco taxes as a revenue source. First, since tobacco contains highly addictive nicotine, demand for tobacco is fairly price inelastic. That is, people are less likely to reduce or eliminate their purchases if the price goes up due to taxation by modest amounts. Data from the U.S. Department of Health and Human Services (USDHHS) for the United States show that a 10 percent increase in price leads to a 3 to 5 percent decrease in consumption of cigarettes. Therefore, cigarette consumption will fall by smaller proportions than any other kind of price increases. Carefully taxing tobacco will not seriously depress tobacco sales, instead preserving those sales so as to increase state revenue. Second, for political reasons, politicians can more easily increase excise taxes than general taxes, like income or property taxes. Cigarette taxes are paid only by smokers, who are increasingly stigmatized, while income taxes are paid by everyone who works. Therefore, if a modest amount of revenue is needed, taxes on tobacco are often a politically feasible solution. Third, taxes on tobacco can be justified on the grounds of public health, even if they are being used for revenue. For most low- to middle-income countries, a tobacco tax increase of 10 percent is about as cost effective as many other government-financed public health interventions, such as child immunization programs. Low-income countries include Pakistan, Ethiopia, and Haiti, while middle-income countries include Mexico, Brazil, and Turkey.

Tobacco Taxes as Regulation

In the mid-1950s several published reports linked smoking with cancer, resulting in a widespread tobacco health scare that caught the attention of governments around the world. In 1962 the United Kingdom's Royal College of Physicians issued an official report linking cancer and smoking. Canada and the United States followed suit, issuing their reports in 1963 and 1964, respectively. These reports opened the door for the regulation of tobacco consumption. In the 1980s taxes became a predominant form of smoking regulation, and in the 2000s cigarette taxes are a significant feature of most government efforts to reduce smoking.

Evidence of their use as regulation is obvious when one compares the average tax rate for U.S. states that grow tobacco with states that do not: In May 2000 the average tax on a pack of cigarettes in the six largest tobacco-growing states was 7.1 cents, while the average tax in the rest of the country was 46.5 cents, according to USDHHS. Using tobacco taxation as regulation is difficult in the face of pro-tobacco interests. In his *Tobacco Control: Comparative Politics in the United States and Canada* (2002), Donley Studlar points out that between 1978 and 2000 six top tobacco growing states never raised their cigarette taxes, while the forty-four other states and Washington, D.C., raised cigarette taxes a total of 160 times. More than half of these other states increased their cigarette taxes at least three times during this time span.

Tobacco taxes work as regulation primarily because they raise the price of tobacco products. Due to the addictiveness of nicotine, responses to price increases may be slow. Still, information from the United States shows that a 10 percent increase in the price of cigarettes reduces consumption by 3 to 5 percent in the short term. In the long term, a permanent price increase can have an even larger effect on consumption (about twice as great of an impact). Simply stated, people smoke less

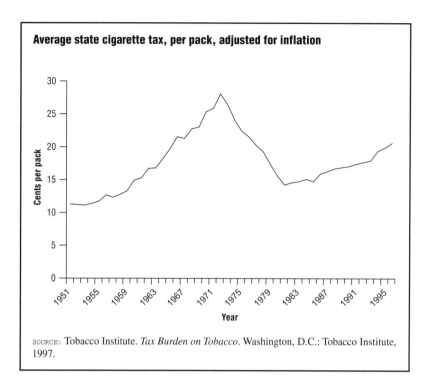

SOURCE: Tobacco Institute. *Tax Burden on Tobacco*. Washington, D.C.: Tobacco Institute, 1997.

after a large tax increase, particularly in the long run since the effects of a price increase may not occur immediately due to addiction. Further, tobacco taxes can work as regulation because they send a signal to smokers that the government is concerned about public health. By raising tobacco taxes with some fanfare about public health, the government simultaneously sends information about the dangers of tobacco use, which can cause some people to try to cut back or quit. Governments may also choose to launch public awareness campaigns at the same time they increase tobacco taxes.

Tobacco Taxes as User Fees

Since health problems related to tobacco use place demands on governments, such as increased hospital costs or increased unemployment compensation due to smoking-related sickness, it is natural that those using tobacco be required to help offset those demands. By viewing tobacco taxes as user fees, and setting aside tobacco tax revenue for public health and smoking reduction programs, smokers in effect help "pay for" these government activities and social costs. For example, state governments spend billions of dollars per year on smoking-related medical expenses alone. As of 2004, sixteen state governments set aside some tobacco tax revenues to be spent directly on programs to reduce tobacco use and improve public health.

Although some scholars argue that cigarette taxes more than adequately cover any social costs, others are a bit more cautious. At a minimum, cigarette taxes would need to be frequently adjusted to keep up with inflation and health care costs if they were being used consistently as user fees. Since this is not the case, it is doubtful that cigarette taxes offset social costs. For example, the figure above shows that the average state tax rate in the United States has fluctuated greatly, once adjusted for inflation. Furthermore, to employ cigarette taxes as both user fees and revenue policy would require raising tax rates so that the revenue from them would not be completely consumed by public health and social costs due to smoking.

The Problems of Tobacco Taxes

The continued use of tobacco taxes for revenue, regulation, and user fees faces some important obstacles. For one, tax avoidance and bootlegging pose a problem, particularly when government jurisdictions share a common border. Avoidance of a tobacco tax involves traveling to purchase tobacco products, for personal consumption, in a government jurisdiction with a low tax. Variations in tax can be large enough so that smokers may be able to save a substantial amount of money simply by crossing into a different jurisdiction to purchase cigarettes. While legal, this tax avoidance will undermine the objectives of the taxation policy. Revenues will be lost, tobacco use will not drop as much, and user fees will be paid to the "wrong" government. This effect can also occur internationally. For example, many French, Germans, and Belgians purchase cigarettes in Luxembourg in order to save money. Estimates have shown that only about 15 percent of tobacco purchased in Luxembourg is consumed there.

Bootlegging (or smuggling), on the other hand, is typically the purchase of tobacco products in a low tax jurisdiction and illegally reselling them in another jurisdiction with a higher tax. Smuggling is a worldwide problem. Global tobacco trade statistics indicate that there are large discrepancies between cigarette imports and exports, and that perhaps a third of exported cigarettes are smuggled. Although there is some evidence that smuggling is not caused simply by price differences between government jurisdictions, the Canadian example is evidence that those differences can act as an incentive. During the 1980s Canada taxed cigarettes aggressively as part of that government's tobacco control policy. By 1991 the difference in the price of cigarettes between Canada and the United States was the largest in the world, at about $3 U.S. per pack. As a result, the amount of cigarettes smuggled increased more than tenfold from 1990 to 1993, and in 1994 the Canadian government reduced its tax rate.

Tobacco taxes are also under scrutiny because there is some evidence that they are unfair. Certainly they are paid by only a relatively small portion of the population. Furthermore, research shows that tobacco taxes take up a larger portion of income among poorer families, making them a regressive tax. Despite evidence that poorer smokers are more sensitive to price increases, and thus tax *increases* are actually progressive, concerns of fairness may be a reason for why tobacco taxes have not been raised to levels consistent with government goals of revenue or regulation. Certainly, issues of equity and smuggling are limiting factors on the usefulness of tobacco taxes.

See Also Brazil; British Empire; Caribbean; Cuba; Dutch Empire; French Empire; Portuguese Empire; Smuggling and Contraband; Spanish Empire.

∎ MICHAEL LICARI

BIBLIOGRAPHY

Centers for Disease Control and Prevention. *Tobacco Control State Highlights 2002: Impact and Opportunity.* Atlanta: Department of Health and Human Services, 2002.

Corina, Maurice. *Trust in Tobacco: The Anglo-American Struggle for Power.* London: Michael Joseph Ltd., 1975.

Goodman, Jordan. *Tobacco in History.* London: Routledge, 1993.

Haas, Mark. "Michigan's Experience with School Finance Reform." Presented at the 56th Annual Federation of Tax Administrators Revenue Estimation and Tax Research Conference, Minneapolis, Minn., 2001.

Joosens, Luk, and Martin Raw. "Smuggling and Cross Border Shopping of Tobacco in Europe." *British Medical Journal* 310 (1995): 1393–1398.

———. "Cigarette Smuggling in Europe: Who Really Benefits?" *Tobacco Control* 7 (1998): 66–71.

Licari, Michael J., and Kenneth J. Meier. "Regulatory Politics When Behavior Is Addictive: Smoking, Cigarette Taxes, and Bootlegging." *Political Research Quarterly* 50 (1997): 5–24.

———. "Regulation and Signaling: When a Tax Is Not Just a Tax." *Journal of Politics* 62 (2000): 875–885.

Lyon, Andrew, and Robert Schwab. "Consumption Taxes in a Life-cycle Framework: Are Sin Taxes Regressive?" *Review of Economics and Statistics* 77 (1995): 389–406.

Manning, Willard et al. *The Costs of Poor Health Habits.* Cambridge, Mass.: Harvard University Press, 1991.

Studlar, Donley. *Tobacco Control: Comparative Politics in the United States and Canada.* Peterborough, Ontario: Broadview Press, 2002.

Taylor, Peter. *Smoke Ring.* London: The Bodley Head, 1984.

Tobacco Institute. *Tax Burden On Tobacco.* Washington, D.C.: Author, 1997.

Townsend, Joy, Paul Roderick, and Jacqueline Cooper. "Cigarette Smoking by Socioeconomic Group, Sex, and Age: Effects of Price, Income, and Health Publicity." *British Medical Journal* 309 (1994): 923–926.

United States Department of Health and Human Services. *Reducing Tobacco Use: A Report of the Surgeon General.* Atlanta: Author, 2000.

Viscusi, Kip. *Smoke-Filled Rooms: A Postmortem on the Tobacco Deal.* Chicago: University of Chicago Press, 2002.

Whelan, Elizabeth. *A Smoking Gun: How the Tobacco Industry Gets Away with Murder.* Philadelphia: Stickley, 1984.

World Bank. *Curbing the Epidemic: Governments and the Economics of Tobacco Control.* Washington, D.C.: Author, 1999.

Therapeutic Uses

In the sixteenth and seventeenth centuries, diverse cultures around the world used tobacco for medicinal purposes in similar ways, although each culture set tobacco's imagined healing properties within a distinctive cosmological framework. Amerindians, the first to use tobacco therapeutically, employed it for a wide range of ailments—as a painkiller for earaches, toothaches, or the pains of childbirth; as a plaster to treat open wounds; as a pesticide to ward off snake and insect bites; and as a remedy against asthma, rheumatism, fevers, convulsions, eye sores, intestinal disorders, various skin disorders, and poisonings. ◆These medical treatments emerged out of complex cultural and religious systems that imbued tobacco with sacred significance.

◆ *See "Social and Cultural Uses" for a depiction of Amerindian treatments involving tobacco.*

THERAPEUTIC USES

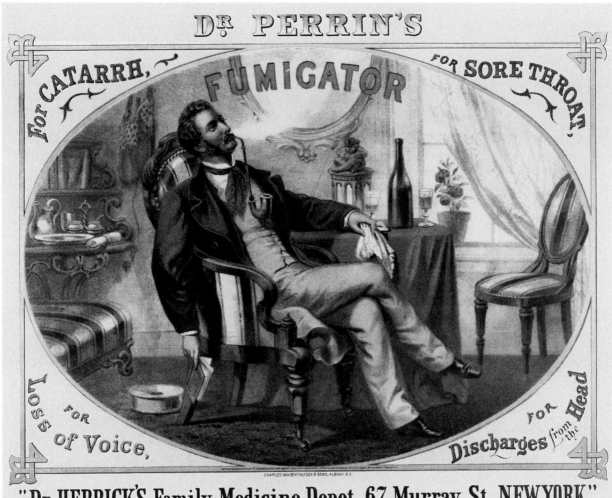

This 1869 advertisement for Dr. Perrin's "Fumigator" seems to be promoting the smoking of tobacco as a remedy for catarrh, sore throat, loss of voice, and discharges from the head. Throughout the 1800s, many Europeans and Americans continued to be swayed by medical claims that today would be considered quackery. © CORBIS

humoral theory the idea, first advanced by the ancient Greeks, that health and temperament were determined by the balance of the four humors, or bodily fluids: blood, phlegm, choler (yellow bile), and melancholy (black bile). This idea dominated medicine until the late 1700s.

Europeans, upon first encountering religious healing rituals involving tobacco in the early sixteenth century, regarded the plant as highly exotic and its ceremonial and medical use as debauched and idolatrous. Over time, European doctors came to regard tobacco as a medicinal herb, but only after they assimilated it into the classical Galenist medical model that had reemerged in Europe in the fifteenth and sixteenth centuries (see the sidebar "Galen and the Four Humors").

Nicolás Mondardes (1519–1588), a well-known author and doctor living in Seville, was the first European physician to unabashedly promote tobacco as a medicine. In 1571, in the second part of his widely read compendium of New World plants, Mondardes highlighted tobacco's healing properties. Basing his assessment of its therapeutic value on the ancient Greek humoral theories of medicine first articulated by Hippocrates (460–377 B.C.E.) and Galen (129–199 C.E.), Mondardes established tobacco's essential properties as hot and dry in the second degree. According to the **humoral theory** of opposites, this meant that tobacco was useful for treating disequilibrium caused by excessive cold or damp by virtue of its hot purgative powers. For example, the hot and drying attributes of tobacco were thought to heal or at least prevent the

Galen and the Four Humors

Prior to the Scientific Revolution that swept through Europe primarily during the seventeenth and eighteenth centuries, tobacco fit easily into a classical conceptual model in which the four primary qualities of hot, cold, wet, and dry were ascribed to all aspects of the living and nonliving world. The four primary qualities expressed aspects of the four elements of air, earth, fire, and water. Together they described the contents of the four humors—blood, black bile, phlegm, and yellow bile— that circulated in all animal and human bodies. Astrologers aligned the four humors with the four cardinal points in the heavens and assigned each humor three constellations.

When tobacco entered the European picture in the sixteenth century, medical theory of the time held that individuals possessed their own *complexio*, or unique combination of humors and environmental factors, such as one's astrological sign and geographic location. For example, if a person with a hot dry *complexio* were to travel to a cold and wet place under a cold and wet astrological sign, she or he might wish to consult a physician in advance for advice about adjusting diet, exercise, rest, evacuation patterns, and passions so as to minimize the tendency to imbalance, and, hence, illness. This is what the Greek physician Galen (129–199 C.E.), who codified humoral theory, termed the "non-naturals."

Given the common assumption that the brain and nerves were cold and phlegmy, it is not surprising that early modern Europeans often turned to tobacco for ailments they located in the head. In 1561, for example, the French queen Catherine de Medici reportedly used snuff to treat her son Francis II's recurrent headaches. Pleased with its apparent success, she later pronounced tobacco to be *herba regina*, or queen's herb. Yet as the seventeenth century progressed, medical opinion shifted, and Europeans debated tobacco's therapeutic values.

▮ ROBERT L. MARTENSEN

---

development of conditions, such as asthma, that were associated with cold and wet situations.

Mondardes went on to outline twenty curative uses of the plant including the topical application of green tobacco leaves to relieve headaches or to soothe snake and insect bites. When inhaled or chewed, it could resolve stomachaches, asthma, parasites, rheumatism, bloating, toothaches, poisonings, various skin diseases, and female hysteria. Many of these prescriptions overlapped with those common among Mesoamericans, leading some scholars to conclude that Mondardes largely appropriated indigenous knowledge and practice into his treatise (Norton 2000).

European Debates Over Tobacco's Therapeutic Use

European physicians debated tobacco's medical efficacy throughout the seventeenth century. Many authors, following the lead of Mondardes, continued to describe tobacco as a wonderful panacea for a host of ailments. Others criticized such claims as **hyperbolic** or even fraudulent and deceptive. While acknowledging tobacco's purgative effects on the human body, critics questioned whether such therapy was effective.

hyperbolic exaggerated; overstated.

Over time, tobacco's reputation as a miraculous drug diminished, and by the eighteenth century tobacco use in Europe came to be recognized as primarily recreational. Nonetheless, **nicotian therapy** continued in Europe well into the nineteenth century. In 1800 European physicians were still using tobacco as an antispasmodic for asthma, an enema for

nicotian therapy a smoking-cessation regimen where other, non-addictive substances are substituted for nicotine.

THERAPEUTIC USES

intestinal obstructions, and as a diuretic for dropsy and similar disorders. Some even continued to advocate its use as a prophylactic against infectious diseases such as cholera well into the 1890s.

Seventeenth-Century Chinese Medicine

Tobacco spread to East Asia on the same wave of globalization that carried it to Europe in the sixteenth century. By the late 1500s, tobacco was being widely cultivated and consumed in many areas of China. China's long tradition of medical botany, stretching back over two thousand years, allowed doctors to easily incorporate the new herb into their **materia medica.** The first to do so were the physicians Zhang Jiebin (1563–1640) and Ni Zhumo (c. 1600). Both mention tobacco in the pharmacopoeia section of texts published simultaneously in 1624. Their descriptions of its presumed medical benefits are quite similar both to one another and to those published in sixteenth and seventeenth-century Europe, although these overlapping conclusions were arrived at quite independently. They begin as they do with any other herb, by noting tobacco's taste (bitter and pungent), its essential properties (warm or hot), and by placing it in the categorization scheme of classical pharmaceutical knowledge. (Zhang Jiebin listed it with plants that grow in marshy conditions, while Ni Zhumo set it within his section on "toxic" herbs). Both described tobacco as having considerable toxicity (*du*). By this they meant not that tobacco was poisonous but that it resembled other powerful drugs that had to be prescribed carefully because they had potentially detrimental and lasting effects.

materia medica a Latin term for the substances and components of medicine.

Bust of Greek physician Galen (129–199 C.E.), who codified humoral theory. LIBRARY OF CONGRESS

These seventeenth-century Chinese authors understood tobacco's effect on the human body within the classical Chinese cosmology of yin (wetness, coolness) and yang (dryness, heat). According to Chinese medical theory, yin and yang regulated the movement of bodily qi (the generative energy essential for life and health) within the human body. Disease arose when the delicate balance between yin and yang was disrupted in some way: by external pathogens, by an excess of activity or emotion, improper or gluttonous eating or drinking, or by excessive lust or overindulgence in sex. Tobacco was pure yang and as such could be used to counterbalance those with an excess of yin by warming the organs of the body and protecting it from dampness and cold. Its warming properties aided digestion, speeded circulation, stopped vomiting, and removed congestion. Its toxicity could kill intestinal parasites in children, reduce masses or lumps in the abdomens of women, and keep poisonous snakes away. If pounded into a paste and applied to the hair, it could kill head lice. Most importantly, perhaps, these two authors identified smoking tobacco and the topical application of tobacco as prophylactics against epidemics generally, or more specifically against **miasmas** (*zhangqi* or malaria) in the south and "cold damage" (*shanghan*) in the north.

miasma a polluted, unhealthy atmosphere.

Drawing upon a strikingly different medical tradition from that invoked by Nicolás Mondardes, Zhang Jiebin and Ni Zhumo came to similar conclusions about tobacco's presumed essential properties and its purgative powers. Where they differed from Mondardes was in their assessment of the herb's potential for harm. Tobacco's toxicity (*du*) made it a potent drug that had to be used with care. Zhang Jiebin warned that tobacco could lead to habitual use and that it could be intoxicating after only a few puffs. To counteract the sensation of becoming tipsy, he recommended the cooling effects of water or refined sugar. Those with strong yang should not smoke at all. Ni Zhumo similarly cautioned those with dryness in their lungs or those spitting up blood to avoid smoking at all costs.

Eighteenth and Nineteenth-Century Chinese Materia Medica

By outlining both the detrimental and beneficial effects they believed tobacco had on the human body, Zhang Jiebin and Ni Zhumo set the tone for subsequent medical writing about the substance in China. About ninety materia medica written during the Qing dynasty (1644–1911) include discussions of tobacco. Many of these, especially those published in the eighteenth and nineteenth centuries, are popularized versions of more sophisticated medical texts, written in simplified language with the layperson in mind. Stripped down to only the most essential content, these pharmacological primers served as introductory textbooks for beginning medical students or as popular family almanacs to be used for basic medical treatment at home. As such, they provide insight into what ordinary eighteenth and nineteenth-century Chinese thought they were doing when they medicated themselves with tobacco.

Popularizers such as Wang Ang (1615–c. 1695) and Wu Yiluo (eighteenth century) stressed the health benefits of tobacco over its dangers. In these highly accessible texts tobacco was identified as a "hot and dry" drug belonging to the fire category, one that had a pungent taste and a strong heat toxicity. In line with Zhang Jiebin and Ni Zhumo, Wang Ang and Wu Yiluo regarded tobacco as a strong heating agent that could replenish yang. These qualities made it useful in the home for treating illnesses caused by pathogenic cold factors (chills, fevers, colds), for reducing swelling and pain in the joints caused by dampness (such as arthritis), for curing malaria, or for warding off poisonous snakes. **Snuff,** which became quite fashionable among the cultural and political elite in the eighteenth century, was thought to prevent all manner of epidemics because the user, by constantly sniffing tobacco through the nose, was less likely to absorb the miasmas that were thought to lead to disease.

snuff a form of powdered tobacco, usually flavored, either sniffed into the nose or "dipped," packed between cheek and gum. Snuff was popular in the eighteenth century but had faded to obscurity by the twentieth century.

By and large, the majority of these popular texts stressed the positive benefits of tobacco over its deleterious effects (the only negative frequently mentioned was the one first brought up by Zhang Jiebin—that it could make people tipsy). This positive pharmacological understanding of tobacco, published in books widely used throughout the eighteenth and nineteenth centuries can partially explain tobacco's **ubiquitous** use by people of all ages, genders, and social classes within China.

ubiquitous being everywhere; commonplace; widespread.

Chinese Debates Over Tobacco's Therapeutic Use

In Qing China, as in seventeenth-century Europe, there were physicians who argued forcefully that tobacco was highly dangerous and potentially harmful to human health. One such critic was Zhang Lu, a well-known physician from Jiangsu Province. In a treatise published in 1715, Zhang Lu conceded that tobacco could be used "with caution" in the treatment of eye diseases but he quickly went on to list

THERAPEUTIC USES

the drawbacks of smoking. He argued that the heat of tobacco smoke suffocated and burned the internal organs and that its toxicity circulated through the body's channels, doing vital harm. Moreover, Zhang Lu pointed out, contrary to the earlier claims of Zhang Jiebin or Ni Zhumo and their followers, tobacco could not prevent malaria in tropical climates nor could it overcome cold pathogenic factors in the north. Zhang Lu believed that tobacco was highly poisonous and that those who received its toxicity over a long period of time were likely to have lungs that "were not clear" to the point that they might spit up "yellow water" and die.

Despite the ubiquity of tobacco smoking throughout Chinese society and the frequency with which it was mentioned in Qing *materia medica*, there is little evidence that doctors actually used it as a drug in clinical practice. Tobacco is rarely listed in formularies (*fangshu*) and medical case records (*yi an*), the two sections of Qing medical texts that reflect what physicians actually did. Those who do refer to tobacco in their formularies, such as Ye Tianshi (1667–1746), discuss only the external application of tobacco paste to sore joints or open sores, not tobacco inhaled as smoke or snuff.

For the most part, by the nineteenth century tobacco appears to have become primarily a folk remedy used in the home rather than one commonly prescribed by classically trained and highly skilled physicians. In this, developments in China replicated those in Europe. While tobacco gradually lost favor with nineteenth-century European doctors after the poisonous quality of nicotine was identified in 1828, it remained popular among laymen who used it to treat toothaches and to ward off infectious diseases such as influenza.

Tobacco in Twentieth-Century China

In twentieth-century China tobacco continues to hold considerable sway as a folk remedy, particularly in the countryside. Although it is listed as a drug in modern pharmacopoeia, for example the *Dictionary of Chinese Drugs*, it is no longer prescribed by physicians. Contemporary doctors, those who practice both Western medicine and traditional Chinese medicine, are well aware of the dangers of nicotine, and there is a vibrant, medically informed antitobacco movement within China. Nonetheless, folklore persists about its presumed health benefits.

A survey conducted in the 1990s by the publication *China Tobacco Work* found that many Chinese still believe that smoking one or two cigarettes a day can prevent malaria or illness from dampness and cold. Puffing on a cigarette is thought to reduce bloating of the stomach, to cure food poisoning, or to settle digestive troubles brought on by nervousness or anxiety. Tobacco continues to be used as a pesticide: smoking in the summer is said to reduce the number of insects that buzz around one's face, washing one's hair with water laced with tobacco juice is said to kill lice, and the oil from a tobacco pipe is still thought to be useful in the treatment of snake bites.

The Chinese, no less than Europeans or those in other cultural areas, long regarded tobacco as a beneficial and efficacious medicine. The presumed therapeutic uses of tobacco undoubtedly contributed to its rapid diffusion around the globe, as consumers believed they were ingesting a substance that was good for their health, or at the very least, not harmful. Nonetheless, medical justifications in China, as in Europe, are perhaps less significant in the end for explaining why people smoked than the fact that doing so was highly pleasurable and ultimately habit-forming.

See Also Disease and Mortality; Doctors; Medical Evidence (Cause and Effect); Social and Cultural Uses; South Asia.

■ CAROL BENEDICT

BIBLIOGRAPHY

Du Yong. "Ming-qing shiqi zhongguoren dui chuiyan yu jiankang guanxi de renshi" (Chinese recognition of the relationship between smoking and health in the Ming-Qing dynasties). *Zhonghua yishi zazhi* (Chinese Journal of Medical History) 30, no. 3 (2000): 148–150.

Hao Jinda. "Dui yancao chuanru ji yaoyong lishi de kaozheng" (Textural research on the history of tobacco's diffusion into China and its medical use). *Zhonghua yishi zazhi* (Chinese Journal of Medical History) 17, no. 4 (1987): 225–228.

Harley, David. "The Beginnings of the Tobacco Controversy: Puritanism, James I, and the Royal Physicians." *Bulletin of the History of Medicine* 67, no. 1 (1993): 28–51.

Jiangsu xin yixue yuan (Jiangsu Province, Institute of New Medicine). *Zhongyao da cidian* (Dictionary of Chinese Medicine). 1912. Reprint, Shanghai, China: Renmin chubanshe (People's Publishing House), 1977.

Ni Zhumo. *Huitu bencao gangmu huiyan* (Illustrated Materia Medica, Arranged by Drug Descriptions and According to Technical Aspects). 1624. Reprint, Jiantang, China: Publisher unknown, 1694.

Norton, Marcia Susan. "New World of Goods: A History of Tobacco and Chocolate in the Spanish Empire, 1492–1700." Ph.D. diss., University of California, 2000.

Stewart, Grace. "A History of the Medical Use of Tobacco, 1492–1860." *Medical History* 11, no. 3 (1967): 228–268.

Walker, Robin. "Medical Aspects of Tobacco Smoking and the Anti-Tobacco Movement in Britain in the Nineteenth Century." *Medical History* 24, no. 3 (1980): 391–402.

Wang Ang. *Bencao beiyao* (Essentials of Materia Medica). 1682. Reprint, Beijing, China: Zhongguo zhongyiyao chubanshe (Traditional Chinese Medicine Press), 1998.

Wu Yiluo. *Bencao congxin* (Restoration of Pharmaceutics). 1757. Reprint, Hong Kong: Shanghai yinshuguan (Shanghai Press), 1960.

Zhang Jianyue. "Yancao de yaoyong jiazhi chutan" (A preliminary discussion of tobacco's medicinal value). In *Zhongguo yancao shihua* (A History of Chinese Tobacco). Edited by *Zhongguo yancao gongzuo* bianjibu (Editorial Department of *China Tobacco Work*). Beijing, China: Zhongguo qinggongye chubanshe (China Light Industry Publishing House), 1993.

Zhang Jiebin. *Jingyue quanshu* (Jing Yue's Complete Works). 1624. Reprint, Beijing, China: Zhongguo zhongyiyao chubanshe (Traditional Chinese Medicine Press), 1994.

Zhang Lu. *Benjing fengyuan* (Elucidation of the Meaning of the Original Classic). 1715. Reprint, Beijing, China: Zhongguo zhongyiyao chubanshe (Traditional Chinese Medicine Press), 1996.

Tobacco as an Ornamental Plant

Tobacco plants, classified by the scientific name of *Nicotiana*, have 67 different species or varieties. *Nicotiana* is endemic to Namibia, parts of Australia, Polynesia, and North America, specifically from California to Southern Oregon, as well as from Texas to Mexico. It is also found in the South American countries of Argentina, Bolivia, Brazil, Chile, and Peru. While many people are undoubtedly familiar with the development of the plant as a commercial crop after its so-called discovery by Europeans, very few appreciate the fact that tobacco has long held a place as an ornamental or decorative plant used in the garden.

Tobacco, whether the commercial or ornamental species, has been grown as an annual—a plant that completes its life cycle in one growing season. The value of tobacco as an ornamental plant was not immediately realized at the time of its introduction to the royal courts of France and Portugal by Jean Nicot for purposes other than smoking and medicine. While late-seventeenth-century naturalists, like John Ray, began to catalog several tobacco species, such as *Nictotiana tabacum, N. rustica, N. latifolium, N. angustifolium, N. minima,* and *N. minor,* it was not until the middle of the eighteenth century that gardeners began to plant tobacco as an ornamental.

In the eighteenth century, gardening boomed among England's wealthy as gardening dictionaries and compendiums were published in greater numbers, chronicling, among other things, the rapidly increasing know-how of growing tobacco. Among the tobacco species planted for ornamental purposes by British gardeners at the time were *Nicotiana paniculata, N. angustifolia, N. fruticosa, N. glutinosa, and N. pusilla.* Gardening "kalendars" instructed that March was for sowing tobacco seed in hothouses and May was for transplanting starts into themed pleasure, flower, or medicinal gardens where the plants would bloom in England's gardens from June to October. British gardeners relied on hothouses to ensure tobacco plants bloomed by the end of their typically cool summers.

In the early nineteenth century, Italian gardeners capitalized on the usage of ornamental tobacco species in clay pots. As early as 1840, Italians were bringing potted *Nicotiana plumbaginifolia* plants to sell at local markets. Apparently around the same time, British gardening authorities remained preoccupied with developing new garden uses for ornamental tobacco species. Certain fragrant, flowering tobaccos, *Nicotiana sylvestris* and *N. alata,* were introduced for the first time into British gardens. In the late 1800s, London gardening circles encouraged English ladies to plant *Nicotiana rotunifolia, N. alata, N. undulata, N. axillaries, N. tristis,* and *N. rugosa,* but withheld approval of *N. rustica* because of its "dirty green flowers" (Loudon 1846). In France, by the end of the nineteenth century, the ornamental tobacco species *Nicotiana tomentosa* had ignited a French attraction in subtropical garden effects. Interestingly, it took another fifty years before these beautiful and fragrant tobacco species caught on among U.S. gardeners.

With the dawn of the twentieth century, more savvy gardeners began to look for increased variety in their plants. In 1901 *Nicotiana forgetiana* was introduced into cultivation. Yet it was valued primarily

TOBACCO AS AN ORNAMENTAL PLANT

for its 1903 creation of a new tobacco species born in England through a cross between *N. alata* and *N. forgetiana*, producing *Nicotiana x sanderae*, which in turn was used to father other ornamental varieties. Five years later, in 1908, Italy introduced two additional ornamental varieties to the U.S. market, *Nicotiana noctiflora* and *N. glutinosa*.

In the 1930s, tobacco was not only being used in gardens, but it also began to appear as an indoor houseplant among Massachusetts Garden Club members. By the late 1930s, *Nicotiana* had gained notoriety as summer porch decorations, probably an extension of the nineteenth-century Italians' earlier success in growing potted tobacco. At the same time, gardeners continued to display *Nicotiana glauca*, *N. tomentosa*, *N. tabacum*, *N. sanderae*, and *N. alata* in subtropical displays.

The practice of **hybridization** reached its peak in the latter part of the twentieth century and early twenty-first century, with additional plant hybrids emerging from *Nicotiana alata* and *N. x sanderae*. With each new hybrid or series, gardeners were able to highlight sought-after characteristics of *Nicotiana*. *Nicotiana alata* offers the semi-dwarf yet fragrant Nikki series, and *Nicotiana x sanderae* first produced a velvety-red Crimson King in the 1930s and shortly thereafter Sensation, which was followed by others like the Domino series with upward facing flowers: the compact-growing Havana, the dwarf Merlin (a perfect container variety), and the more weather-tolerant Starship.

Understanding the historical use of tobacco as an ornamental adds splendid detail to the plant's historical tapestry. Though much maligned

Initially grown in seventeenth-century British gardens as a medicinal herb, tobacco quickly became recognized for its value as an ornamental. This modern-day British garden includes fragrant white flowering tobacco, lobelia, and impatiens. © ERIC CRICHTON/CORBIS

hybridization the practice of cross-breeding different varieties of plants or animals to produce offspring with desired characteristics.

TOBACCO AS AN ORNAMENTAL PLANT

Domino Pink and White is a hybrid species of tobacco known for its upturned petals and greater summer heat tolerance. Gardeners favor hybrids because their flowers are not sensitive to the length of daylight, whereas the flowers of "unimproved" varieties open only on cloudy days and at night. © TANIA MIDGLEY/CORBIS

for its more common usage, ornamental varieties of tobacco hold a key place in our gardening heritage. From seventeenth-century England to modern-day America, the intoxicating perfume of the tobacco plant's flowers ensures its legacy as an ornamental plant.

See Also Botany (History); Native Americans; Origin and Diffusion; Social and Cultural Uses; Therapeutic Uses.

▮ CHRISTINA A. S. PERRON

BIBLIOGRAPHY

Abercrombie, John. *The Garden Vade Mecum, or Compendium of General Gardening*. Dublin: J. Jones, 1790.

Bailey, L. H. *The Standard Cyclopedia of Horticulture*. New York: The Macmillan Company, 1937.

Huxley, Anthony, ed. *The New Royal Horticultural Society Dictionary of Gardening.* New York: Stockton Press, 1992.

Loudon, Jane. *Gardening for Ladies: and Companion to the Flower Garden.* New York: Wiley & Putnam, 1846.

Meller, Henry J. *Nicotiana.* London: E. Wilson, 1833.

Tobacco Control in Australia

Australia ranks high among nations that have made effort to reduce the burden of tobacco-caused death and disease. Between 1945 and 2001, adult male smoking prevalence fell mostly continuously from 72 percent to 22 percent. Female rates have fallen as well, from a peak of 31 percent in 1983 to 22 percent in 2001 (Woodward et al., . . . c01t1; Siahpush). Annual adult per capita consumption has fallen 61 percent from 3.54 kilograms in 1961, to 1.37 kilograms in 1999, reflecting falling smoking prevalence and reduced consumption (Woodward et al., . . . c01t2). While a greater proportion of people with less education and income smoke than their more educated and affluent counterparts, the growing proportion of ex-smokers varies little across socioeconomic status, meaning that cessation efforts have impacted evenly across the population. An exception here is Australia's Aborigines, whose smoking rates often exceed 60 percent. Since the late 1980s, male lung cancer rates have been falling dramatically, while female rates have reached a plateau.

Antismoking sentiment has existed in Australia since the end of the nineteenth century. However, efforts to explain rapidly declining tobacco use have generally pointed to milestones that have been in place since the 1970s, including various tobacco-control policies, tobacco-control laws, and prominent antismoking campaigns. However, such explanations tend to credit nothing to those enabling factors that cause these visible and recorded "events" to happen. The advocacy of individual citizens and grassroots organizations that precedes the introduction of a law or the factors responsible for the evolving enthusiasm of a politician for tobacco control are seldom recorded, "counted," or considered in evaluative research. Yet when basic questions are asked—like "How did Australia manage to get tobacco advertising banned?" or "Why has tolerance of smoking in public indoor areas reduced so much?"—no valid account of the process could fail to place advocacy at center stage.

A 1992 Philip Morris Corporate Affairs planning document said the following about Australia: "Australia has one of the best organized, best financed, most politically savvy and well-connected antismoking movements in the world. They are aggressive and have been able to use the levers of power very effectively to propose and pass draconian legislation. . . . The implications of Australian antismoking activity are significant outside Australia because Australia is a seedbed for antismoking programs around the world."

Achievements

In the past three decades, Australian tobacco control advocates have made significant gains in the following areas:

- **Harm Reduction.** Australian advocates were among the first to arrange for the tar and nicotine content of cigarettes to be tested and to advocate for the potential significance of tobacco yield (tar, nicotine, and carbon monoxide) data for harm reduction. This work can be seen as an early chapter in the evolution

TOBACCO CONTROL IN AUSTRALIA

An example of the proposed new "High Impact" graphic images to replace warnings on cigarette packets. Parliamentary Secretary for Health Trish Worth said the government was going through the process under the Trade Practices Act to replace the current warnings on tobacco products with 14 high-impact graphic images. AP/WIDE WORLD PHOTOS

of recent international interest in harm minimization and the international momentum for tobacco products to be regulated in ways that parallel food and drug regulation.

- **Advertising Bans.** Australia was one of the first democracies to ban all tobacco advertising and sponsorship.
- **Pack Warnings.** Australia has one of the world's largest pack warnings and pioneered research into warnings so as to have maximum impact on youth. Plans to implement these warnings saw prolonged periods of fierce counterlobbying by the industry.
- **Mass-Reach Campaigns.** From the late 1970s, governmental and nongovernmental organizations in Australia were among the first to run mass-reach antismoking campaigns. These were sometimes attacked by the tobacco industry and removed by an industry-dominated self-regulatory process.
- **Civil Disobedience.** Australia was the first nation to experience widespread civil disobedience against the tobacco industry through a campaign where health and community activists sprayed graffiti on tobacco billboards. This effort dramatically reframed tobacco advertising from something most would have seen as an unremarkable, normal part of the commercial

environment into a phenomenon that focused community discourse around irresponsible industry and collusive government policy unwilling to restrain it.

- **Smokeless Tobacco.** In 1986, South Australia became the first government in the world to ban smokeless tobacco. This ban subsequently went national.

- **Small Packs Banned.** In 1986, South Australia was the first government in the world to ban small "kiddie" packs (less than twenty sticks). Again, this ban subsequently went national.

- **Tax.** Australia has a relatively high tobacco tax by international standards, being about the fourth highest in the world.

- **Replacement of Sponsorship.** Victoria pioneered the use of a dedicated 5 percent rise in tobacco tax (hypothecation), which goes to a specific use, not into the general revenue, to enable the buy-out of tobacco sponsorships.

- **Clean Indoor Air.** Australia has among the world's highest rates of smoke-free workplaces and, progressively, domestic environments. Smoking is banned on all public transport with violations being uncommon. All states have restaurant smoking bans.

■ SIMON CHAPMAN

BIBLIOGRAPHY

Anderson P., and J. R. Hughes. "Policy Interventions to Reduce the Harm from Smoking." *Addiction* 95 (2000): Supp. 1:S9–11.

Barnsley K, and M. Jacobs. "World's Best Practice in Tobacco Control: Tobacco Advertising and Display of Tobacco Products at Point of Sale: Tasmania, Australia." *Tobacco Control* 9 (2000): 230–232.

Borland R., M. Morand, and R. Mullins. "Prevalence of Workplace Smoking Bans in Victoria." *Aust N Z J Public Health* 21 (1997): 694–698.

Borland R. et al. "Trends in Environmental Tobacco Smoke Restrictions in the Home in Victoria, Australia." *Tobacco Control* 9 (1999): 266–271.

Centre for Behavioural Research in Cancer, Anti-Cancer Council of Victoria. *Centre for Behavioural Research in Cancer: Health Warnings and Content Labelling on Tobacco Products*, 1992.

Chapman S. "Civil Disobedience and Tobacco Control: The Case of BUGA UP." *Tobacco Control* 5 (1996): 179–185.

Chesterfield-Evans, A. "BUGA-UP (Billboard Utilizing Graffitists Against Unhealthy Promotions): An Australian Movement to End Cigarette Advertising." *New York State Journal of Medicine* 83 (1983): 1333–1334.

Egger G. et al. "Results of Large Scale Media Antismoking Campaign in Australia: North Coast 'Quit For Life' Programme." *BMJ Clinical Research Ed.* 287 (1983): 1,125–1,128.

Gray N., and D. Hill. "Cigarette Smoking, Tar Content, and Death-Rates from Lung Cancer in Australian Men." *Lancet* 1 (1975): 1,252–1,253.

Hill, D. J., V. M. White, and M. M. Scollo. "Smoking Behaviours of Australian Adults in 1995: Trends and Concerns." *Med J Aust* 168 (1998): 209–213.

Holman, C. D. et al. "Banning Tobacco Sponsorship: Replacing Tobacco with Health Messages and Creating Health-Promoting Environments." *Tobacco Control* 6 (1997): 115–121.

Morgan, L.C. et al. "Lung Cancer in New South Wales: Current Trends and the Influence of Age and Sex." *Med J Aust* 172 (2000): 578–580.

Pierce, J.P. et al. "Evaluation of the Sydney 'Quit.for.Life' Anti-smoking Campaign. Part 1: Achievement of Intermediate Goals." *Med J Aust* 144 (1986): 341–344.

Scollo, M., and D. Sweanor "Cigarette Taxes." *Tobacco Control* 8 (1999): 110–111.

Siahpush, M. "Prevalence of Smoking in Australia." *Aust NZ J Pub Health* 27 (2003): 556.

Tyrell, I. *Deadly Enemies: Tobacco and Its Opponents in Australia*. Sydney: University of New South Wales Press, 1999.

Wilson D.H., et al. "15's: They Fit Anywhere—Especially in a School Bag: A Survey of Purchases of Packets of 15 Cigarettes by 14 and 15 Year Olds in South Australia." *Supplement to Community Health Studies* 9, no. 1 (1986): 16–20.

Woodward, S., M. Winstanley, and N. Walker. "Tobacco in Australia." Facts and Issues. Available: <http://www.quit.org.au/FandI/fandi/c01s1.htm#c01t1>.

Woodward, S., M. Winstanley, and N. Walker. "Tobacco in Australia." Facts and Issues. Available: <http://www.quit.org.au/FandI/fandi/c02s1.htm#c02t1>.

Tobacco Control in the United Kingdom

At the turn of the twentieth century, many people in British society, including some medical authorities, believed tobacco had medical uses and health benefits, including relieving stress. The advent of the cigarette in the 1880s made tobacco use more entrenched in British society. By the end of the following century, however, its use had been undermined by science and hedged about with restrictions. Smoking was increasingly a lower-class rather than a cross-class activity; its ubiquity among young, single mothers was a matter of concern for policy makers. Smoking had undergone a process of social repositioning. What were the key factors that brought about this change in the United Kingdom?

Early Research

In the 1930s, statisticians employed by insurance companies took note of rising lung cancer death rates in the population. Researchers suggested that the rising popularity of cigarette smoking—along with worsening environmental pollution from motor cars, industrial plants, and tarred roads—possibly contributed to an increase in lung cancer. This hypothesis led statistician Sir Austin Bradford Hill and his research assistant Richard Doll to study the smoking habits of hospital patients with lung cancer and to compare them with patients without lung cancer. Their paper, published in the *British Medical Journal* in 1950, reported that lung cancer patients were more likely to be smokers, suggesting that smoking could be a cause of lung cancer. However, their

A "no smoking" sign placed at the side of a bar in central London. Campaigners said on 9 March 2004 that more people's lives would be saved by banning smoking in public places than are lost every year in road accidents. To mark No Smoking Day, the campaign's director, Ben Youdan, said banning tobacco in pubs, bars, and all workplaces would save 4,800 lives a year in England, Wales, and Scotland. AP/WIDE WORLD PHOTOS

findings were not immediately accepted, and some scientists and physicians were critical of their methods. For example, legendary British statistician R. A. Fisher suggested that there could be a hereditary basis for the association; that is, smokers might carry a gene that made them prone to smoke and also made them more likely to develop cancer. The epidemiologic methods employed by Doll and Hill were novel at the time and were not yet widely accepted as a valid method of scientific proof.

Later Findings

A key turning point was the publication on 7 March 1962 of the Royal College of Physicians' report *Smoking and Health*. Sir George Godber, the deputy chief medical officer in the Ministry of Health, had encouraged the publication of this report in order to increase pressure for governmental action. The report concluded that "Cigarette smoking is a cause of lung cancer, and bronchitis and probably contributes to the development of coronary heart disease and various other less common diseases" (Royal College of Physicians 1962). The report also recommended a series of actions the government could take to address the problem, including education of the public concerning the hazards of smoking, more effective restrictions on the sale of tobacco to children, restrictions on tobacco advertising, wider restriction of smoking in public places, and an increase in taxes on cigarettes.

The report's findings had a substantial impact in the United Kingdom, largely because of the media attention it received. Rather than simply summarizing their findings in medical journals or at scientific conferences, Royal College researchers held press conferences and communicated their findings to the general public. Government authorities did not immediately adopt the recommended actions of the report, but they did initiate a health education campaign to inform the public about the health impact of smoking. A ban on cigarette advertising on television followed in 1967, and health warnings on cigarette packets appeared in 1971 after another Royal College report.

In addition to advising people to quit smoking, medical professionals in the 1960s and 1970s sought strategies to reduce harm from smoking in people who would not or could not quit smoking entirely. For example, they proposed that cigarette smokers switch to pipes or cigars and advocated research efforts to identify and remove harmful cancer-causing substances in tobacco and smoke. The main vehicle for these efforts was the Independent Scientific Committee on Smoking and Health (ISCSH), a Department of Health committee made up of independent public health scientists. Through a series of voluntary agreements with the tobacco industry, the committee supported research into "less dangerous smoking," evaluated cigarette additives and synthetic tobacco substitutes, and established industrywide targets for reducing the **tar** and nicotine content of cigarettes.

tar a residue of tobacco smoke, composed of many chemical substances that are collectively known by this term.

However, in the late 1970s the medical and public health communities began to take a stronger activist stand against tobacco, this time stressing the need for abstention rather than reduction of harm. A key organization was ASH (Action on Smoking and Health) founded in 1971 by doctors active in the Royal College. ASH was a new type of health pressure group, an alliance between doctors and health publicists that used the media as a primary arena for its activities.

A Turning Point

In the late twentieth century, public health efforts were limited by the fact that many people still viewed smoking as an act of free will, an activity that harmed only the smoker and not others. However, that argument was undermined from two directions in the 1980s and 1990s, and a turning point came in tobacco control in the United Kingdom. First, the publication of Japanese epidemiologist Takeshi Hirayama's paper in the *British Medical Journal* in 1981 about the health impacts of smoking on the nonsmoking wives of smokers brought the debate on passive smoking to the fore. Smoking was no longer an activity that just harmed the smoker, but one that involved innocent victims, a powerful argument for policy change. Thus, what had been a moral issue was redefined as a scientific one. Second, as scientists began to understand the addictive properties of nicotine, these findings clashed with the tobacco industry's traditional defense that smoking was a matter of personal choice. Tobacco control activists and policy makers drew parallels between the use of nicotine and that of illicit drugs. But rather than attempting to eliminate nicotine, public health advocates increasingly focused on ensuring access to nicotine replacement therapy for smokers, particularly for smokers in the lower social classes, who could not afford to pay for treatment.

■ VIRGINIA BERRIDGE

BIBLIOGRAPHY

Berridge, Virginia. "Post War Smoking Policy in the UK and the Redefinition of Public Health." *Twentieth Century British History* 14, no. 1 (2003): 61–82.

Doll, Richard, and Austin Bradford Hill. "Smoking and Carcinoma of the Lung. Preliminary Report." *British Medical Journal* 2 (1950): 739–748.

Hirayama, Takeshi. "Nonsmoking Wives of Heavy Smokers Have a Higher Risk of Lung Cancer : A Study from Japan." *British Medical Journal* 282 (1981): 183–185.

Royal College of Physicians. *Smoking and Health*. London: Pitman Medical Publishing Company, 1962.

Tobacco Industry Science

For decades, the tobacco industry has used several strategies to attempt to generate controversy about the health risks of its products. These strategies are: 1) fund and publish research that supports the tobacco industry's position; 2) suppress and criticize research that does not support the industry's position; and 3) disseminate tobacco industry–generated data in the lay press and directly to policy makers. The strategies used by the tobacco industry have remained remarkably constant since the early 1950s when the industry focused on refuting data on the harmful effects of active smoking. Beginning in the 1970s and through the 1990s, the industry became more concerned with refuting data on the harmful effects of secondhand smoke. The tobacco industry funded and manipulated research on topics ranging from nicotine and addiction, adverse health effects of active smoking and secondhand smoke, exposure assessments, and product design studies.

Previously secret internal tobacco industry documents that have been made public through litigation have given the public health community unprecedented insight into the tobacco industry's internal and external scientific research programs. The documents reveal that industry research programs have been pursued primarily to protect the industry's economic interests rather than to generate independent scientific knowledge. In addition, the tobacco industry has also funded research that is not directly related to its products in order to distract from tobacco as a health problem, to generate good publicity, and to enhance its image. For example, one of the criteria that Philip Morris' Worldwide Scientific Affairs Program considered when deciding whether to fund a research application was whether the research would enhance the credibility of the company.

The tobacco industry funded research through a number of mechanisms: 1) its trade association, The Tobacco Institute; 2) tobacco companies; 3) legal firms with tobacco companies as clients; and 4) research organizations funded by tobacco companies. These research organizations, the Council for Tobacco Research (CTR) founded in 1954 and the Center for Indoor Air Research (CIAR) founded in 1988, often funded projects that were reviewed by tobacco industry lawyers and executives, rather than scientists. Lawyers were not only involved in selecting

TOBACCO INDUSTRY SCIENCE

A chemist examines tobacco residue in test tubes during a 1956 experiment to discover the effects of smoking on the lungs. © HULTON-DEUTSCH COLLECTION/CORBIS

projects for funding, but also in designing the research and disseminating the results of the selected projects. Although CTR and CIAR were disbanded as part of the 1998 master settlement agreement, in 2000, Philip Morris reinitiated an external research grants program that is almost identical in terms of grant review procedures and scope to CIAR.

The tobacco industry documents have revealed the extent to which industry lawyers were involved in decisions in every research arena. Correspondence among industry lawyers shows how scientific research, marketing, public relations, and almost all industry activities were directed by tobacco industry lawyers' efforts to protect the industry from litigation. Lawyers monitored in-house and externally funded research. Tobacco industry lawyers edited scientific papers produced by company scientists and externally funded scientists prior to their publication in the peer-reviewed literature. This editing included deleting references, changing wording, and deleting acknowledgements of tobacco industry sponsorship. Lawyers at Brown & Williamson developed methods for handling documents in order to protect them from discovery during lawsuits. Documents were circulated on restricted lists and some documents were labeled, sometimes inappropriately, as "work product" or "privileged" in order to prevent them from being used in court. Documents that dealt with the toxicity of tobacco smoke or the pharmacological properties of nicotine often fell into these protected categories. The documents that were considered by lawyers to have the most potential for damage to the industry were shipped from U.S. tobacco companies to non-U.S. companies to avoid discovery during litigation.

A defining characteristic of the tobacco industry's research program was the secrecy that surrounded the involvement of industry lawyers, executives, and scientists in the design, conduct, and dissemination of research, particularly externally funded research. Lawyers not only edited research reports, but also sometimes prevented their release. For example, by the early 1960s, the tobacco industry learned that nicotine was addictive. Studies on nicotine addiction appeared years later in the general scientific literature and the surgeon general did not conclude that nicotine was addictive until 1979. The industry kept their knowledge of nicotine from the public. The documents show that the tobacco industry has been engaged in deceiving policy makers and the public for decades about the harmful health effects of their products.

The tobacco industry's development of the Japanese Spousal Smoking Study provides a case example of the industry's involvement in the design, conduct, and dissemination of research and its efforts to hide this involvement. The tobacco industry created the Japanese Spousal Smoking Study in order to refute an influential independent study by researcher Takeshi Hirayama showing that there was an association of secondhand smoke exposure and lung cancer. Although the Japanese Spousal Smoking Study had named academic investigators, project management was conducted by Covington and Burling (a tobacco industry law firm), the research was supervised by a tobacco industry scientist, and a tobacco industry consultant assisted in reviewing the study design and interpretation of data. The tobacco companies that funded the study did not want any of these individuals named as co-authors on any of the resulting scientific publications. Although the tobacco companies considered using CIAR as "a cover" to fund the study, three companies agreed to fund the study directly. Progress reports for the study were prepared on Covington and Burling stationery. When the study was prepared for publication, the tobacco industry consultant was the sole author. The publication acknowledged "financial support from several companies of the tobacco industry." This acknowledgement tells the reader little about who was actually involved in the design, conduct and publication of the study. The hidden roles of the tobacco company lawyers and scientist raise questions about who is accountable for the research.

The tobacco industry has also funded scientific publications as a way to disseminate their research. These publications, often in the form of symposium proceedings, were generally not peer-reviewed by scientists, presented an unbalanced view of the health risks of tobacco, and were of poor research quality. The tobacco industry also funded review articles on the adverse effects of secondhand smoke and these reviews are a favored source of information for policy makers. However, tobacco industry funded review articles on the health effects of secondhand smoke are about 90 times more likely to conclude that passive smoking was not harmful compared to reviews funded by any other source, even when controlling for a variety of other factors that might be associated with the outcome of the research.

Tobacco industry lawyers also played a major role in organizing scientific symposia. They arranged for funding, screened speakers, and sometimes arranged for sponsorship through a third party so that the conference would be perceived as "independent" of the tobacco industry. To suggest that the research it funds meets scientific standards and that there is substantial support for its position, the tobacco industry often cites its industry-funded, non–peer reviewed publications in the lay press, as well as scientific and regulatory settings.

Symposia were often attended and presented by scientific consultants who were hired by the tobacco industry. A major goal of the tobacco industry's scientific consultancy program was to refute data about the harmful effects of tobacco. The industry consultants were paid to criticize independent research on tobacco and secondhand smoke in a variety of forums, including participation in scientific conferences; publication in the form of conference proceedings, journal articles, and books; media appearances; testimony at tobacco litigation trial; and the preparation of statements for government committees. The industry consultant programs were international and used to discredit research conducted by non-industry scientists around the world.

Tobacco companies have coordinated their efforts at an international level (U.S. tobacco companies did not just work with U.S. tobacco companies but with companies all over the world to do research and promote tobacco). In addition, tobacco companies also coordinated their activities with other corporate interests. Financial ties between tobacco companies and the chemical, pharmaceutical and food industries have given the tobacco companies leverage to influence policies that could affect these other industries. For example, the tobacco industry used its financial tie with pharmaceutical companies to pressure them to weaken their marketing of nicotine replacement therapies.

See Also Chemistry of Tobacco and Tobacco Smoke; Medical Evidence (Cause and Effect); Public Relations.

▌LISA BERO

BIBLIOGRAPHY

Barnes, Deborah, and Lisa Bero. "Industry-Funded Research and Conflict of Interest: An Analysis of Research Sponsored by the Tobacco Industry through the Center for Indoor Air Research." *Journal of Health Politics, Policy, and Law* 21 (1996): 515–542.

Barnoya, Joaquin, and Stanton Glantz. "Scientific Quality of Original Research Articles on Environmental Tobacco Smoke." *Tobacco Control* 6, no. 1 (1997): 19–26.

———. "Why Review Articles on the Health Effects of Passive Smoking Reach Different Conclusions." *Journal of the American Medical Association* 279 (1998): 1566–1570.

———. "Tobacco Industry Success in Preventing Regulation of Secondhand Smoke in Latin America: The 'Latin Project.'" *Tobacco Control* 11, no. 4 (2002): 305–314.

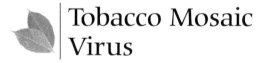

Tobacco Mosaic Virus

Tobacco Mosaic Virus (TMV) derives its name from the mosaic pattern it causes on the leaves of infected tobacco plants. The importance of TMV, however, goes far beyond agriculture. It is one of the best-studied viruses. Over a century of TMV research has repeatedly extended the boundaries of fundamental biological knowledge.

TOBACCO MOSAIC VIRUS

In 1886 the German bacteriologist Adolf Mayer named a mottled leaf pathology tobacco mosaic disease after showing it was infectious. Six years later, the Russian scientist Dmitri Iosifovich discovered that surprisingly, unlike other known infectious agents, the infectivity of the sap from diseased plants could pass through a bacteria-retarding filter. The discipline of **virology** effectively began in 1898 when the Dutch microbiologist Martinus Beijerinck argued that the agent of tobacco mosaic disease was caused not by a very small bacterium but rather by something fundamentally different, something he called a virus. TMV was the first demonstration of a nonbacterial infective agent.

virology the study of viruses and viral diseases.

To great fanfare, the American scientist Wendell Stanley announced that he had crystallized TMV in 1935. It was the first virus and the first living organism crystallized. His needle-shaped para-crystals of TMV particles blurred the distinction between biology and chemistry and raised questions about the nature of life. Prior to 1935, scientists thought that only nonliving things could be crystallized. But if the crystals of TMV were dissolved, the released TMV particles remained infectious, or still alive, so to speak. Perhaps there was no sharp distinction between living and nonliving things.

With more precision than Stanley, the British scientists Fred Bawden and Bill Pirie demonstrated in 1936 that TMV contained 5 percent **ribonucleic acid (RNA)** as well as protein. Further knowledge of the nature of TMV came from applications of the physical techniques of electron microscopy and x-ray diffraction. The invention of electron microscopy allowed scientists to visualize the rod shape of individual TMV particles. In the late 1930s Isadore Fankuchen and John Bernal analyzed the structure of TMV using x-ray diffraction and proposed that the rod-shaped TMV particle itself consisted of an ordered arrangement of small subunits.

ribonucleic acid (RNA) an essential component of all living matter, present in the cytoplasm of all cells.

Partially interrupted by World War II, research on the substructure of TMV began to flourish in the 1950s. James Watson, the codiscoverer of the structure of deoxyribonucleic acid (DNA), speculated that the rod-shaped TMV particle was in fact a long helix of protein subunits and RNA, a prediction verified and refined by Donald Caspar and Rosalind Franklin in the 1950s using x-ray diffraction. From their work, a picture emerges of a helical virus 300 nanometers (nm) long, with a hollow core, consisting of approximately 2,100 identical protein subunits assembled much like stairs in a spiral staircase and a single strand of RNA interlaced through the protein subunits. At the time, this description of TMV was the most detailed of any virus. There is no significant connection between the structure of TMV and that of DNA, other than both being helical.

A further milestone in molecular biology involving TMV occurred in 1955. In the United States, Heinz Fraenkel-Conrat and Robley Williams and independently in Germany, Alfred Gierer and Gerhard Schramm successfully dismantled TMV into its constituent protein and RNA components and then reconstituted them in vitro to produce infective particles. They in effect reassembled the virus from its parts. When different strains of TMV were taken apart, their parts exchanged and reconstituted, progeny virus resembled the strain that contributed the RNA, not the strain that contributed the protein. These reconstitution experiments were important in demonstrating that nucleic acid and not protein was the genetic material of life.

In the second half of the twentieth century, TMV served as a model for the self-assembly of large biological molecules and the interplay between RNA and protein. Although some controversy remains over the initiation of the process, the assembly pathway of TMV is still one of the best understood of all viruses. Research continues on TMV, but its heyday as a cutting-edge research model appears to have passed.

See Also Cigarettes; "Light" and Filtered Cigarettes; Nicotine; Product Design; Toxins.

■ GREGORY J. MORGAN

BIBLIOGRAPHY

Creager, Angela N. H. *The Life of a Virus: Tobacco Mosaic Virus as an Experimental Model, 1930–1965*. Chicago: University of Chicago Press, 2002.

Creager, Angela N. H., et al. "Tobacco Mosaic Virus: Pioneering Research for a Century." *Plant Cell* 11, no. 3 (March 1999): 301–308.

Scholthof, Karen-Beth G., John G. Shaw, and Milton Zaitlin, eds. *Tobacco Mosaic Virus: One Hundred Years of Contributions to Virology*. St Paul, Minn.: American Phytopathological Society Press, 1999.

Toxins

Natural tobacco contains more than 3,000 separate chemical compounds. In commercial products, the composition of tobacco is altered by the use of pesticides, curing, processing, and the addition of flavorants. In smoked tobacco products, further changes are produced by partial combustion of the tobacco and the transfer of chemicals to smoke. Toxic compounds present in manufactured tobacco and smoke are the basis for the many negative health effects associated with tobacco use. A number of specific compounds contributing to increased cancer and other disease risks have been identified. Government attempts to regulate tobacco product toxicity have focused on the need for reduction of these harm-producing compounds. This has led manufacturers to adopt changes in product design and delivery, including the development of **low-yield cigarettes** and, more recently, reduced nitrosamine tobaccos and other potential reduced-exposure products. Because of the complexity of tobacco products and product compounds, the total impact of these changes on exposure may never be fully understood.

low-yield cigarettes a cigarette designed to deliver less nicotine to the smoker, and, theoretically, less harmful. Actual results seem to dispute this contention.

Historical Assessment of Tobacco Toxicity

Although the perceived negative effects of tobacco have been a subject of controversy at least since its introduction into Europe in the early part of the sixteenth century, modern investigation of smoking and health began around 1900. At that time, observed statistical increases

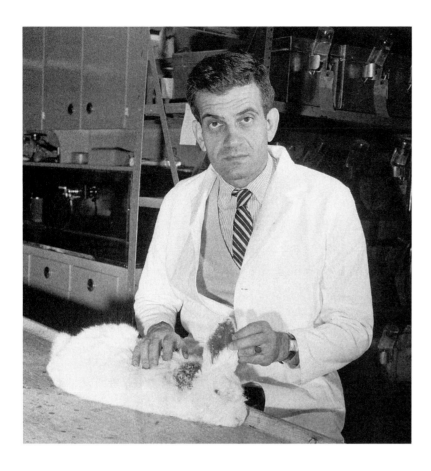

Scientist Ernst Ludwig Wynder (1922–1999) in 1957 shows the cancerous lesions on a laboratory rabbit. The rabbit's skin had been coated with tobacco tars (particulate matter). © BETTMANN/CORBIS

in lung cancer rates in the United States and England led to studies seeking to identify the relationship between tobacco use and lung cancer and other disease.

The earliest common method for estimating the cancer-causing potency of tobacco smoke and specific tobacco toxins involved painting collected tobacco particulate matter, or **tar,** onto the shaven backs of mice in order to observe the possible development of tumors and lesions. Skin painting studies in the 1950s and 1960s confirmed a clear relationship between the amount of tar applied and the percentage of observed cancers. Inhalation-based animal studies were later used to assess the development of cancers in the respiratory system, specifically in the larynx, nasal cavity, and lung. A clear relationship between dose and response has been established with other noncancerous diseases, including heart and respiratory disease.

tar a residue of tobacco smoke, composed of many chemical substances that are collectively known by this term.

Measures of human exposure to tobacco smoke and isolated smoke compounds have generally been obtained by use of a smoking machine that mimics human smoking of a cigarette using a standard set of puffing parameters, including puff volume, duration, and time between puffs. The American Tobacco Company began using a standard machine smoking measurement in 1936, which was subsequently adopted in 1969 by the Federal Trade Commission for routine measurement of tar and nicotine in tobacco smoke with only slight modifications. But these early efforts only measured overall tar and nicotine content and could not measure individual compounds or determine which were most toxic.

TOXINS

Chemistry of Tobacco Toxins

Tobacco smoke is extraordinarily complex—scientists estimate that at least 4,800 compounds are present in burned tobacco, even before the use of additives or other commercially introduced changes (National Cancer Institute 2001). The smoke is generally divided into two phases: the particulate, consisting of larger smoke particles, and more highly volatile gas or vapor compounds. Measures of smoke tar do not describe a particular smoke compound but are rather a sum of the thousands of chemicals in the total smoke particulate (usually excluding water and nicotine).

Particulate smoke demonstrates the greatest carcinogenic (cancer-causing) activity in cigarette smoke, including 55 separate compounds that have been identified as possible human carcinogens. Of this group, polynuclear aromatic hydrocarbons (PAH) and tobacco-specific nitrosamines (TSNA) have received particular attention. PAH are produced by the incomplete combustion of organic material, and can be found in a number of sources including vehicle exhaust, wood smoke, and grilled meats. TSNA are formed from nicotine and other tobacco **alkaloids** and are present in tobacco, tobacco smoke, and secondhand smoke. These compounds are potent cancer-causing agents.

Smoke gases include carbon monoxide, believed to play a major role in the development of heart disease, and **ciliatoxic** agents such as hydrogen cyanide and acrolein, which are likely contributors to respiratory disease. Overall, smoking tobacco results in exposure to hundreds of known carcinogens and other toxic agents with different potencies and effects. Other compounds not yet identified are also likely to be present.

At least 28 tumor-producing agents have been identified in smokeless tobacco products such as chewing tobacco and **snuff.** The most abundant carcinogens are volatile aldehydes, which although weak carcinogens, likely contribute to carcinogenic potential. Nitrosamines, including TSNA, are believed to be a major contributor to smokeless tobacco risk.

Tremendous effort has been expended to characterize tobacco and to identify disease-producing compounds in tobacco and smoke. The identification of carcinogens, ciliatoxins, and other harmful compounds represent an important advance for understanding the risks of cigarette smoking. However, the data to assess whether some compounds play a greater role in determining overall health effects of tobacco use, and the significance of their interaction with each other, is still incomplete.

Product Regulation and Changes

In response to growing health concerns, tobacco manufacturers introduced major cigarette modifications between 1950 and 1975. Product changes included the introduction of filters and ventilation to reduce the delivery of smoke toxins, increased use of processed tobaccos (which lowered smoke delivery as well as product costs), and the use of chemical additives, largely to offset ensuing losses in taste and sensory perception. Together, these changes have resulted in significantly reduced tar and carbon monoxide levels as measured by smoking machines; however, there is no conclusive evidence that they have made tobacco products less toxic.

alkaloid an organic compound made out of carbon, hydrogen, nitrogen, and sometimes oxygen. Alkaloids have potent effects on the human body. The primary alkaloid in tobacco is nicotine.

ciliatoxic being harmful to the cilia of the lungs. Cilia are tiny, hair-like, structures on the inner surfaces of the lung.

snuff a form of powdered tobacco, usually flavored, either sniffed into the nose or "dipped," packed between cheek and gum. Snuff was popular in the eighteenth century but had faded to obscurity by the twentieth century.

First, common measures of total smoke delivery (such as tar) do not reflect the complexity of possible changes in the composition of smoke. It has been demonstrated that reductions over time in the smoke levels of some toxins, such as benzo(a)pyrene, an important PAH, have at the same time produced increases in others, such as NNK, a potent TSNA. Reliance on smoke machine measures fails to account for possible shifts in one or a number of classes of harmful compounds contained in the smoke.

A second problem results from changes in human smoking patterns that have developed in response to product changes. Smokers have adapted behaviors to inhale more smoke from a cigarette and to inhale more deeply in order to maintain levels of smoke delivery. The amount of toxins a smoker is exposed to depends not only on how many cigarettes are smoked but also on how they are smoked. Because smoke machines cannot adapt their behavior to cigarette changes, they are likely to misrepresent actual smoke delivery and, therefore, the smoker's exposure. These behavioral changes may particularly affect the delivery of compounds, such as carbon monoxide, for which absorption is dependent on the depth of inhalation.

A significant reduction in nitrosamines and TSNA formation appears to be feasible based on manufacturing processes currently used for smokeless tobacco (Snus) produced in Sweden, and now being considered experimentally in the United States and elsewhere. Other potentially harmful agents, particularly harm-producing additives, have been considered for regulation internationally, including in the European Union, Canada, and Australia. However, because of the complexity of tobacco and tobacco toxins, the actual impact of these changes on health effects requires more careful study.

See Also Cigarettes; "Light" and Filtered Cigarettes; Menthol Cigarettes; Nicotine; Product Design.

▌GEOFFREY FERRIS WAYNE

BIBLIOGRAPHY

Brunnemann, Klaus D., and Dietrich Hoffmann. "Chemical Composition of Smokeless Tobacco Products." In *Smokeless Tobacco or Health: An International Perspective.* Smoking and Tobacco Control Monograph 2. Bethesda, Md.: U.S. Department of Health and Human Services, National Institutes of Health, 1992. Pp. 96–108.

Hoffmann, D., and I. Hoffmann "The Changing Cigarette, 1950–1995." *Journal of Toxicology and Environmental Health* 50, no. 4 (March 1997): 307–364.

National Cancer Institute. *Risks Associated with Smoking Cigarettes with Low Machine-Measured Yields of Tar and Nicotine.* Smoking and Tobacco Control Monograph 13. Bethesda, Md.: U.S. Department of Health and Human Services, [Public Health Service,] National Institutes of Health, National Cancer Institute, 2001. Full report Available: <http://cancercontrol.cancer.gov/tcrb/monographs/13/index.html>.

Stratton, Kathleen, et al., eds. *Clearing the Smoke: Assessing the Science Base for Tobacco Harm Reduction.* Washington, D.C.: Institute of Medicine, National Academy Press, 2001. Full report available: <http://www.nap.edu>.

U.S. Department of Health and Human Services. *The Health Consequences of Smoking: The Changing Cigarette.* A Report of the Surgeon General. Rockville, Md.: U.S. Department of Health and Human Services, Office on Smoking and Health, 1981. Full report available: <http://www.cdc.gov/tobacco/sgr/sgr_1981/index.htm>.

Trade

In the fifteenth century, the tobacco plant (*Nicotiana rustica* and *Nicotiana tabacum*) was dispersed throughout the Americas by the native people. This diffusion process extended beyond the continent to embrace the islands of the Caribbean where the first recorded European encounter with tobacco took place. Christopher Columbus's journal records that, on 15 October 1492, the explorer intercepted a single Native American, paddling a canoe, whose small cargo of goods included some dried tobacco leaves. Columbus observed that tobacco "must be a thing highly valued by them, for they bartered with it at the island of San Salvador" (Goodman 1993).

Knowledge of tobacco was transmitted to Europe in a multistage process. Printed reports, based on the accounts of Columbus and other early explorers, appeared comparatively quickly. The chronicler Peter Martyr published an account of tobacco usage among Taino Indians on Hispaniola in 1511 based on an earlier testimony written by Friar Ramon Pané—a Catalan priest who had accompanied Columbus on his second voyage to the New World in 1493. In view of the novelty of printing technology itself, it is probable that manuscript and verbal accounts similar to Pané's circulated even earlier than printed texts. European consumption of tobacco in the New World was increasingly well documented from the 1530s in Spanish, Portuguese, and subsequently French colonial settlements. Some scholars have also suggested that Brazilian colonists had begun growing small quantities of tobacco by 1534. Transcontinental experiments in planting followed soon afterward as seedlings were introduced into Portugal around 1548. By the 1570s, successful attempts at cultivation had been made in Northern Europe, Africa, and Asia. For the first time in history, tobacco had become a globally produced commodity.

The first distinct set of European consumers of nicotine consisted of the direct participants in colonial ventures, particularly merchants and mariners of Spain and Portugal. Spanish sailors, for example, are reported to have consumed tobacco as a regular part of their diet during the 1550s. During the last quarter of the sixteenth century, the health properties of tobacco began to be discussed in Europe's medicinal literature as consumption spread more widely. The first medical treatise devoted exclusively to tobacco was published by Anthony Chute in 1595. At approximately the same time, clay replaced walnut and silver as the staple material used to manufacture tobacco pipes in England, while in 1597 the first tax on tobacco was imposed by Elizabeth I. Viewed collectively, this evidence suggests strongly that tobacco consumption had become an established part of European consumers habits by the later sixteenth century.

Surviving sources documenting tobacco's diffusion suggest a rapid transmission of reports of the commodity, a delay of approximately half a century between the first encounter and experiments in cultivating and consuming tobacco in Europe, and a further gap (of approximately fifty years) until trade in tobacco expanded to the point where it became worthwhile for governments to tax it.

It is instructive to compare tobacco's chronology of dissemination with that of coffee, using England as a test case. Tobacco and coffee possessed similarities: both were entirely new commodities with stimulating properties. Moreover, in each case successful introduction was associated with important product innovations (clay pipes, ceramic cups with handles, roasting machinery, expansion of taverns, the establishment of coffee houses). Circulation of printed reports of coffee occurred within a decade of the travellers Leonard Rauwolfius and Constantino Garzoni's experience of the drink in Aleppo and Constantinople in 1573; the first experiments of consumption in England, however, occurred after a lag of six decades; the first tax was levied approximately eighty-seven years after the first documented European encounter with coffee.

Tobacco and coffee's trajectories of diffusion are similar but there were also important differences between the two commodities. Commercial coffee cultivation existed prior to the establishment of European trade and demand in Europe initially absorbed only a small proportion of output. Despite problems of periodic disruption and distance, in the case of coffee a land alternative existed to maritime supply. Indeed, the caravan route was initially used to bring the drink westward. In contrast, Amerindian agriculture lacked a tobacco surplus sufficient to meet European demand as it expanded. For a trade to develop, therefore, direct investment in production was required.

Tobacco Colonies and New World Settlement

In areas of Europe where agriculture was specialised and highly commercialized, land could be diverted to tobacco and the plant cultivated as a cash crop. Holland proved the most prolific European region of production, though successful attempts were also made to grow the plant in England, in the French territories of Guyenne and Alsace, and in the German **Palatinate.** The share held by European cultivators in total commercial output was, however, small and Dutch tobacco was primarily used to supplement imports into Europe from the Americas. In Amsterdam and Rotterdam, dealers mixed low-quality Dutch and German leaf with imported supplies to produce a blend that was consumed primarily in the form of **snuff.**

Despite Dutch success in tobacco growing, the bulk of tobacco consumed by Europe's smokers and snuff takers during the seventeenth and eighteenth centuries was supplied by New World producers. In Figure 1, which excludes Caribbean production), the dominant position of the Americas is apparent. Tobacco growing in the Americas was based initially on the assimilation of Amerindian agricultural practices. European innovation, however, was responsible for transforming cultivation into a major export industry and tobacco formed a staple element of Britain's colonization of North America. Commercialization of tobacco growing was achieved, however, only at the expense of environmental damage and high mortality among both white servants and black slaves shipped across the Atlantic (Menard and Walsh 1974).

palatinate the territory of a feudal lord having sovereign power within his domain.

snuff a form of powdered tobacco, usually flavored, either sniffed into the nose or "dipped," packed between cheek and gum. Snuff was popular in the eighteenth century but had faded to obscurity by the twentieth century.

TRADE

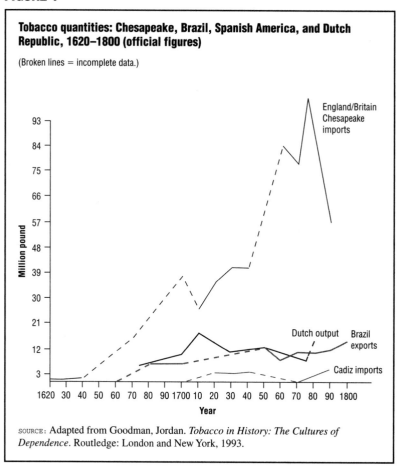

FIGURE 1

Tobacco quantities: Chesapeake, Brazil, Spanish America, and Dutch Republic, 1620–1800 (official figures)

SOURCE: Adapted from Goodman, Jordan. *Tobacco in History: The Cultures of Dependence*. Routledge: London and New York, 1993.

Initially, the willingness of Native Americans to trade small quantities of tobacco enabled Europeans to experiment with smoking and snuff taking. Unlike the commodities coffee and tea, however, tobacco did not form part of a large-scale commercial network within the New World. Spanish and Portuguese colonists were the first to expand tobacco's supply by cultivating and trading tobacco leaf. Tobacco first became available in England through illicit trade with Spanish settlers, particularly in Trinidad and Guiana, supplemented by the capture of prize cargoes by privateers. By the end of the sixteenth century, direct supply of tobacco grown in both North America and England itself began to replace contraband leaf. Political factors played an important part in this shift as deteriorating relations between England and Spain provoked a crack down on smuggling by the Spanish imperial authorities while also providing a stimulus to English colonizing ventures.

Tobacco leaf was not cultivated by British settlers in either the location or form in which it was first encountered by colonial settlers. Sweet-scented varieties of the plant were preferred to the astringent-tasting tobacco grown by the majority of Amerindians. Production was also characterized rather by shifting comparative advantage as Europeans transplanted tobacco within the Americas. First Brazil and then the Caribbean became centers of production, before the focus of trade switched to the Chesapeake colonies of Virginia and Maryland during the second half of the seventeenth century. Like all staple crops entering into colonial trade, tobacco was grown in regions with favorable ratios of land mass to coastline in order to exploit the low cost of transportation by water relative to land. For this reason, tobacco was grown on Bermuda and the smaller Caribbean islands of Barbados, Saint Kitts, Martinique, and Guadeloupe during an early stage of each colony's settlement.

The introduction of sugar into the Caribbean during the 1640s and 1650s was followed by a reorganization of colonial trade. Tropical islands increasingly specialized in sugar and rum, while tobacco migrated to Virginia and Maryland colonies, characterized by a vast network of rivers and creeks draining into the Chesapeake Bay.

European colonizers confronted an immense task in establishing viable settlements in the Americas capable of producing a cash crop, shipping tobacco across the Atlantic, and then marketing the commodity successfully in a society with no prior experience of smoking or snuff taking. The potential of New World commerce was recognized by syndicates of London merchants who attempted to develop trade with the Chesapeake and the West Indies. Elite city businesspeople financed the Virginia and Roanoke settlements during the 1580s, established the Virginia Company (in 1606) and allied themselves with the Earl of Carlisle, who had obtained a proprietary patent to settle Barbados in 1627. These early ventures proved disastrous for investors and many of the participants. Captain John Smith's *General History of Virginia* (published in 1624) describes the "starving time" experienced by the Jamestown colonists in 1609, which reduced the Virginia population from 500 to barely sixty survivors. Yet despite generating few profits, the experiments yielded valuable information about the economic realities of New World settlement, including knowledge about tobacco cultivation and use.

After the enormous practical difficulties of establishing settlements was grasped, most early investors withdrew their support. As a consequence, the large quantities of capital and labor required to meet the substantial development costs of colonization was not forthcoming from established commercial sources. The confidence and commitment required to sustain the long-term investment process that colonization required was instead supplied by entrepreneurs hitherto located on the fringes of England's commercial community. Provincial gentry, west coast merchants (initially in Bristol and later Liverpool), and representatives of middling groups (shopkeepers and tradesmen of England's county towns) were instrumental in financing trade and settlement in what became the tobacco colonies. The shift in the profile of investors was accompanied by a change in commercial policy. Company monopolies and proprietary rights in colonies were replaced by a more open-access policy, whereby settlers occupied land as freeholders and merchants were permitted to trade with the colonies free of guild restrictions.

The physical barrier of the Atlantic and the disease environment in the New World placed formidable obstacles in the way of successful tobacco planting. The former problem was overcome by adapting the ancient system of European apprenticeship, enabling prospective colonists to finance their crossing by becoming indentured servants for terms of up to seven years. In effect, migrants took out a loan secured by their own person, which they subsequently paid off by working for a colonial master. Limited medical knowledge prevented any easy solutions to the problem of identifying hostile diseases, such as malaria and water pollution, contributing to the rise in the death rate. New arrivals to the West Indies and the Chesapeake, as a consequence, faced a much greater risk of death during their first years in the New World than their counterparts in Europe. Children born to settlers likewise lacked immunity and paid a grim mortality penalty. Delayed marriage until after the completion of terms of indentured servitude, coupled with a shortage of females (a majority of servants were males), compounded the reproduction problem.

Tobacco was an ideal start-up crop for a colony because it required limited amounts of capital (an iron hoe and a wooden dry house comprising the most important inputs) and also lacked economies of scale. A smallholder, working less than twenty acres alongside a couple of servants, could be as productive as a large plantation owner. In order to sustain output, however, constant flow of labor was needed to overcome the high death rate and low fertility in the colony. During the second half of the seventeenth century, the supply of servants was diminished because of competition from other areas of settlement (particularly the Middle Colonies, most notably Pennsylvania), slower population growth in England, and higher wages for workers in Europe. Therefore, enslaved Africans were introduced into the Chesapeake in large numbers for the first time during the 1680s and 1690s. This dramatic shift in the composition of the labor force was assisted by a fall in the price of slaves associated with the ending of the Royal African Company's monopoly of supply.

A trade depression, aggravated by disruption to commerce during the War of the Spanish Succession (1702–1713) brought great hardship to the tobacco colonies. The temporary fall in numbers of servants and slaves shipped to the Chesapeake, however, assisted the region in the long term because it helped equalize the sex ratios and speeded the emergence of a Creole (native-born) majority among both the white and black population. **Creoles** possessed greater natural immunity to disease and were usually in a position to marry earlier than new servant arrivals. From the second decade of the eighteenth century onward, natural increase (a surplus of births over deaths) supplied the greater part of the Chesapeake's labor requirements.

Creole originally, a person of European descent born in the Spanish colonies. Later, the term was applied to persons of mixed European and African descent. As an adjective, it can describe admixtures of European and African cultural components such as language, cookery, and religion.

One of the best-known charts in colonial economic history is the "scissor chart," as shown in Figure 2, which contrasts the falling farm price of Maryland tobacco with the rising volume of leaf exports to Britain shipped from the Chesapeake (McCusker and Menard 1985). British imports of tobacco grew exponentially from less than 10,000 pounds (4,549 kilo) in 1620 to just under 40 million pounds (18 million kilo) by 1700; at the same time the farm price of tobacco in the Chesapeake fell from 22 old pence per pound (roughly 80 cents per kilo) to less than 4 cents by the 1680s. These spectacular developments reflected the striking gains in productivity achieved by planters.

Knowledge gained from practical experience revealed the best soil locations and seasons for planting tobacco. The substitution of cheaper enslaved workers for increasingly expensive servants led to further cost savings. While few economies of scale existed on the production side, distribution was another matter. The early practice employed by British merchants of chartering a ship and sending it to the Chesapeake speculatively to sell a cargo of manufactures and purchase tobacco was replaced by the establishment of permanent factors in the colony to

FIGURE 2

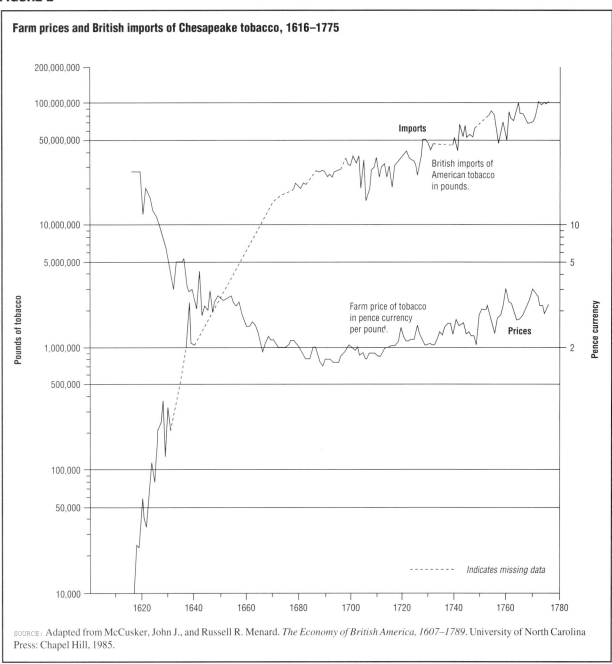

Farm prices and British imports of Chesapeake tobacco, 1616–1775

SOURCE: Adapted from McCusker, John J., and Russell R. Menard. *The Economy of British America, 1607–1789*. University of North Carolina Press: Chapel Hill, 1985.

coordinate business. In some areas, planters established direct correspondences with merchants in London or the outports, enabling credit relations to form. This system of trade was known as the consignment system. During the course of the eighteenth century, Scottish merchants were instrumental in creating networks of stores in the Chesapeake in a further attempt to reduce risk and uncertainty.

The improved productivity of shipping in the tobacco trade can be measured directly in a reduction in turnaround times by vessels sailing to the colonies. Whereas between 1694 and 1701 vessels calling at Maryland and Virginia spent an average of nearly 100 days in port, by

TRADE

hogshead a large wooden barrel formerly used to store and transport cured leaf tobacco. A hogshead typically held approximately 800 to 1000 pounds (350 to 450 kg) of tobacco.

the 1760s they spent less than 50 days tied up loading and unloading merchandise. More efficient use of cargo space was also achieved by designing improved containers for tobacco, called **hogsheads,** and packing them ever more densely with tobacco leaf. Because there was considerable local variation, it is difficult for historians to be exact about trends in the mean dimensions of the hogsheads used to pack tobacco, but they have suggested that capacity increased from approximately 13 cubic feet in 1674 to 22 cubic feet by 1747. The development of sophisticated insurance markets permitted losses to be spread among the trading community as a whole, reducing the risk of bankruptcy. Sailing ships were extremely costly capital assets with high variable costs due to the large crew required to operate them. Improved techniques that saved resources, therefore, had profound implications for the trade.

Overdependency on Tobacco in the Chesapeake

Tobacco dominated economic activity in the Chesapeake for much of the seventeenth and eighteenth centuries and probably provided the major portion of income earned by upward of 80 percent of families. While the crop held many attractions for a newly settled region, over the longer-term significant problems of dependency emerged. Though spectacular, the productivity gains made in the tobacco industry were of a once-and-for-all nature and eventually a limit was reached. The point where the price of tobacco ceased falling on trend and instead started to fluctuate in cycles marks the beginning of difficulties for producers. While economies continued to be made on the distribution side, cultivators found it difficult to realize further significant productivity gains; therefore, earnings were linked increasingly to the state of the market in Europe. The atomized structure of tobacco farming made planning difficult to accomplish within the Chesapeake. Therefore, high prices in one season led planters to overreact by increasing their output, leading to overproduction and subsequent falls in price during the next season.

The limited skill and capital requirements of tobacco also meant that the crop possessed few linkages with other sectors. Agricultural techniques (leaving tree stumps to rot in the ground and failure to make extensive use of manure) were also wasteful of land and favored extensive cultivation with large fallow patches. Hoe cultivation generated a smaller demand for purchases of iron manufactures than agricultural systems based on plough cultivation. The processing of tobacco leaf itself took place in simple wooden dry houses and, unlike cereal crops, did not require the construction of complex mill machinery.

diversification in agriculture, avoiding overdependence upon one crop by producing several different crops.

Tobacco's status as a cash crop within the colony (accounting in 1727 for 87% of export earnings) inhibited the development of local manufacturing as producers imported goods from Britain to meet the bulk of their needs. The absence of large towns in the Chesapeake is a striking symptom of tobacco's minimal potential as an engine of economic **diversification.** Approximately 30 percent of the population of British North America lived in the Chesapeake in 1770 yet the region lacked a major urban center. Baltimore, the largest city with approximately 6,000 inhabitants, was tiny in comparison to Philadelphia, New York, and Boston; moreover, Baltimore's growth occurred late in the colonial period and was driven by grain cultivation rather than tobacco.

Europe's Expanding Market for Tobacco

The statistics of the English and Scottish tobacco trade, which dominated European supplies, can be divided into two phases. For most of the seventeenth century, most of the tobacco imported from the Chesapeake was probably retained for domestic consumption. For example, during the years 1693 to 1697 (the earliest for which complete data have been found), 12.7 million pounds (5.7 million kilo) of tobacco was retained for the home market in Britain, whereas 14.2 million pounds (6.5 million kilo) was re-exported. Thereafter, re-exports of tobacco grew faster than retained imports.

Following Anglo-Scots Union in 1707, Scottish traders played an increasingly prominent role in the organisation of the tobacco trade. During the second quarter of the eighteenth century, the port of Glasgow established itself as the second port after London in the British tobacco trade; during the third quarter, Glasgow overtook the metropolis, accounting for 45 percent of imports during the years 1771 to 1773. Profits from tobacco trading helped finance Scottish industries and land improvement in Glasgow's hinterland. Scots merchants also established commission houses and mercantile partnerships in London that further strengthened their grip over Chesapeake tobacco distribution, particularly with respect to the Virginia trade and the contract to supply the French tobacco farmers general.

The growth of smoking in Britain itself is obscured by the haze of smuggling (particularly re-export frauds designed to evade the high level of **duty** levied on imports). The best attempts to adjust for the effects of smuggling suggest that up until the beginning of the eighteenth century, tobacco usage in Britain rose from virtually nothing in the early 1630s to approximately 1 pound per head during the 1660s; during the next thirty years, consumption grew further and had attained a level of 2 pounds per head by the early 1690s. Thereafter, only modest increases in the level of consumption per head were recorded in the trade data prior to the introduction of cigarettes during the early twentieth century. It is likely, however, that improvements in the efficiency in distribution raised the amount of tobacco reaching consumers. A shift from pipe smoking to snuff taking probably also contributed to a rise in tobacco usage that is masked by the available statistics.

duty a tax, usually on certain products by type or origin. A tariff.

Britain's consumption trends were similar to those of other European markets where importation of tobacco per head also grew rapidly before leveling off. There are a number of possible explanations of this phenomenon. Tobacco possesses addictive properties and it may simply be that once consumers satisfied their cravings they ceased to demand further quantities of leaf. (Yet across Europe a significant proportion of tobacco was taken as snuff and the addictiveness of tobacco ingested in this form is weaker than if it is smoked so other explanations of consumer behavior must be considered.)

European states were quick to introduce taxation once imports of tobacco reached a level where this was viable. The form of taxation varied. In Britain, tobacco was taxed by levying import duties. In other countries, beginning with Spain and including France and the Papal States, governments created a monopoly franchise and sold or "farmed" the right to distribute tobacco to a private company in order to raise revenue. As a result of these impositions, the price fall recorded in Figure 1 was not fully reflected in the retail price of tobacco, notwithstanding

TRADE

the encouragement that the tax system gave to revenue fraud and smuggling. Confusion between farm prices received by planters and retail prices paid by consumers has led to the misconception that demand for tobacco increased in Europe primarily because Chesapeake production lowered prices. Without innovation and productivity gains a large-scale trade in tobacco could not have developed, but only a limited reduction in retail prices occurred owing to the fact that government increased taxes as farm prices declined. High duties provided a stimulus to smugglers, but contraband was neither a riskless nor costless activity and the gains accruing to smuggling gangs made at the expense of the state did not necessarily provide consumers with a cheaper product. Taxation policies also contributed, therefore, to the failure of tobacco consumption to broach a ceiling level.

The social context of tobacco usage also appears to have set limits to the growth of demand. Studies by both cultural and economic historians have emphasized how smoking and snuff taking were social activities that were influenced by contemporary conceptions of masculinity and status. The form research has taken differs; cultural historians have examined contemporary references to tobacco in literature and pamphlets while economists have considered the role of income distribution on demand. Nevertheless, the conclusions reached by researchers are similar: tobacco usage was influenced by the institutions associated with smoking, by distinctions of status and gender, and by the income levels and consumer preferences of different groups within society.

Static market conditions at home meant that for most of the eighteenth century well in excess of 60 percent of tobacco grown in the Chesapeake and shipped to British ports was re-exported to Continental Europe, most notably to the markets of France, Flanders, and Holland which accounted for approximately two-thirds of the trade, as shown in the table below, which displays British tobacco trade from 1721–1770 (Price 1973). The re-export trade to France was dominated by the supply of tobacco to the French farmers general, who held the contract to supply France, whereas the Dutch ports of Rotterdam and Amsterdam mixed together blends of tobacco for distribution in the German territories and shipment to other markets.

Years	Imports	Retained Imports	Re-Exports	Re-Exports: France and Flanders	Re-Exports: Holland
1721–1730	39.23	15.32	23.91	6.62	8.42
1731–1740	45.97	13.55	32.42	12.21	9.18
1741–1750	61.54	21.10	40.44	12.71	13.34
1751–1760	67.75	19.93	47.82	15.91	15.65
1761–1770	76.20	9.28	66.92	22.19	24.18

Impressive Achievements but Gained at Heavy Cost

Within two centuries of the first documented encounter with tobacco, European merchants and planters had succeeded in transforming tobacco into a global commodity and forging a long-distance supply chain to meet the demands of European consumers. This task was not accomplished easily. It cost the lives of many Amerindians and white settlers, while also condemning generations of Afro-Americans to chattel slavery. Reviewing the growth of the tobacco trade before 1800, however, it is difficult not to be impressed by the invention and ingenuity with which private individuals on both sides of the Atlantic employed a range of interdependent technologies and innovated across a broad front to bring tobacco from colonial plantation to clay pipe and snuff box.

Tobacco Trade from the Revolutionary War into the Nineteenth Century

Tobacco trade of the United States was curtailed during the Revolutionary War. The U.S. government cultivated French support in the war through tobacco. Benjamin Franklin obtained a loan by using 5 million pounds of tobacco as collateral. As a result, the British concentrated their armies in southern colonies in the last years of the war and destroyed **plantations** and considerable warehouse holdings of tobacco. Virginia planters were indebted to British merchants for some 2 million pounds sterling, but planters got no relief since the Treaty of Paris and subsequent court cases provided for full recovery of debts. Consequently, members of many families left for a new life in the virgin soil in the West.

plantation historically, a large agricultural estate dedicated to producing a cash crop worked by laborers living on the property. Before 1865, plantations in the American South were usually worked by slaves.

Agriculture in the United States was stimulated by the war, but tobacco remained the leading U.S. export, worth about $4 million in 1790. About one-half of the southern population, including most of the black slaves, was engaged in or dependent on tobacco growing. After 1800, the relative importance of tobacco exports declined rapidly because of the expansion of cotton production, which was revolutionized by the invention of the cotton gin in 1793. With numerous inlets on tidewater suitable for ocean-going ships, Maryland and Virginia growers dominated the tobacco export trade until preference shifted around 1850 to the milder-flavored **flue-cured tobacco** of North Carolina.

flue-cured tobacco also called Bright Leaf, a variety of leaf tobacco dried (or cured) in air-tight barns using artificial heat. Heat is distributed through a network of pipes, or flues, near the barn floor.

From 1790 to 1814, U.S. exports of tobacco (reported officially as unmanufactured tobacco) ranged widely, from 118 million pounds in 1790 to only 3 million pounds (War of 1812). It was not until 1851 that exports permanently exceeded 100 million pounds. At the beginning of the nineteenth century, with British trade connections reconnected, U.S. tobacco exports represented close to 90 percent of U.S. production and may have accounted for around 90 percent of world tobacco export trade.

Several factors helped to undermine the position of United States tobacco in the world market during the first half of the nineteenth century: the disturbed trade conditions resulting from the Napoleonic wars; the post-1815 attempt of England to stimulate domestic tobacco production or West Indian importation; the high duties imposed by countries of continental Europe anxious for revenue; and the competition of Cuba, Sumatra, and Columbia markets.

Beginning in the late 1700s, planters in Kentucky and Tennessee annually floated thousands of hogsheads of tobacco down the Mississippi River to New Orleans, but they periodically encountered disputes with inspectors and other port officials. France had acquired Louisiana from Spain in 1800, and then in 1803 United States purchased Louisiana to clear up these disputes. Because early tobacco growing laid waste to lands, settlers sought new lands in the West rather than struggle with worn-out fields.

In Europe, the Napoleonic wars meant that soldiers from Britain, France, and other countries came into contact with different tobacco habits, especially cigar-smoking. The Spanish upper classes enjoyed smoking cigars, a habit adopted by British cavalry officers. In the 1830s, French travelers observed Spanish women smoking *papalettes*, or shredded tobacco wrapped in paper, and the French called the smoking devices cigarettes. A smoking revival occurred in Britain (pipes as well as cigars), and snuffing went into decline. Demand for cigars meant more tobacco production outside the United States and new factories in the Spanish colony of Cuba and in Spain itself.

Worldwide Trade in the Early Twentieth Century

By the beginning of the twentieth century, at least fifteen countries reported exporting 688 million pounds of tobacco, or about one-third of the total world production. Exports ranged from none from Japan and only 6 percent of India's tobacco crop to nearly 100 percent for Brazil and Dutch East Indies (Java and Sumatra). The United States remained the leading exporter, with nearly one-half of the world's tobacco trade. With the growth of the U.S. domestic market, exports came to represent less than one-half of U.S. tobacco production. Both India and Japan had a considerable historical record of producing tobacco for local consumption, whereas Brazil and Dutch East Indies tobacco trade was primarily in cigar leaf tobaccos and was fostered by tobacco dealers and exporters. Cuba, the Philippines, Greece, and Turkey were major suppliers in the world tobacco market.

International trade in tobacco at the beginning of the twentieth century was fostered largely by buyers representing European countries. Cigarette production was still a minor outlet for tobacco and smoking tobacco, and cigars dominated the tobacco market in most countries. Germany represented 22 percent of world imports; Germany, Austria-Hungary, Belgium, France, Italy, Netherlands, Spain, and the United Kingdom accounted for three-fourths of tobacco-import trading. A government tobacco monopoly operated in France, Italy, and Spain, while the American Tobacco Company acquired control of major tobacco companies elsewhere in Europe and in the British Colonies. A few countries in Western Europe grew almost no tobacco commercially (Denmark, Finland, Norway, Sweden, and United Kingdom) and purchased from other countries the kinds of leaf they required. Some of those countries may have recorded exports because they acted as trans-shipping points (receiving and holding tobacco for later shipment).

Most producing countries grew more than one type of tobacco and supplemented their own production by imports of other types required by the domestic industry. By the 1920s, with expanding cigarette consumption pulling up the tobacco markets, the United States remained the premier exporting country but it also imported oriental or **Turkish tobacco,** which represented about 10 percent of a typical American-style cigarette blend. (Domestic-grown flue-cured, burley, and Maryland

Turkish tobacco a variety of mild, aromatic tobacco. Ironically, Turkish tobacco is not native to Turkey but was imported from North America. Turkish tobacco leaves are smaller and more delicate than American varieties. It is usually blended with Bright Leaf (Virginia) and Burley in cigarettes.

tobaccos represented about 90 percent.) Similarly, China's tobacco production was close to that of the United States; China also needed sizable imports for blending and ranked third among importing countries, with about one-half the quantities going to Germany and United Kingdom.

By contrast, some countries in the 1920s were nearly self-sufficient in tobacco: Greece, Turkey, Bulgaria, Cuba, and the Soviet Union. Typically, this situation was due to government import control to regulate foreign exchange or protect domestic tobacco growers.

Exports in the pre–World War II period (1935–39) were below that of a decade earlier because of the decline in cigar consumption, which meant reduced shipments for Indonesia, the Philippines, Cuba, and the Dominican Republic. U.S. exports were also lower, primarily because of a sharp fall-off in shipments of fire-cured and dark **air-cured** types (primarily used for cigars and snuff). Accounting for 4 percent of the world total, Canada, Nyasaland (now Malawi), and Rhodesia (now Zimbabwe) could provide tobacco under the umbrella of the U.S. agricultural price support program. Some U.S. growers moved to Canada to avoid U.S. production restrictions. Growers in the British Africa colonies followed the pattern of U.S. growers of two centuries earlier with a labor-intensive cash crop and sales arrangements with international tobacco dealers.

air-cured tobacco leaf tobacco that has been dried naturally without artificial heat.

After three decades as the leading tobacco import destination, Germany fell behind United Kingdom in the 1930s in part because of the sharp rise in sales of UK cigarettes. They were made entirely of flue-cured tobacco in contrast to German cigarettes, which consisted of the American-style blend. Also, the German government exercised considerable control over trade matters and began an antismoking campaign during that period. European countries accounted for three-fourths of tobacco imports. The United States and China imported the most tobacco in this period. For both the United Kingdom and China, the United States was the principal supplier.

Changes after World War II

World War II brought great hardships to many around the world, including relocations of populations, independence to many former colonial countries, and interrupted trade routes. From 1935 to 1959, tobacco exports rose 25 percent, and shifting shares among countries continued as cigarettes took the lead in tobacco manufacturing. U.S. exports continued to lead, with 35 percent of the total. The most striking shift was the slump in exports from Indonesia (formerly Netherlands India) to only one-third of the prewar volume and to a mere 2 percent share of total exports. As the local industry became more oriented to satisfying the local population, surpluses for export diminished. Canada, Rhodesia, and Malawi now accounted for 12.5 percent of export trade. With the overall market expanding, these newer suppliers could more easily enter the trade. Tobacco auction markets, patterned after those in the United States, facilitated sales to foreign manufacturers.

Among import destinations in the period from 1955 to 1959, the United Kingdom remained the leader, with one-fourth of the total; the European continent accounted for 70 percent. Data from China and the Soviet Union in this period are not available, but their trade was considered to be largely internal and strictly controlled by the central

TRADE

The ship "Eibe Oldendorff Monrovia" sitting at a loading dock in Seattle, Washington. © MIKE ZENS/CORBIS

authorities. The United States took a 10-percent share of imports as cigarette output rose and U.S. cigar manufacturers shifted to overseas sources.

At about this time, European countries still remembered food shortages during World War II, and several came together for new trade arrangements that became the European Common Market (later the European Union). Also included was a Common Agricultural Policy (CAP) that provided support for domestic agriculture without production controls and sales of any surpluses to areas outside the Common Market. Many observers at that time commented that such a system would be unsatisfactory. Another major government policy change was a unilateral declaration of independence in 1965 by Rhodesia from the British Commonwealth. The United Nations declared sanctions and asked other nations not to accept Rhodesia's exports. Initially, U.S. exports filled the void, but eventually Rhodesia shippers were able to move their tobacco through intermediary countries. In 1980, normal trade resumed following the election of a new government.

Between 1976 and 1978, tobacco trade had risen to 2.9 billion pounds. U.S. shipments led the way at a peak of 642 million pounds, or a 22 percent share. Brazil had jumped to become the second largest exporter, with 238 million pounds and an 8 percent share. In earlier years, Brazil produced and exported a considerable amount of cigar leaf tobacco, so a network of dealers and exporters was in place. The jump occurred in cigarette leaf production located in the southern state of Rio Grande do Sol. As in many other countries, tobacco, as an export crop was an ideal source of cash income for diversified crop farms.

With emphasis on domestic food production, Cuba's tobacco trade languished and fell to only 1 percent of the world total by the 1970s. Tobacco production in the Mediterranean area is extremely labor-intensive but exports have held their own; nonetheless, Greece and Turkey's share of world trade has fallen. Italy's exports (3.5 percent of total) represent surplus production that the CAP was required to sell outside the European Union.

Trade in the Twenty-First Century

By the start of the twenty-first century, tobacco trade had expanded by over one-half in twenty-five years. But half of the increase was due to the inclusion of data from the newly independent countries that had been part of the Soviet Union, whose trade had previously been internal. Brazil has become the leading exporting country, with 21 percent of the total. The United States and Zimbabwe were in a virtual tie for second place, with 9 percent each. Under the U.S. production-control and price-support program, the U.S. Department of Agriculture was required to reduce the tobacco **marketing quotas,** and so less tobacco was available for purchase. Exporters claimed that U.S. price levels were noncompetitive with comparable tobaccos from other countries, costing about one-half of U.S.-grown crops. Even by the end of the 1980s, U.S. tobacco imports were exceeding leaf tobacco exports.

Greece and Turkey retained about a 10 percent share as the market for oriental tobacco continued to thrive and with cigarette production shifting to American-style blends. Malawi is probably the country most heavily dependent on tobacco exports, with 5 percent of the world leaf trade. India's exports vary somewhat, but the share of around 5 percent indicates that the crop retains many economic advantages for producers. Italy's surplus continues to cascade onto the world tobacco market, at 5 percent in 2001. The surplus from the EU countries (except Greece) represents 14 percent of world exports.

Tobacco buyers in Zimbabwe check the aroma of the leaf on the opening day of the selling season, April 2000. Tobacco, the country's leading foreign currency earner, started trading at one-third normal capacity caused by an overvalued local currency and the disruption caused by the invasion of farms by war veterans and villagers © REUTERS NEWMEDIA INC./CORBIS

marketing quota the amount of cured leaf tobacco (usually expressed in pounds) that a tobacco grower may produce and sell in a given year. The United States Department of Agriculture began setting marketing quotas for tobacco in 1933.

TRADE

Although the greater portion of tobacco production of the world is consumed in the producing countries, a considerable amount, about one-third in recent years, enters international trade, in the form of leaf or products, supplements for deficient supplies in some countries, or total requirements in others. World exports of leaf tobacco amounted to 2.1 billion metric tons in 2002, compared with .4 billion in the years 1935–1939. This was an increase of over 400 percent, in contrast to a 125 percent increase in world production of leaf tobacco. This development reflects the rapid rise in consumption almost worldwide, with many populous countries needing proportionately larger imports than could be acquired from domestic production to meet the rising leaf requirements of manufacturers. The principal importing countries are the Peoples Republic of China, the Russian Federation, the United States, Germany, the United Kingdom, and the Netherlands.

With the antitobacco efforts gathering force in the developed countries, major cigarette manufacturers have established joint ventures with former state manufacturers in Eastern Europe. They have also stepped up cigarette sales to Third World populations and have pushed into other business lines, such as food and beverages. As markets shift, many growers and marketing firms find almost no useful alternatives for their production and processing facilities.

See Also Africa; Brazil; British Empire; Caribbean; Chesapeake Region; China; Cuba; Dutch Empire; French Empire; Mexico; Middle East; Oceania; Philippines; Portuguese Empire; South and Central America; South Asia; South East Asia; Spanish Empire; Zimbabwe.

■ SIMON SMITH
■ ROBERT MILLER

BIBLIOGRAPHY

American Tobacco Company. *"Sold American!" The First Fifty Years*. Privately printed, 1954.

Batie, Robert Carlyle. "Why Sugar? Economic Cycles and the Changing of Staples in the English and French Antilles, 1624–1654." In *Caribbean Slave Society and Economy*. Edited by Hilary Beckles and Verene Shepherd. New York: New Press, 1991.

Brenner, Robert. *Merchants and Revolution: Commercial Change, Political Conflict, and London's Overseas Traders, 1550–1653*. Cambridge and New York: Cambridge University Press, 1993.

Brigham, Janet. *Dying to Quit: Why We Smoke and How We Stop*. Washington: John Henry Press, 1998.

Burrough, Bryan, and John Helyar. *Barbarians at the Gate: The Fall of RJR*. New York: Harper and Row, 1998.

Capehart, Thomas. *Tobacco Situation and Outlook Yearbook. Economic Research Service*. Washington: U.S. Department of Agriculture, 2002.

Carr, Lois Green, and Russell R. Menard. "Land, Labor, and Economies of Scale in Early Maryland: Some Limits to Growth in the Chesapeake System of Husbandry." *Journal of Economic History* 49 (1989): 407–418.

Chute, Anthony. *Tabaco: the Distinct and Severall Opinions of the Late and Best Phisitions that have Written of the Divers Natures and Qualities Thereof*. London, 1595.

Cooper, Mary H. "Regulating Tobacco." *The CQ Researcher.* Congressional Quarterly (1994): 841–864.

Devine, T. M. *The Tobacco Lords: A Study of the Tobacco Merchants of Glasgow and Their Trading Activities, c.1740–1790.* Edinburgh, Scotland: John Donald Publishers, 1975.

———, ed. *A Scottish Firm in Virginia, 1767–1777: William Cunninghame and Company.* Edinburgh, Scotland: Scottish History Society Publications, 1984.

Dickson, S. A. *Panacea or Precious Bane: Tobacco in Sixteenth Century Literature.* New York: New York Public Library, 1954.

Edwards, Everett E. "American Agriculture: The First 300 Years." *Yearbook of Agriculture: Farmers in a Changing World.* Washington: U.S. Department of Agriculture, 1940.

Friedman, Kenneth Michael. *Public Policy and the Smoking Health Controversy.* Lexington: Lexington Books, 1975.

Fritschler, A. Lee, and James M. Hoefler. *Smoking and Politics: Policymaking and the Federal Bureaucracy.* New York: Appleton-Century-Crofts, 1969.

Gage, Charles E. *American Tobacco Types: Uses and Markets.* Washington: U.S. Department of Agriculture Circular 249, 1942.

Galenson, David. "White Servitude and the Growth of Black Slavery in Colonial America." *Journal of Economic History* 41 (1981): 39–47.

Gately, Iain. *Tobacco: A Cultural History of How an Exotic Plant Seduced Civilization.* New York: Grove Wiedenfeld, 2001.

Goodman, Jordan. *Tobacco in History: The Cultures of Dependence.* London and New York: Routledge, 1993.

Grise, Verner. *Structural Characteristics of Flue-Cured Tobacco Farms and Prospects for Mechanization*, Ag. Economic Report 277. Washington: U.S. Department of Agriculture Economic Research Service, 1975.

———. *Tobacco: Background for 1990 Farm Legislation.* Washington: U.S. Department of Agriculture Economic Research, 1989.

Historical Statistics of the United States, Colonial Times to 1970. Washington: United States Department of Commerce, Bureau of the Census, 1970.

Jost, Kenneth. "Closing In on Tobacco". *The CQ Researcher.* Congressional Quarterly (1999): 977–1000.

Kulikoff, Allan. *Tobacco and Slaves: The Development of Southern Cultures in the Chesapeake, 1680–1800.* Chapel Hill: University of North Carolina Press, 1986.

Lorimer, J. "The English Contraband Tobacco Trade in Trinidad and Guiana, 1590–1617." In *The Westward Enterprise: English Activities in Ireland, the Atlantic, and America, 1460–1650.* Edited by K. R. Andrews et al. Liverpool, U.K.: Liverpool University Press, 1978.

Mann, Charles Kellogg. *Tobacco: The Ants and the Elephants.* Salt Lake City: Olympus, 1975.

Mann, Jitendar. *Dynamics of the U.S. Tobacco Economy.* Washington: U.S. Department of Agriculture Economic Research Tech. Bulletin 1499, 1974.

McCusker, John J., and Russell R. Menard. *The Economy of British America, 1607–1789.* Chapel Hill: University of North Carolina Press, 1985.

Menard, Russell R. "From Servants to Slaves: The Transformation of the Chesapeake Labor System." *Southern Studies* xvi (1977): 355–390.

———. "Immigrants and Their Increase: The Process of Population Growth in Early Colonial Maryland." In *Law, Society, and Politics in Early Maryland.* Edited by Aubrey C. Land, Lois Green Carr, and Edward C. Papenfuse. Baltimore and London: Johns Hopkins Press, 1977.

———. "Population, Economy, and Society in Seventeenth-Century Maryland." *Maryland Historical Magazine* 79 (1984): 71–92.

Menard, Russell, and Lorena Walsh, "Death in the Chesapeake: Two Life Tables for Men in Early Colonial Maryland." *Maryland Historical Magazine* 69 (1974): 211–227.

Mollenkamp, Carrick, Adam Levy, and Joseph Menn. *The People vs. Big Tobacco: How the States Took on the Cigarette Giants*. Princeton: Bloomberg Press, 1998.

Nash, R. C. "The English and Scottish Tobacco Trades in the Seventeenth and Eighteenth Centuries: Legal and Illegal Trade." *Economic History Review* 35 (1982): 354–372.

North, Douglass C. "Sources of Productivity Change in Ocean Shipping, 1600–1850." *Journal of Political Economy* 76 (1968): 953–970.

O'Rourke, Kevin, and Jeffrey G. Williamson. "After Columbus: Explaining the Global Trade Boom, 1500–1800." *Journal of Economic History* 62 (2002): 417–456.

Osler, Gerry, Graham A. Colditz, and Nancy L. Kelly. *The Economic Costs of Smoking and Benefits of Quitting*. Lexington: Lexington Books, 1984.

Price, Jacob M. "The Rise of Glasgow in the Chesapeake Tobacco Trade, 1707–1775." *William and Mary Quarterly* 43 (1954): 179–199.

———. *France and the Chesapeake: A History of the French Tobacco Monopoly, 1674–1791, and of Its Relationship to the British and American Tobacco Trades*, 2 vols. Ann Arbor: University of Michigan Press, 1973.

———, ed. *Joshua Johnson's Letterbook, 1771–4: Letters from a Merchant in London to his Partners in Maryland*. London: Record Society Publications, 1979.

———. "The Last Phase of the Virginia-London Consignment Trade: James Buchanan & Co., 1758–68." *William and Mary Quarterly* 43 (1986): 64–98.

Robert, Joseph C. *The Story of Tobacco in America*. New York: A. Knopf, 1949.

Roessingh, H. K. "Tobacco Growing in Holland in the Seventeenth and Eighteenth Centuries: A Case Study of the Innovative Spirit of Dutch Peasants." *The Low Countries History Yearbook* 11 (1978): 18–54.

Shammas, Carole. "American Colonization and English Commercial Development, 1560–1620." In *The Westward Enterprise: English Activity in Ireland, the Atlantic and America, 1500–1650*. Edited by K. R. Andrews et al. Liverpool, U.K.: Liverpool University Press, 1978.

Shepherd, J. F., and G. M. Walton. *Shipping, Maritime Trade, and the Economic Development of Colonial North America*. Cambridge and New York: Cambridge University Press, 1972.

Smith, S. D. "The Market for Manufactures in the Thirteen Continental Colonies, 1698–1776." *Economic History Review* 51 (1998): 676–708.

———. "The Early Diffusion of Coffee Drinking in England" In *Le Commerce du Café avant l'ère des Plantations Coloniales: Espaces, Réseaux, Sociétés, xve-xixe Siècle*. Edited by Michel Tuscherer. Cairo: Institute Français d'archéologie orientale, cahier des annales islamologiques, 2001.

Soltow, J. H. "Scottish Traders in Virginia, 1750–1775." *Economic History Review* 12 (1959): 83–98.

Taylor, Peter. *The Smoke Ring: Tobacco, Money, and Multinational Politics*. New York: Pantheon Books, 1984.

Thirsk, Joan. "New Crops and their Diffusion: Tobacco-Growing in Seventeenth-Century England." In *Rural Change and Urban Growth, 1500–1800*. Edited by C. W. Chalkin and M.A. Havinden. London and New York: Longman, 1974.

Tobacco: Deeply Rooted in America's Heritage. Tobacco Institute. Privately printed, 1985.

White, Larry C. *Merchants of Death: The American Tobacco Industry.* New York, Beech Tree Books, 1988.

World Production and Trade. Circulars 2–4, 10–30. Washington: United States Department of Agriculture, Foreign Agricultural Service, 2003.

United States Agriculture

For better or worse, tobacco is thoroughly American. Native Americans grew and smoked tobacco long before Europeans and Africans came to the New World. Tobacco farming is America's oldest commercial enterprise. For almost four centuries, American farmers have sent the weed in its many forms throughout the world. And while other nations grow tobacco, American products enjoy a popularity in world markets not accorded any other producer. Historically, vast areas of the American South were devoted to producing tobacco, and entire rural populations marched to the cadence of tobacco culture. The living standards of millions of Americans have risen and fallen with the fortunes of the crop. This chapter of the tobacco story begins with American independence.

Geographic Expansion, 1780s–1800s

By the end of the Revolutionary War (1783), tobacco was well established as America's leading export. From Europe to China, smokers wanted American tobacco, and by 1792 exports were up 36 percent over prewar levels. Demand for American leaf drove its expansion far beyond its original home in Tidewater Virginia and Maryland. Predictably, tobacco culture followed patterns of settlement. As farm families migrated west and south, they carried tobacco seeds and knowledge with them to their new homes. Thus, tobacco farming spread into the Virginia/North Carolina border region, Tennessee, Kentucky, and the South Carolina Piedmont. By the early 1800s, tobacco culture had crossed into Ohio and Missouri as well.

At first, tobacco played a minor role in the frontier economy as pioneer families divided their time between raising food crops and hunting game. Tobacco was grown in small plots for home consumption and barter. Soon, however, tobacco evolved into a cash crop as backcountry settlers grew greater quantities, not only to smoke but also to sell. The new tobacco regions had advantages over the historic culture areas. After many decades of growing tobacco, eastern Virginia soils had lost

UNITED STATES AGRICULTURE

Aerial photo of striped tobacco fields, Blue Ridge Mountains, North Carolina, 1991.
© RICHARD A. COOKE/CORBIS

much of their fertility, and Western growers could sell leaf profitably at prices below the East's cost of production. Poor transportation and remoteness from markets hindered backcountry farmers for a while, but by the 1840s expanding rail service was linking the backcountry with the coast. Other Western tobacco growers shipped their leaf down the Ohio, Missouri, and Mississippi Rivers to New Orleans markets. Western competition ultimately forced Tidewater Virginia farmers to plant other, more profitable crops.

Tobacco 1800–1865

Tobacco farming was labor intensive. Indeed, tobacco was virtually a handmade crop as each leaf was handled several times during cultivation and processing. Seedbeds were sown in winter. Farmers typically located seedbeds at the edge of a wooded area to shelter them from wind. After sowing, growers covered the beds with straw to protect the young plants. By spring, the plants were several inches high and ready for transplanting. Growers pulled the plants from the seedbeds and transplanted them into fields of a few acres. Sometimes, a child would ladle water from a wooden bucket onto each young plant.

As the plants grew, farmers plowed or hoed the fields to remove weeds. After six weeks or so, the plants reached adolescence and flowers began forming at the top. Seeking to concentrate the plants' energies on leaf production, growers "topped" the plants, removing the flowers by hand. Also removed were "suckers," leaf-like outgrowths at the axils—the juncture of leaf and stalk—that diverted nourishment from the leaves. Growers pruned the suckers by hand, starting at the top and working down. Tobacco worms—fat, green caterpillars—were picked off at the same time. "Suckering" was tedious work, hard on the arms and back. Sometimes, growers removed the "lugs"—the bottom leaves—to direct nourishment to the more valuable middle and upper leaves. The plants ripened through the summer, the leaves gradually turning from green to yellow. When the plants were judged fully ripe, farmers harvested the tobacco, severing the stalks at the bottom.

After harvesting, the tobacco farmers' work was only half done: Curing, **stripping,** bulking, grading, and tying the crop were still to come. Tobacco leaves contain moisture that must be removed. Thus, tobacco growers were required not only to nurture and cultivate the plants but to dry or "cure" them as well. Typically, harvested plants were hung in well-ventilated barns to air-cure. **Air-curing** could take several weeks. Sometimes, farmers lit small fires on the dirt floors of the barns to speed curing. Fire-cured tobacco acquired a heavy, smoky flavor. In late autumn, farmers stripped the dry, brown leaves from the stalks and "bulked" them in piles to improve quality.

Usually, farmers sorted the leaves into four or five grades based on color and size. The cured leaves were then "tied." A folded leaf was wrapped around the stems of several leaves forming a small bundle called a "hand." The hands were packed in large, wooden barrels called "**hogsheads**" and brought to market. A hogshead of tobacco often weighed 1,000 pounds (450kg) or more. By the mid-nineteenth century, many farmers abandoned hogsheads and carried their tobacco to market in large baskets.

For much of the nineteenth century, Europe was the most important market for American tobacco. By 1800, the European passion for **snuff** had about run its course, and Europeans were turning more to the cigar. Demand for cigar tobaccos, especially the flavorful wrapper leaves, was met by tobacco growers in the Connecticut Valley. Fine leaf had been produced there since the 1760s, and by the early nineteenth century Connecticut Valley growers were specializing in producing cigar tobaccos for Europe and a growing American market. Later in the century, Connecticut Valley growers learned to imitate Asian climatic conditions by growing tobacco under porous cloth. The attendant rise in temperature and humidity together with the region's unique soils produced superior cigar tobaccos. Cigars were the smoke of choice for much of the nineteenth century on both sides of the Atlantic.

As the nation grew, a strong domestic market developed for all types of American tobaccos. At first, only tobacco judged too poor for export was consumed locally. By the early 1800s, however, more and better leaf was manufactured and consumed by Americans. Besides Connecticut Valley wrappers, Americans smoked cigars rolled from dark Tennessee and Kentucky leaf. Much tobacco was consumed in pipes as well. Frontier women commonly smoked, and even children sometimes puffed small pipes. But the most popular tobacco product in the United States in the early nineteenth century was chewing tobacco. "**Plug,**" "quid," or "chaw," as it was variously known, was commonplace by the 1830s. Cured leaves were coated with molasses, licorice, rum, fruit brandies, or other flavorings and molded into blocks—plugs—about the size of a deck of playing cards. Users cut off the desired amount with a pen knife and packed the "chaw" between cheek and gum. Copious amounts of saliva were produced and expelled from the mouth in brown, malodorous

stripping in the Burley and fire-cured tobacco cultures, cured leaves must be separated from the dead stalk. This is called "stripping."

air-curing the process of drying leaf tobacco without artificial heat. Harvested plants are hung in well-ventilated barns allowing the free circulation of air throughout the leaves. Air-curing can take several weeks. Burley tobacco is air-cured.

hogshead a large wooden barrel formerly used to store and transport cured leaf tobacco. A hogshead typically held approximately 800 to 1000 pounds (350 to 450 kg) of tobacco.

snuff a form of powdered tobacco, usually flavored, either sniffed into the nose or "dipped," packed between cheek and gum. Snuff was popular in the eighteenth century but had faded to obscurity by the twentieth century.

plug a small, compressed cake of flavored tobacco usually cut into pieces for chewing.

UNITED STATES AGRICULTURE

A 1944 advertisement for Lucky Strike cigarettes features a detail of "Grading Leaf" by Peter Hurd. © BETTMANN/CORBIS

streams. As the tobacco was not burned, chewing was less of a fire hazard than smoking. But the health risks of sidewalks and lawns covered with human expectorations were often as hazardous as fire.

Cigarettes were rare before the 1870s. They were hand rolled in New York sweatshops, and the price was steep. Rolling one's own was cheaper but inconvenient. Moreover, the strong cigarette tobaccos were harsh-tasting, and attempts to inhale the smoke often led to a coughing spasm. Therefore, most tobacco users contented themselves with pipes, cigars, snuff, and chewing.

The cotton boom of the early 1800s was a momentous event in American agriculture, and its effects were felt in the tobacco industry. With the invention of the cotton gin, large-scale cotton production became feasible for the first time. Driven by strong demand for cloth, cotton prices rose well above prices for leaf tobacco. Responding to higher prices, many farmers reduced their tobacco acreage in favor of cotton.

In some areas, cotton drove tobacco culture virtually to extinction. For example, by the 1820s tobacco shipments through Savannah, Georgia, had fallen 89 percent from 1,500 hogsheads per year to only 170. In Charleston, South Carolina, former tobacco processing facilities were rededicated to cotton. For two centuries, tobacco had reigned as America's leading export. In the early 1800s, however, cotton succeeded tobacco as king of American agriculture.

From the beginning, tobacco and slavery were bound together. The first African slaves brought to America toiled in tobacco fields. Tobacco, therefore, set slavery in motion and provided an eager slave market for two centuries. Some **plantations** were small with only a few slaves; some held more than one hundred people in bondage and were essentially independent little villages. By the 1800s, most slave-based tobacco farms—especially the larger plantations—had adopted the task system. Under the task system, slaves were given individual job assignments rather than laboring in gangs. Tasks included jobs related to every stage of cultivating and processing tobacco. For example, a hoeing or suckering task might allot a certain area of a field. Sometimes, wooden stakes would mark the boundary of the task area. A stripping task could specify a certain poundage of leaf to be stripped. Once the task was completed, the slave was released from labor for the day. Most slaves could finish their tasks by four or five o'clock, but very industrious workers might finish earlier.

plantation historically, a large agricultural estate dedicated to producing a cash crop worked by laborers living on the property. Before 1865, plantations in the American South were usually worked by slaves.

Unfortunately, slavery also moved west with tobacco culture. Western lands were cheap while slaves were expensive. Moreover, human property was mobile while real estate was not. Thus, many planters found it expedient to move to new lands every few years, taking their slaves with them. The impact of slavery was felt throughout the white community as well. Slavery placed small, independent farmers at a competitive disadvantage since they had to pay for labor. Slavery also undermined the free labor market by lowering wages for free blacks and white workers. In Tidewater Virginia, many former tobacco planters sold their slaves to toil in the newer cotton and tobacco regions of the West and South.

Post-Slavery Changing Labor Relations

The victory of the Union in the Civil War (1861–1865) ended slavery in the United States. But resolving old questions raised new ones. What would replace slavery as the dominant labor system in Southern agriculture? Most freedpeople wanted land of their own, but while a fortunate few acquired land, most continued to work for white landowners. **Sharecropping** did not originate in the post–Civil War period. The custom of landowner and laborer dividing a crop was well established before the war, but both parties had typically been white. With emancipation, however, sharecropping increasingly involved white landowners and black laborers. Although details varied, the basic sharecrop arrangement was fairly simple. Landowners provided land, horses or mules, implements, and a house. The cropper family furnished all the labor, and the parties equally divided the proceeds from the crop. That housing was provided was, perhaps, the most tangible benefit to the cropper. A large tobacco farm could have several cropper families living and working on assigned parcels of land.

sharecropping a form of agricultural labor that gained popularity after the Civil War. Laborers, usually families, lived and worked on land belonging to a proprietor. They grew staple crops like tobacco and cotton. Rather than regular cash wages, they were paid with shares of the crop at harvest time.

Farmers harvesting tobacco leaves on a Virginia plantation, c. 1612. HULTON ARCHIVE/GETTY IMAGES

In the beginning, most freedpeople viewed sharecropping as a temporary transition to owning land. But as the years passed, few realized their dream of economic independence. Sometimes croppers obtained food and other necessities on credit, but their earnings were insufficient to settle their debts and they slid into hopeless poverty. Low tobacco prices worsened the problem. In years of low prices—and there were many—there was little profit for either the cropper or the landlord. For many, therefore, sharecropping became a trap into which they were born, lived, and died.

Another form of land tenure in tobacco farming was tenantry. Basically, tenantry was a type of rental. Tenants typically owned a team of horses or mules, plows and other implements, and had established credit with local merchants. Sometimes, tenants paid landowners a flat cash rental for the land and dwelling. More commonly, tenants paid a quarter of the crop. Although the distinctions may seem subtle, tenantry was considered a step above sharecropping in the agrarian hierarchy of the rural South. Many hands were needed to ready a crop for market, and a sizeable farm would be home to several tenant families. The landowner often reserved a "home place," a parcel around his dwelling to cultivate himself. Over time, tenantry gradually replaced sharecropping on Southern tobacco farms, and by the 1940s probably eighty percent of the Southern tobacco crop was produced by tenants and small landowners. Tenantry declined rapidly in the 1960s as machines replaced human and animal labor. Few persons born after 1945 ever held tenant status as adults.

Tobacco crops ranged in size from 3 acres to more than 100, but the most common unit was the small family farm raising 6 to 10 acres of tobacco. The family lived on the property and provided their own labor with perhaps a few hired hands in busy seasons. The entire family contributed to making the crop. Children as young as six or seven were given simple tasks like picking up dropped leaves; old men and women, no longer up to field work, could grade and tie the cured leaves.

Tobacco farmers raised other crops and livestock for food and extra income. Most learned to rotate their crops, including soil-building plants like clover in the rotation. A balanced farming plan included corn, wheat, beef, pork, and poultry as well as a vegetable garden. Women and children usually tended the garden and fed the chickens; older boys and men tended the horses, cattle, and hogs. Farmers pastured cattle on fallow fields to renew the fertility of the soil.

Bright Leaf Tobacco

The development of Bright leaf or **flue-cured tobacco** was a critical event in the tobacco story. Farmers learned to cure tobacco with artificial heat employing a furnace outside the curing barn. Heat was distributed through a system of stove pipes, or "flues," running parallel to and a few inches above the floor. Harvested plants were hung overhead. The furnace chimney carried exhaust outside the barn so plants were untainted by smoke and soot. With the higher temperatures, a barn of tobacco cured in about five days rather than the five weeks needed for air-curing. Not only did flue-cured tobacco cure faster, but it was lighter, milder, and more aromatic than **air-cured** tobacco. Its characteristic yellow color gave it the name Bright Leaf. Consumers liked Bright Leaf from the start, and by the 1870s many farmers in the North Carolina-Virginia border region were flue-curing.

flue-cured tobacco also called Bright Leaf, a variety of leaf tobacco dried (or cured) in air-tight barns using artificial heat. Heat is distributed through a network of pipes, or flues, near the barn floor.

air-cured tobacco leaf tobacco that has been dried naturally without artificial heat.

To obtain the best possible color and flavor, Bright leaf growers replaced stalk harvesting with leaf harvesting. Tobacco leaves ripen in sequence beginning at the bottom of the stalk, then the next higher leaves, and so on to the uppermost leaves. Waiting for the upper leaves to ripen insured that the lower leaves were overripe and less valuable. Harvesting the leaves as they ripened guaranteed the highest value for every leaf. Farmers learned to harvest their crop four times, taking one-fourth of the leaves with each priming. Stalks never left the field but were simply plowed under after the final harvest.

Before the 1870s, cigarettes were a novelty as few consumers enjoyed the harsh-tasting, pungent smoke. Bright leaf tobacco changed that. The milder, lighter tobacco was ideal for cigarette smoking. Moreover, the milder smoke was more easily inhaled, enhancing the smoker's satisfaction. Inhaling also enhanced and accelerated the smoker's addiction to the product. At the same time smokers were embracing Bright leaf, technological breakthroughs in cigarette manufacture made mass production possible for the first time. Economies of scale enabled manufacturers to cut cigarette prices in half, and the price of the ten-unit pack fell from 10 cents to 5 cents. Buttressed by strong advertising support, sales soared. Cigarette consumption in the United States increased sharply, rising from 42 million in 1875 to 500 million in 1880, a rate of increase approaching 1,200 percent in five years. Cigarette consumption passed 1 billion in 1884 and 2 billion in 1887.

Why did cigarette smoking become so popular? Several factors are responsible. Smokers liked the mild, aromatic Bright leaf tobacco, and lower prices placed cigarettes within reach of all social classes. Cigarettes were convenient. Unlike cigars and pipes, cigarettes required no special paraphernalia—only a match. And while a cigar or pipe required considerable time to enjoy, a cigarette could be consumed in five minutes. Of course, addiction underlay everything. Occasional smokers soon became steady users.

Geographic Expansion, 1880s–1930s

The dramatic increase in cigarette smoking required a corresponding increase in the supply of leaf tobacco. By the 1890s, Bright leaf tobacco culture had spread into eastern North Carolina and eastern South Carolina. By the early twentieth century, farmers in southern Georgia and northern Florida were learning to grow and cure tobacco. Burley production expanded also. Nearly every county in Kentucky raised that variety, and production expanded in Missouri and Ohio as well. Historic culture areas of Maryland and Virginia saw a resurgence of tobacco culture for the first time in living memory.

By the 1920s, supply was outpacing demand. Moreover, tobacco manufacturers exploited the farmers' weak marketing position to keep leaf prices low. In the Great Depression of the 1930s, prices fell below the cost of production, and all tobacco regions faced misery and want. Through the Agricultural Adjustment Act of 1933, the national government limited supplies of leaf tobacco through allotments and quotas. Expansion was frozen in place; no new regions could produce tobacco. Existing tobacco farms were assigned a leaf quota, controlling the amount of tobacco they could produce and sell. The quota was adjusted every year to reflect changes in demand. By limiting supply, leaf prices rose dramatically and tobacco farming became profitable again. The tobacco program has proven very popular and durable. It has survived, with some changes, since the 1930s.

Mechanization, 1940s–2000s

Mechanization—replacing human and animal labor with machines—provided another important source of change on American tobacco farms in the twentieth century. With greater profitability, tobacco growers could invest in labor-saving and highly efficient farm machinery. The farm tractor multiplied the horsepower on a typical tobacco farm by a factor of ten. Through the years, tobacco-specific implements were developed to mechanize certain tasks while leaving others unchanged. Chemical technologies were important also. Scientifically developed fertilizers reinforced by nitrogen and potassium supplements pushed yields above 2,000 pounds (900 kg) per acre. Herbicides and pesticides controlled insects and undesirable weeds. In the 1970s, mechanized harvesting and curing equipment automated those labor-intensive operations. By the 1980s, sharply reduced labor requirements were increasingly met by migrant labor. Greater capital requirements made tobacco farming more of a commercial activity, and smaller farmers found it difficult to compete. Many sold or rented their quotas to larger and better capitalized growers. Between 1970 and 2000, the size of the average tobacco farm more than tripled. In the 2000s, tobacco farming is larger, more mechanized, and employs fewer people than at any time in its history.

See Also Labor; Plantations.

∎ELDRED E. "WINK" PRINCE JR.

BIBLIOGRAPHY

Badger, Anthony J. *Prosperity Road: The New Deal, Tobacco, and North Carolina.* Chapel Hill: University of North Carolina Press, 1980.

Daniel, Pete. *Breaking the Land: The Transformation of Cotton, Tobacco, and Rice Cultures Since 1880.* Urbana: University of Illinois Press, 1985.

Fite, Gilbert C. *Cotton Fields No More: Southern Agriculture, 1865–1980.* Lexington, Ky.: University Press of Kentucky, 1984.

Gately, Iain. *Tobacco: The Story of How Tobacco Seduced the World.* New York: Grove Press, 2001.

Goodman, Jordan. *Tobacco in History: The Cultures of Dependence.* London: Routledge Press, 1993.

Hirschfelder, Arlene B. *The Encyclopedia of Smoking and Tobacco.* Phoenix, Ariz.: Oryx Press, 1999.

Kulikoff, Allan. *Tobacco and Slaves: The Development of Southern Cultures in the Chesapeake, 1680–1800.* Chapel Hill and London: University of North Carolina Press, 1986.

Prince, Eldred E., Jr., and Robert Simpson. *Long Green: The Rise and Fall of Tobacco in South Carolina.* Athens and London: University of Georgia Press, 2000.

Tilley, Nannie May. *The Bright-Tobacco Industry, 1860–1929.* Chapel Hill: University of North Carolina Press, 1948.

Virginia Slims

Throughout the 1960s, Philip Morris (PM) was a rapidly growing company, its expansion fueled primarily by the explosive growth of its Marlboro brand. With PM's success based largely on one brand, the company sought to broaden its customer base by establishing other brands, so as not to be dependent solely on Marlboro for its profitability.

As a growing, hungry company, PM sought out new **marketing niches** in a market that was fragmenting rapidly, because PM's many competitors were launching numerous new brands. Ironically, one of the niches it exploited most successfully was the same one it had abandoned in 1955 when the original Marlboro was discontinued: women.

American Tobacco Company's Silva Thins brand was first introduced in 1967. In 1968 Virginia Slims appeared on the scene. Although several other brands for women were available, Virginia Slims was the most successful, with a marketing campaign that hit its target dead-on. Conceptually, the marketing strategy was to use a feminine name that also connoted beauty and thinness. The cigarette itself was extremely slim—markedly thinner than the industry standard—and, correspondingly, the package was slender and appealingly stylish and was considered by many a fashion accessory.

By capitalizing on the emerging women's movement of the 1960s, PM built a successful advertising campaign that established Virginia Slims as a best-seller in its niche. Ads featuring the slogan "You've come a long way, Baby" simulated old photographs that both contrasted and compared Victorian era women to modern women. The result was a successful new product that, by 1971, was selling 5.5 billion units a year, a little over one percent of the market. While Virginia Slims did not hold a large share of the market, it was enough to be a highly profitable brand.

Starting in 1970, PM began sponsoring the Virginia Slims Tennis Tour, originally because PM's president, Joseph Cullman III, was a tennis fan. It turned out to be a good move from PM's standpoint, as

marketing niche the area of commerce most appropriate for a particular product; for example, sugar-free soda would fill the marketing niche for dieting persons.

VIRGINIA SLIMS

Popular advertisements for Virginia Slims cigarettes contrasted modern, liberated women with those of the Victorian era. In this magazine ad from the 1970s, the image of a stylish model is juxtaposed against a fictitious historical newspaper that announces "Alaskan Town Allows Woman to Smoke." THE ADVERTISING ARCHIVE LTD./VIRGINIA SLIMS

irony an event or circumstance that is the opposite of what might be expected, for example, athletes endorsing cigarettes.

the players were anxious not only because female players wanted to earn what their male counterparts were making, but also for the star status that came with the publicity. In return, the tour provided good publicity for Virginia Slims. This relationship, however, became a lightning rod for women's health advocates who publicized the **irony** of athletes promoting cigarettes. After the University of Illinois at Chicago hosted the Virginia Slims Tournament in 1993 and 1994, protests by health advocates forced the University to cancel its contract with PM.

Over the years, PM has targeted the female smoker by promoting a consistent image for Virginia Slims. Humorous advertisements told women how far they had come and congratulated them on having their own cigarette. Promotional items included the Virginia Slims Book of Dates, so that such independent women could keep track of all their men. In the early 1990s, PM also introduced a line of clothing and accessories worn by attractive, young, and slim models to promote the brand.

By the early 1990s, Virginia Slims had captured roughly 3 percent of the American market and was the leading women's cigarette. By the start of the twenty-first century, in a highly fragmented market where only one brand, Marlboro, captured more than 10 percent of the market, Virginia Slims retained a **market share** of approximately 2.4 percent. This figure represented approximately 10 billion cigarettes, which made Virginia Slims the tenth largest-selling cigarette in the United States in 2002.

See Also Camel; Cigarettes; Gitanes/Gauloises; Lucky Strike; Marlboro; Philip Morris.

▪ JOSEPH PARKER

market share the fraction, usually expressed as a percentage, of total commerce for a given product controlled by a single brand; the consumer patronage for a given brand or style of product.

BIBLIOGRAPHY

Jones, Raymond M. *Strategic Management in a Hostile Environment: Lessons From the Tobacco Industry.* Westport, Conn.: Quorum Books, 1997.

Kluger, Richard. *Ashes to Ashes: America's Hundred-Year Cigarette War, the Public Health, and the Unabashed Triumph of Philip Morris.* New York: Alfred A Knopf, 1996.

Sobel, Robert. *They Satisfy: The Cigarette in American Life.* New York: Anchor Press/Doubleday Publishing, 1978.

Visual Arts

In the twentieth century tobacco companies spent millions on advertising campaigns that used glamour to lure people into smoking particular brands of cigarettes and cigars. Highly paid art directors and photographers created images of pleasure and power that featured celebrity smokers, including actors and sports heroes, who already were fantasy subjects for admirers. As noted by Patrick Carroll, founder of Freedom Tobacco International, in a 7 August 2003 article for *The Miami Herald*, "To be honest, celebrities make or break your brands. If you look at who drinks what or that sort of thing, celebrity endorsements have always meant a lot."

Historical Background

Since its sixteenth-century rise to prominence in western culture, smoking has possessed a degree of glamour as a mood-altering substance and as an exotic practice. Its effect on health was debated, with supporters (including tobacco merchants) claiming it was a panacea (cure-all) for depression and other maladies. Opponents, especially religious figures, considered smoking a new form of sensual indulgence and condemned it as a threat to health, morals, and the social order. These controversies have only added an aura of sinful pleasure to smoking.

Centuries before modern ad campaigns, painters and printmakers found smoking an appealing subject for formal and symbolic reasons. Smoke, a fluid element without stable proportions—rather like water or

VISUAL ARTS

Before the eighteenth century, European portrayals of tobacco use typically were negative. This detail from Jan Steen's 1663 painting *Topsy-Turvy World* gives an unglamorous portrayal of people governed by carnal desires, including wine, women, and song. One of the neglected children in the background holds a pipe to his lips. © ARCHIVO ICONOGRAFICO, S.A./CORBIS

clouds—allows an artist to compose freely, using spirals, wisps, or puffs, and veils of varying transparencies. Smoke leads the viewer's eye through a canvas and establishes links between characters and objects. A pipe can call attention to a character's hands, or mouth, and the spark from a flint lends drama to a canvas. Imitations of fire and smoke demonstrated the sort of virtuosity prized by collectors.

The authority of religious leaders in western culture, before the eighteenth century, was strong enough to curb glamorous representations of smoking. Seventeenth-century Protestant Holland, for example, was the nucleus for *vanitas*—still-life paintings that emphasize the foolishness of indulging the senses at the eternal soul's expense. In such a still life an abandoned pipe could refer to a departed owner, smoke to the brevity of life, and ashes to penance or death.

A number of paintings from this period depicted market or tavern scenes that showed the lower classes as governed by their appetites. The seventeenth-century Dutch painter Jan Steen, in *The Effects of Intemperance*, shows a woman lost in alcoholic slumber with a pipe dangling in front of her legs. Similar disorder is portrayed in Steen's *Topsy-Turvy World*, in which an unattended child in the background appears to be smoking a pipe. Such portrayals of women smokers and drinkers as neglectful of their children and their personal virtue are decidedly lacking in glamour. However, glamorous representations began to appear in the eighteenth century, especially in France, as philosophers supported critical attitudes toward religion and social habits.

VISUAL ARTS

Members of the eighteenth-century French aristocracy considered the pursuit of pleasure a privilege of their class. As trendsetters to the western world, they lent glamour to tobacco ingestion. During days filled with flirting, gambling, eating, and drinking, snuff-taking became as ritualized as tea-drinking in Japan. A courtier would open a gold **rococo** snuffbox with a flourish of a lace cuff, insert snuff in a nostril, and unfurl a lavishly embroidered handkerchief for a sneeze. Highly skilled craftsmen catered to the desire for glamorous tobacco accessories; these paved the way for elegant nineteenth- and twentieth-century **Art Nouveau** and **Art Deco** cigarette cases.

Tobacco use grew enormously in the nineteenth century thanks, in part, to the industrial revolution, which introduced rail transport to better distribute products and mass media to better publicize them. Major wars glamorized smoking, and self-assertive women began to acquire what had been a largely a male habit. Western travel literature linked smoking habits in the Near East with sensual fantasies. Moreover, as Paris became the art and pleasure capital of the world, its artists and writers frequently treated smoking as an attribute of modernity.

rococo an artistic and architectural style of the eighteenth century (1700s) characterized by elaborate ornamentation.

Art Nouveau the leading ornamental style from the 1890s through World War I. Art Nouveau employs idealized human figures together with natural elements like vines, acorns, and palms. Art Nouveau advertising copy and packaging was very popular with tobacco products, especially cigars.

Art Deco the most fashionable style of design in the 1920s and 1930s. Art Deco is usually characterized by geometric lines and shapes. Smoking tobacco tins, cigarette packages of this period were often rendered in the Art Deco style.

War and Smoking

As a repetitive, addictive act, which provides oral satisfaction and encourages camaraderie, smoking became endemic among the military during the nineteenth century. In the Crimean War (1853–1856) British soldiers learned from their Turkish allies to smoke cigarettes for their cheapness and convenience. They brought the habit home to England, where civilians began to associate smoking with Near Eastern exoticism and the imagined adventures of war.

Wars also served to strengthen the link between smoking and male virility. Caricaturists and pornographers, such as the Belgian painter and graphic artist Félicien Rops, often capitalized on the resemblance of pipes, cigars, or cigarettes to penises and of expelled smoke to ejaculations. In ads for cigars, the continued sexual vigor of a wealthy, elderly man could be implied by the upward tilt of his cigar or the vigorous ejection of smoke. The Austrian neurologist and founder of psychoanalysis Sigmund Freud, in his *Three Essays on Sexuality and Other Writings* (1901–1905), noted that "There is a constitutional intensification of the erotogenic significance of the labial region. If that significance persists, these same children, when they are grown up, will become epicures in kissing, will be inclined to perverse kissing, or, if males, will have a powerful motive for drinking and smoking."

The New Woman

In nineteenth-century fine and commercial art women were seldom shown in the act of smoking, unless they were such outsiders as actresses, prostitutes, degenerate society women, or so-called new women, who acted like men. Beautiful women were frequently shown in posters or on cigar box labels, but—with few exceptions—not with a smoking implement in their mouths. The blonde-haired woman depicted by the Art Nouveau Czech illustrator Alphonse Mucha in his

VISUAL ARTS

Decorative tendrils of hair swirl about the shoulders of a woman as she smokes a Job cigarette and slips into reverie. This 1896 Art Nouveau–style poster was created by Alphonse Mucha. © HISTORICAL PICTURE ARCHIVE/CORBIS

1896 Job cigarette poster swoons in a cigarette-induced orgasm, yet the cigarette does not touch her lips.

Antismoking campaigns, connected with temperance movements, railed against women smokers. The physician J. H. Cohausen argued that "Freedom in smoking and drinking goes with freedom in morals." Yet increasing numbers of women who wanted greater political and social privileges took on the male habit of smoking. The French novelist George Sand (the male pseudonym of Armandine Aurore Lucille Dupin) was known as a notorious smoker, and British actress Lily Langtry, perhaps the first actress to be photographed smoking, insisted on being included in after-dinner smoking parties, where men traded ribald stories over their cigars.

The British novelist Ouida also demanded to stay in the dining room when men lit up, pronouncing: "Smoke and drink as if you were at the club; talk as if you were in the smoking-room there." The heroine of Ouida's 1867 novel *Under Two Flags* is nicknamed Cigarette because she smokes, takes lovers, and is as brave as any of her foreign legion comrades.

In the 1890s Frances Benjamin Johnston, the first commercially successful American woman photographer, created a witty bohemian self-portrait in which she holds a cigarette in one hand, a beer stein in the other, and displays her petticoats. Her faint smile suggests that she enjoys flaunting social norms. These nineteenth-century female celebrities added an aura of glamour to smoking by women, thus foreshadowing the twentieth-century Virginia Slims ad campaigns.

Impressionism and Pleasure

The French painter Renoir was one of several impressionist artists who celebrated the pleasures of the senses. In his *Luncheon of the Boating Party* amorous, healthy-looking men and women communicate the joy of being alive. The man closest to the viewer is smoking a cigarette. Such art had a tremendous impact throughout Europe and America.

The lives of artists and writers struck many who lived through the industrial revolution as last bastions of freedom. In the vast majority of nineteenth-century magazine illustrations of artists at work, they are smoking cigarettes. Smoking as an aid to inspiration and as a sign of a glamorous, liberated lifestyle soon became set in the public mind. In the nineteenth-century English novelist Oscar Wilde's *The Picture of Dorian Gray* (1891) the protagonist Dorian Gray claims "A cigarette is the perfect type of a perfect pleasure. It is exquisite, and it leaves one unsatisfied."

Exoticism

During the 1880s, thousands of small cigar factories began to use labels to attract buyers. Cigar box labels, illustrated insert cards, and posters were often collected in albums for entertainment prior to radio and television. Such art was designed mainly for white manufacturers by white male artists for a white male audience.

Frequently, exotic women from countries associated with tobacco production were featured on these labels, and their images often implied that they were offering themselves along with the smoke. At this time it was a cliché for a man to speak of his cigarette or cigar as his lover. The English author Rudyard Kipling referred to his beloved Cuban cigars as "a harem of dusky beauties tied fifty in a string." The exotic woman expressed a sensuality that became identified with smoking products, which are held in the mouth to kiss and suck.

The women featured were most often Turkish, Spanish (including Cuban and Gypsy), Native American, or African. Of these, only the African woman is not usually shown as glamorous. A Turkish woman, calling to mind Scheherazade of *Thousand and One Nights*, typically lounged within a harem and had a nearby hookah to smoke. She appeared as a pleasure commodity to be bought and sold to satisfy male appetites, as were cigars.

Native American women were often shown offering tobacco leaves to European explorers or settlers, supporting the popular notion that uncivilized people were to provide raw materials to civilized ones. Native American women were also idealized as being in harmony with nature and were seldom shown in overtly sexual roles.

VISUAL ARTS

Spanish women could be femmes fatales reminiscent of the literary protagonist Carmen, who in Prosper Mérimée's 1845 novel worked in a cigarette factory and took pleasure in the masculine practices of sex and smoking. When Georges Bizet's opera of the same name opened in Paris in 1875, the audience was shocked to see cigarette girls smoking on stage as a male chorus sang "Look at them! Bold looks, flirting ways."

In contrast, blacks in nineteenth-century tobacco illustrations are typically shown as figures for ridicule, or as distant field workers overseen by a white plantation owner. Harem women, with whom the average European or American male had no contact, encouraged sexual fantasies, but Black women, who frequently had given birth to children sired by their white masters, did not.

Twentieth-Century Images of Glamour and Smoking

Cultivating brand loyalty became a key goal of twentieth-century advertising. Ad agencies developed more sophisticated ways to stimulate desire, drawing upon psychology and consumer surveys. Such agencies were particularly drawn to the movies and organized sports, whose stars were idolized by the public.

WARTIME ADVERTISING. During World War I and World War II tobacco ads featured the glamour of extreme experiences—foreign travel, heroism, and intense camaraderie. War's aura of glamour came largely from media-constructed images rather than from war's grim realities. In 1918 the U.S. War Department purchased the entire output of Bull Durham tobacco, and the company advertised, "When our boys light up, the Huns will light out." In a Camel ad of the 1930s, a middle-aged man smokes while watching his wife gift wrap cigarettes for their overseas soldier son and comments, "Like you sent 'em to me—remember?" The ad continues: "He recalls the times when he received the same gift from his wife. Camel. Yes, they were first with men in the army then—and they are today. Not only in the Army but in the Navy, the Marines, the Coast Guard . . . with the millions who stand behind them . . . for Camel is America's favorite."

THE MOVIES. During the 1920s and into the Depression years, Americans flocked to films that depicted glamorous lifestyles. They purchased magazines that depicted the extravagant homes of favorite actors and actresses. Within films, the glamour of smoking was enhanced by gorgeous costumes and settings. When the American actress Marlene Dietrich toyed with a cigarette set in a long holder she communicated emotions that ranged from boredom to passion.

The Motion Picture Production Code of 1930—which prohibited, among other things, "excessive and lustful kissing, lustful embraces, suggestive postures and gestures"— unwittingly encouraged greater use of smoking in movies. Directors found in smoking powerful ways to imply sexual intimacy. As Paul Henried lights Bette Davis's cigarette in the 1942 film *Now, Voyager*, their hands touch, and their glances are

VISUAL ARTS

In the 1947 film *Out of the Past* Jane Greer lights Robert Mitchum's cigarette. Greer's usage of cigarettes contributed to the seductive and sinister spirit of her film noir character. © THE KOBAL COLLECTION

full of desire. To step up the heat, Henried lights Davis's cigarette from the smoldering one he holds between his lips.

Smoking within the dark, film noir movies of the 1940s and 1950s might involve a startling burst of flame from a match and the snaking of smoke through a cellar. Gangsters, femme fatales, and private eyes all smoked in these films that linked smoking with the danger and glamour of living life on the edge.

Tobacco and Health

During the 1950s, as research established links between smoking and disease, tobacco ads featured health-oriented testimonials from celebrities. The film star Linda Darnel is featured in one such ad where she is shown smoking in a mirrored, neorococo dressing room and wearing a strapless red evening dress and diamond jewelry. When asked why she changed to Camels, she responds: "I found they got along wonderfully with my throat. I especially appreciate Camel mildness when I'm making a picture."

In 1955 Philip Morris introduced the cowboy figure of the Marlboro Man in their ads in an effort to counteract the effeminate stigma attached to filtered cigarettes.✦ Cowboy life in films and ads was glamorized as heroic and individualistic, a popular fantasy at a time when most men worked in urban environments or for large corporations. The campaign made Marlboro the top-selling filtered cigarette in the United States .

While sports heroes have played a role in the marketing of tobacco since the early twentieth century when they appeared on tobacco trading cards, with the proliferation of mass media—newspapers, magazines, radio, movies, and television—their influence became more appealing to

✦ See "Marlboro" for a photograph of Philip Morris advertisements featuring early versions of the rugged cattleman.

Tobacco in History and Culture
AN ENCYCLOPEDIA

cigarette manufacturers. In 1968 Philip Morris introduced its Virginia Slims cigarettes to America, and two years later the brand began sponsoring a women's tennis tournament. The tennis champion Billie Jean King, in 1973, wearing Virginia Slims colors, defeated Bobby Riggs in a highly publicized match, thus associating smoking with the glamour of victory.

The Cigarette Labeling and Advertising Act

In response to the 1964 U.S. surgeon general's *Report on Smoking and Health*, the following year Congress passed the Cigarette Labeling and Advertising Act, which called for a nationwide ban on tobacco advertising on radio and television. It also required the printing of the statement "Caution: Cigarette Smoking May Be Hazardous to Your Health" on all cigarette packages and in advertising.

The act encouraged groups and individuals to attack the aura of glamour that had been constructed around smoking. Some of the most effective efforts were made by celebrities who had smoked on television, such as Edward R. Murrow and William Talman, the prosecuting attorney in the *Perry Mason* series. Both made antismoking pleas, which were especially poignant after their deaths from lung cancer.

In 1992 Marboro shareholders received a visit from Wayne McLaren, who had modeled as their Marlboro Man, at their annual meeting. He told them he was dying of cancer and asked the company to voluntarily limit its advertising. The corporation, then boasting a market worth $32 billion, did not cooperate. McLaren's plea, however, ran in major newspapers. In 1976 a television program titled "Death in the West—the Marlboro Story," created for Britain's Thames Television by Peter Taylor and Martin Smith, interspersed Marlboro ads with interviews with cowboy smokers stricken with lung cancer. Legal problems in Britain led to suspension of initial showings, but in 1982 the film played nationwide in the United States.

Recent Developments

Now that public consciousness has been impressed with the dangers of smoking and the manipulations of tobacco companies, some television and film writers have used smoking to identify the bad guys in their works. In *The X-Files* (1993–2002), a popular television series that investigated paranormal phenomena, the protagonists Fox Mulder and Dana Scully were nonsmokers, but the truly sinister characters in the program—particularly the mysterious character called simply "Cigarette Smoking Man"—were chain smokers.

Writers have also used smoking to establish a character's willingness to take risks. In the 1999 film *Fight Club* the actor Brad Pitt plays Tyler Durden , a character invented, unknowingly, by Jack (played by Edward Norton) to act out Jack's suppressed desires. Jack does not smoke and works in an office where smoking is prohibited. His alter ego, Tyler, does smoke. While smoking, Tyler tells Jack, "I'm free in all the ways you want to be." Tyler organizes increasingly vicious fights among young men and gradually lures Jack into becoming both a smoker and fighter. The actress Helena Bonham Carter, who in the film

---

"Tobacco"

Tobacco is a dirty weed. I like it.

It satisfies no normal need. I like it.

It makes you thin, it makes you lean,

It takes the hair right off your bean.

It's the worst darn stuff I've ever seen.

I like it.

GRAHAM LEE HEMMINGER
(*PENN STATE FROTH*, 1915)

plays Marla, a woman desired by both sides of this split personality, smokes constantly. As she exhales, smoke pours from her open mouth, as if from an internal furnace.

Tobacco companies have developed ways to bypass the Cigarette Act. For example, producers of the film *License to Kill* accepted a $350,000 payment to have James Bond smoke Larks on camera, and the movie stars and role models Arnold Schwartzenegger, Bruce Willis, Demi Moore, and Pierce Brosnan have all appeared on the cover of *Cigar Aficionado* magazine. The cigarette manufacturer Freedom Tobacco International, in introducing its cigarette brand Legal, paid actresses to smoke the brand in Manhattan bars and nightclubs in the spring of 2003 to promote the brand. With so much money at stake, it seems likely that tobacco companies will continue to invent ways to make smoking appear glamorous.

See Also Advertising; Advertising Restrictions; Antismoking Movement From 1950; Arents Collection; Camel; Film; Intellectuals; Literature; Marlboro; Music, Classical; Music, Popular; Social and Cultural Uses; Virginia Slims.

▌DOLORES MITCHELL

BIBLIOGRAPHY

Apperson, G. L. *The Social History of Smoking.* London: M. Secker, 1914.

Arents, George, Jr. *Tobacco: Its History Illustrated by the Books, Manuscripts, and Engravings in the Library of George Arents Jr.*, 5 vols. Edited by Jerome E. Brooks. New York: Rosenbach Co., 1937–1952.

Kluger, Richard. *Ashes to Ashes: America's Hundred-Year Cigarette War, the Public Health, and the Unabashed Triumph of Philip Morris.* New York: Alfred A. Knopf, 1996.

Pearlstone, Zena. "Native American Images in Advertising." *American Indian Art* 20, no. 3 (summer 1995): 36–43 .

Warning Labels

Studies have provided evidence that warning labels on packages and in advertising of tobacco products, particularly if graphic and highly visible, can reduce smoking. Over time, these labels have become more direct and explicit in their warnings and larger in size in an attempt to discourage smoking.

United States

The practice of placing warning labels on tobacco products in the United States was initiated in 1965 as an outcome of the landmark 1964 report *Smoking and Health: Report of the Advisory Committee to the Surgeon General of the Public Health Service* (see sidebar). Although the report provided evidence of the negative effects of smoking on health, it did not make any specific policy recommendations. Within a week after its release, however, the Federal Trade Commission (FTC) proposed that a health warning be placed on cigarette packages and in advertisements.

Convinced that it would fare better in Congress than with the FTC because of the help of prominent tobacco-friendly members of Congress, the tobacco industry encouraged legislative action, and in 1965 the Cigarette Labeling and Advertising Act (PL 89–92) was passed. This law required a health warning on cigarette packages, but suspended the FTC's proposed warnings in advertising. It also prohibited other federal agencies from requiring health warnings in advertising and prohibited state and local governments from enacting requirements for more stringent regulation. Because of these latter provisions, the bill was viewed by critics such as Senator Frank Moss (D-Utah), Federal Trade Commissioner Philip Elman, and the *New York Times* as a protection for the industry.

Beginning in 1966, cigarette packages were required to carry a label that read "Caution: Cigarette smoking may be hazardous to your health." The cautious wording of this statement was replaced in 1970 as the result of the passage of the 1969 Public Health Cigarette Smoking

Dr. Luther L. Terry

After President John F. Kennedy appointed him surgeon general in 1961, Dr. Luther L. Terry spearheaded the first surgeon general's report on smoking and health, which conclusively linked smoking to lung cancer and chronic bronchitis. Terry ensured that the report received substantial media attention by organizing a press conference to announce his committee's highly anticipated findings on 11 January 1964. The report also prompted Congress to take action by passing the *Cigarette Labeling and Advertising Act of 1965*, which mandated the first surgeon general's warning on cigarette packages. Despite stepping down as surgeon general in 1965, Terry maintained a strong crusade against the tobacco industry until his death in 1985. He successfully lobbied for the 1971 end to radio and television cigarette advertising, and later helped lead the way to ban cigarettes from the workplace. The American Cancer Society recognizes individuals worldwide for exemplary work in tobacco control and prevention through their Luther L. Terry Awards.

▮ DONALD LOWE

Act (PL 91–222), which changed the label to the stronger "Warning: The Surgeon General has determined that cigarette smoking is dangerous to your health." That legislation also banned all television advertising of cigarettes as of 1971 and empowered the FTC to consider the inclusion of warning labels in all other forms of advertising within six months after the broadcast ban went into effect.

Preferring health warnings that were vague, the tobacco industry had defeated an effort to add a reference to lung cancer and other diseases in the labels mandated by the Public Health Cigarette Smoking Act. The industry was not as successful on this issue in 1984, however, when new legislation required that cigarette packages carry four rotating messages (all of which were expressed as Surgeon General's warnings) that were to appear in type that was 50 percent larger than before. One of the rotating messages specifically stated that smoking causes lung cancer, heart disease, and emphysema. The other warnings stressed that smoking could have negative effects on a fetus, that cigarette smoke contains carbon monoxide, and that quitting smoking greatly reduces serious health threats. The industry succeeded in blocking efforts to include any reference to addiction to tobacco in the proposed rotating labels. The 1984 legislation also extended the reach of the warning labels to most forms of cigarette advertising.

Under the Comprehensive Smokeless Tobacco Health Education Act (PL 99–252) of 1986, the FTC required that warning labels also appear on all packaging and in all advertising for smokeless tobacco products. This act also banned broadcast advertising of smokeless tobacco products. Although the advertising of cigars and little cigar products was banned from radio and television by amendments in 1971 and 1973 to the Cigarette Labeling and Advertising Act, no warning labels were required under these amendments. In 2000, however, the Federal Trade Commission reached a settlement with seven of the largest American cigar companies requiring health warnings on cigar products and in various types of advertising for these products.

In the 2000s, many tobacco critics are not convinced that the warning labels on cigarettes are effective in the effort to reduce the number of smokers. They want larger and more graphic warnings with pictures that will readily attract the attention of smokers. Several bills to make warning labels more prominent have been introduced into Congress since 2000.

Other Countries

The practice of placing warning labels on tobacco products was adopted by numerous countries from the 1970s to the 1990s. Health warnings were introduced on cigarette packages, for example, in the United Kingdom in 1971 (at first voluntarily by tobacco companies, but later mandated by law), in Australia in 1972, in Thailand in 1974, and in Singapore in 1980. By 1991, 77 countries required such warnings, although in most cases the warnings were not very strong. In Japan, for instance, cigarette packages from the late 1980s until 2003 carried only the vague warning, "Since smoking might injure your health, let's be careful not to smoke too much." In some developing countries, either no warnings or only rudimentary health warnings were required, and American tobacco companies were criticized by

WARNING LABELS

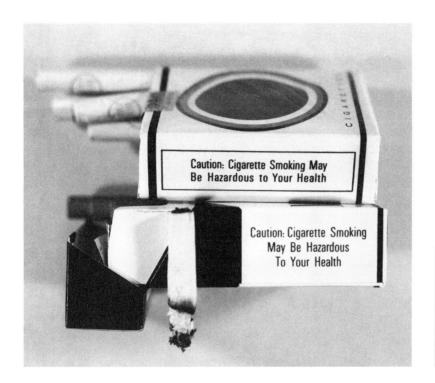

In the United States, early warning labels in the 1960s carried the mandatory warning label, "Caution: Cigarette smoking may be hazardous to your health." By 2002, warning labels in the United States and other countries were larger in size, were more explicit in their message, and sometimes included graphic images.
© AP/WIDE WORLD PHOTOS

groups such as Public Citizen and the Center for Communications, Health, and the Environment for meeting only the minimum local requirements on cigarette packages sold in these nations rather than including the stronger warnings required on cigarettes sold in the United States.

During the 1990s, a number of countries strengthened the warnings on tobacco products so that the United States was no longer the leader in this field. The 2000 Surgeon General's report *Reducing Tobacco Use* concluded that by the time the report was compiled American practices in this area were less restrictive than those of several other countries. The toughest warning labels were mandated in Canada. By 2001, all cigarette packages in that country were required to carry strong rotating warning labels (such as, "Cigarettes are highly addictive") accompanied by images, which in some cases were highly graphic (for example, an image of a human heart with clogged arteries).

In 2001, the European Parliament adopted stringent regulations concerning warning labels for member countries of the European Union. The size of the warnings was increased from a minimum of 4 percent to at least 30 percent of the front and 40 percent of the back surfaces of cigarette packages. The required warning statements were blunt, such as "Smoking kills." Smokeless and oral tobacco products were also required to display stronger and more visible warnings. The European Union Commission later developed guidelines to allow member states to add graphic images to tobacco packaging.

The effort to promote stronger warning labels was extended to a global level with the unanimous passage of the Framework Convention on Tobacco Control by the 192 nations of the World Health Organization in 2003. Along with a variety of provisions, this treaty obliges nations that sign it to adopt and implement large, clear, visible, and rotating

WARNING LABELS

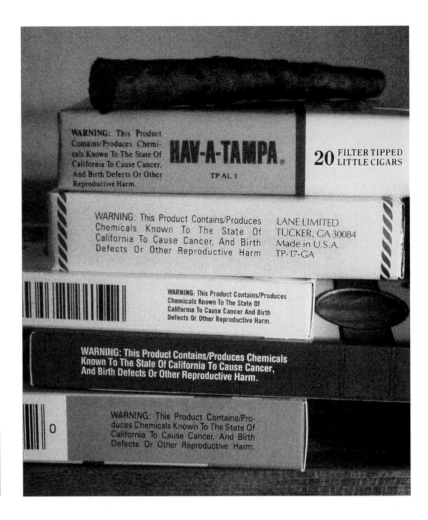

These cigar boxes from California contain warnings of the health hazards associated with the chemicals they contain.
PHOTOGRAPH BY ROBERT J. HUFFMAN.
FIELD MARK PUBLICATIONS

health warnings that occupy at least 30 percent of the display area on a tobacco product package. The United States delegation was the last to accept the treaty. As of 2004, the treaty has not been signed by the president or ratified by Congress. Some observers believe that tobacco companies in the United States and other nations will lobby heavily against the signing of the treaty.

See Also Cigarettes; Disease and Mortality; Regulation of Tobacco Products in the United States

▮ JOHN PARASCANDOLA

BIBLIOGRAPHY

Action on Smoking and Health. Available: <http://www.ash.org.uk/html/regulation/html/warnings.html>.

Centers for Disease Control and Prevention. Available: <http://www.cdc.gov/tobacco/sgr/sgr_2000/factsheets/factsheet_labels.htm>.

Jha, Prabhat, and Frank Chaloupka. *Curbing the Epidemic: Governments and the Economics of Tobacco Control.* Washington, D.C.: World Bank, 1999.

Kluger, Richard. *Ashes to Ashes: America's Hundred-Year Cigarette War, the Public Health, and the Unabashed Triumph of Philip Morris.* New York: Vintage Books, 1997.

U.S. Department of Health and Human Services. *Reducing Tobacco Use: A Report of the Surgeon General*. Washington, D.C.: U.S. Department of Health and Human Services, 2000.

Water Pipes *See* Pipes.

Women

The history of tobacco consumption is gendered and enmeshed in sexual politics. Cultural rules about who can use tobacco, in what form, and in what contexts, have often been related to ideas about masculinity, femininity, and appropriate gender relations. Cigarette smoking in particular has, since the late 1800s, often been used and interpreted as a symbol of women's emancipation. However, from the 1970s, amidst growing fears about the health risks of tobacco use, the global expansion of girls' and women's smoking has been widely viewed as a response to persistent gender inequalities.

Gendered Tobacco Consumption in Europe and North America

In preindustrial Europe and North America, there is no evidence of any proscription of tobacco use by gender; in fact, tobacco was regarded by some Europeans as beneficial for pregnant women. Where tobacco was used by men, it was also used by women. Women from all classes smoked in seventeenth- and eighteenth-century Britain, but it was not common except among "humbler" folk in areas such as Cornwall and the west of England (Apperson 1914). **Snuff**-taking was, however, a luxury enjoyed by women in the upper echelons of eighteenth-century England. Snuff was also popular among the French, including the "noblest ladies" of the French Court and the middle-class women who "imitated them" (Schivelbusch 1992). Dutch and French paintings also suggest that elegant women smoked clay pipes at this time (Goodman 1994). In colonial America it was also not unusual for women to smoke pipes, or to take snuff. In 1686 a French traveler in the English-American colonies noted that women smoked everywhere, including church. Although women were not prevented from smoking there is some evidence of gender-specific restrictions. For example, in colonial America it was acceptable for women to smoke in their homes or on their doorsteps but not in taverns, particularly in the company of strangers. In seventeenth-century London, pipe smoking in public was not considered appropriate for all women: "gentlewomen . . . and virtuous women accustom themselves to take it [tobacco] as medicine, but in secret" (Goodman).

The onset of industrialization in the late eighteenth and nineteenth centuries had far-reaching implications for gender identities and gender relations. Industrialization resulted in a shift of production from the

snuff a form of powdered tobacco, usually flavored, either sniffed into the nose or "dipped," packed between cheek and gum. Snuff was popular in the eighteenth century but had faded to obscurity by the twentieth century.

WOMEN

Actress Helena Bonham Carter (right) lights a cigarette in her role as the independent-minded Kate Croy in the 1997 film *The Wings of the Dove*. The film, based on the novel by author Henry James, takes place in the early 1900s. PHOTOGRAPH BY MARK TILLIE. THE KOBAL COLLECTION

home to the workplace, the establishment of "separate spheres" for men and women, and a pronounced sexual division of labor. These arrangements were buttressed by an ideology of sexual difference whereby women were constructed as guardians of morality, the home, and the family. The pronounced gender structuring of industrializing societies, and the distinctive meanings attached to being male and female, were accompanied by shifts in the gender dimensions of tobacco consumption. Smoking became firmly associated with men and masculinities; respectable women did not smoke, and respectable men did not smoke in the presence of women. Smoking was seen as "unwomanly" as it endangered female reproductive health and was inconsistent with women's roles as mothers and guardians of morality. Concerns about women's smoking were fuelled by assumptions about the greater vulnerability of women than men to both the health risks of smoking and the lure of cigarettes.

Despite these common ideas about smoking, some women continued to smoke but they were usually placed outside the boundaries of respectable womanhood. In Britain, old working-class women sometimes smoked clay pipes; smoking was also common among Irish women street traders in London and women laborers in Northumberland and the Scottish border. In the United States, poor European immigrants smoked cigarettes. While poor women smoked principally for the pleasures it afforded, some radical upper-class women deliberately used tobacco to make a statement. When George Sands, the bisexual French writer, smoked in public and the American "femme fatale" Lola Montez posed in 1851 for a photograph with a cigarette in her gloved hand, both flaunted conventions of "respectable" womanhood.

Although "respectable ladies" did not smoke in the nineteenth century, tobacco use was often linked, at a symbolic level, to male ownership and/or control of women's bodies. Tobacco has long been personified as a woman and in the nineteenth century it was a cliché for a man to refer to cigars or cigarettes as lovers or mistresses. Rudyard Kipling described his cigars as "a harem of dusky beauties." This description

conveys the association between the color of tobacco and that of an "exotic" woman's skin and the captivity of both the cigars and the woman, awaiting the "owner's" needs (Mitchell 1992). These associations were visible in nineteenth-century pornography, commercial art, and high culture.

Seductive women were central to the marketing of tobacco products to men. "Exotic" women—Turkish, Spanish, Gypsy, Native American, and African—were commonly used to decorate American tobacco cartons. Labels depicting seductive gypsy women were designed to appeal to "the buyer who wanted an aura of danger along with his smoke—these were fiery women who only the most virile of males might dare to take on" (Mitchell). The sexual associations of smoking were enhanced by the use of **cigarette cards** which featured photos of scantily clad actresses and other beautiful women.

Links between smoking and sexuality were also established in opera, most notably *Carmen*, which was popular in Britain and the United States in the 1880s and 1890s. In *Carmen*, as the scholar Dolores Mitchell described, the cigarette factory is like a harem, as men cannot enter without a permit because of women's state of undress, and the heroine is a temptress who enslaves her lovers. The English novelist Ouida's *Under Two Flags* (1867), and the female character "Cigarette," similarly aligned smoking with female sexuality.

Two sailors from HMS Fury share a smoke with two women while they are ashore, UK, 1935. © HULTON-DEUTSCH COLLECTION/CORBIS

cigarette cards paper trading cards sometimes featuring sports personalities or movie stars packaged with cigarettes and offered as an incentive for purchase.

WOMEN

Cigarettes and the "Emancipation" of Western Women, Pre-1950

Smoking was principally a male pursuit in industrializing countries at the end of the nineteenth century even though it was increasingly popular among women from the upper and professional middle classes in urban centers of North America, Australia, and Europe. The growth of smoking among women was facilitated by the emergence, and accessibility, of manufactured cigarettes from the 1880s. Also, the availability of mild tobacco blends made cigarettes more palatable. However, wide-reaching social and economic changes in the position of women and in their expectations were the main reasons why women increasingly wanted to, and could, smoke. Upper-middle- and upper-class women used smoking as a visual statement of rebellion against the prevailing domestic ideal and as a critique of gender relations that positioned women as unequal, and subordinate, to men. Cigarette smoking symbolized a woman's emancipation and her claim to privileges previously the preserve of men. One alarmist German newspaper in the 1840s referred to "miniature George Sands" smoking cigars in the streets of Berlin as a sign of their emancipation; caricatures of women smoking in public also appeared in the French press of the 1840s (Schivelbusch). In Britain, from the 1880s, cigarettes were advertised in the suffrage press, and some women's clubs provided smoking rooms for their members.

World War I played a significant role in further shifting gender patterns of tobacco use in Western countries due to the disruption it caused to the public/private distinction and conventional notions of femininity and gender relations. The war introduced women to employment previously deemed as men's work and demonstrated their competence in these areas; fostered women's confidence and spurred a demand for equality and a rejection of prewar gender conventions; increased women's disposable income; and fostered rebellion against bourgeois values and the embrace of a modern world order. In the United States, the war de-stigmatized cigarette smoking for both sexes and aligned it with patriotism. More widely, the war increased women's exposure to tobacco as they collected cigarettes to send to soldiers or, in the course of nursing, lighted cigarettes for the wounded. For women on the home front as well as on the military one, smoking served also as a symbol of their newfound status.

In the 1920s cigarettes were smoked by progressive and professional women, but they were most visibly associated with young and affluent "modern girls" and "flappers." Women's smoking was, however, still a minority practice; in 1924, it constituted only 5 percent of national tobacco consumption in the United States and 1.9 percent in Britain. Although increasingly accepted, women's smoking still attracted criticism. British, American, and Australian women smokers were often ridiculed by men on the grounds that they were unable to smoke properly. In Britain, and possibly elsewhere, women were also criticized when they smoked like men; the thin line between acceptable and nonacceptable smoking was drawn in relation to the perceived threat which smoking posed to gender and gender relations.

Attractive images of women smokers became increasingly visible during the interwar period, following the establishment of cigarette smoking among women of the higher social classes. Advertisements featuring women holding, or smoking, cigarettes began to appear in

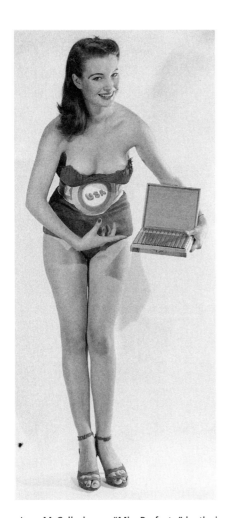

June McCall, chosen "Miss Perfecto" by the California Cigar Retailers Association, displays a box of "stogies," Los Angeles, 1950. She won the title, according to the cigar men, because in this costume she exhibits "quality wrapper, snug binder, and superb filler." © BETTMANN/CORBIS

upper-class British and Australian magazines around the time of World War I and became a staple of middle-class women's magazines by 1930 in Britain, and a few years later in Australia. U.S. advertising was slower to target women directly because the tobacco industry was cautious not to attract the attention of the anticigarette lobby. Advertisements for Milo Violets targeted North American women in 1918, but advertisers of mass produced brands were more cautious in doing this. The first notable advertisement was in 1926 when a woman was depicted accepting a Chesterfield cigarette. This was followed in 1927 by an advertisement for Marlboro which portrayed a woman smoking. By World War II cigarette advertisements appeared in most middle-class North American women's magazines.

Scholars debate the significance of advertising as it relates to the growth of women's smoking in the early 1900s though it is widely accepted that the proliferation of images of women smokers did provide people with favorable ways of understanding this practice. Some commentators have argued that news coverage of early women smokers played a more important role than advertising in popularizing women's smoking. Film has also been identified by some as more significant. In American and British film, the woman smoker became commonplace by 1930 and heroines, more so than female villains, brandished cigarettes.

By 1939 cigarette smoking was established as a respectable feminine practice, at least among the upper and middle classes. However, smoking was not universally respectable. In Britain, smoking among working-class women continued to have strong associations with loose sexual behaviour; not surprisingly, "nice" working-class girls did not smoke (Tinkler 2003). The scholar Michael Koetzle noted that in Germany, and the West more generally, there remained an "ambivalence" about women smoking. A thin line existed between on the one hand, modern, fashionable, elegant smoking and, on the other hand, lasciviousness. Harping back to earlier sexual associations in the history of tobacco, sensuality continued to be part of the symbolic repertoire of cigarette smoking for women.

World War II served as a catalyst for a dramatic increase in women's smoking and by the late 1940s roughly one-third of British and U.S. women smoked cigarettes as did one-quarter of Australian women. Nevertheless, forms of tobacco consumption remained highly gendered throughout the twentieth century. Cigar and pipe smoking remained firmly masculine smoking practices in Britain and the United States. In Australia, where tailor-made cigarettes were not popular with men prior to the 1950s, pipes and roll-your-own cigarettes marked the gender boundaries of smoking before and after 1950. Some women did use "masculine" products, but they were invariably of low social status. Pipes and roll-your-own cigarettes were, for example, smoked by some Australian women, usually aboriginal women or "rough" white women (Tyrrell 1999). In interwar Britain, clay pipes and cigars were smoked by gender and sexual rebels, especially lesbians, for example the artist Gluck, the speed boat racer Joe Carstairs, and the novelist Radclyffe Hall; **briar** pipes and cigars were also commonly smoked by lesbians in the 1950s. While in the West the cigarette was unisex, the consumption of other forms of tobacco, such as cigars, pipes, and chewing tobacco, remained firmly masculine in association and continued to signify sexual difference.

briar a hardwood tree native to southern Europe. The bowls of fine pipes are carved from the burl, or roots, of briar trees.

Cigarette Smoking on a Global Scale, 1970–Present

Until the late twentieth century, high-income countries generally had higher rates of smoking among women than did low- and middle-income countries. A difference in the proportion of males and females that smoke is also more pronounced in low-income countries. Part of the explanation for this pattern lies in different cultural traditions that are sometimes, as in Islamic states, underpinned by religious proscription. Traditional forms of tobacco consumption are, however, still prevalent among women in mainly rural communities in parts of Asia, Africa, and the Middle East. In India, for example, some women smoke **bidis,** chew pan (tobacco mixed with betel nut and lime), or smoke reverse-chutta (the lit end of a long cigar that is inserted into the mouth and puffed on).

bidis thin, hand-rolled cigarettes produced in India. Bidis are often flavored with strawberry or other fruits and are popular with teenagers.

Urbanization and shifts in the position of women after World War II have usually been accompanied by dramatic increases in women's cigarette smoking. In Europe socioeconomic and political changes have been aligned with increases in women's smoking. After the fall of the Generalissimo Franco regime in Spain, the proportion of women smoking increased from 17 percent in 1978 to 27 percent in 1997. In former socialist countries there were also notable changes. In Lithuania the proportion of women smoking doubled over a five-year period in the 1990s. In former East Germany between 1993 and 1997, rates of smoking among twelve- to twenty-five-year-old women rose from 27 percent to 47 percent. Increased tobacco consumption by young women has also been noted among the more developed countries in the West Pacific region. Commentators frequently refer to an "epidemic" of smoking in which women and young people are increasingly implicated.

In general, proportionally more men than women smoke. However, in high-income countries in the 1980s the proportion of women smoking began to converge with that of men. One example illustrates this trend. In Britain in 1948, twice as many men as women smoked (including cigars and pipes as well as cigarettes), and one and a half times as many men as women smoked cigarettes. By 1990, 38 percent of men and 31 percent of British women smoked, and 31 percent of men and 29 percent of women smoked cigarettes. At an international level, the convergence of male and female levels of smoking has been attributed to two key processes. First, levels of smoking in many "developed" countries decreased more quickly among men than women. In the "rich world" in the early 1980s, the proportion of men who smoked was falling in 19 out of 22 countries while the proportion of women who smoked was rising or stable in 11 of them. Second, among young smokers the gender balance was often reversing. Whereas before 1980 more boys than girls smoked, by the 1980s girls were often more likely to be smokers than their male peers. Young women in their teens and early twenties emerged as a fast growing group of smokers. Among Spanish fourteen- to eighteen-year-olds in 2000, 25 percent of males smoked compared to nearly 36 percent of females, and among fifteen-year-olds in Britain in 2003, 19 percent of boys smoked compared to 25 percent of girls.

In the context of well-publicized health risks of smoking in many Western countries, cigarettes were increasingly perceived as "ladykillers" rather than as "torches of freedom"; smoking was now a feminist issue.

The questions addressed were twofold: Why did women smoke when faced with information about the health risks of smoking, and why was smoking more attractive to girls than boys?

Explanations are not straightforward. A common theme was the importance of smoking for girls and women as a means of coping with the material and cultural constraints and pressures that arose from gender inequalities. Gender inequalities are manifest in a range of historically shifting forms and, for some women, this type of inequality was cross-cut by other forms of disadvantages that correlated usually with high rates of smoking such as a low social class and, in some countries, ethnic minority status. Managing a combination of paid work and childcare responsibilities featured as a specifically gendered characteristic of German women smokers in the late 1970s. In Britain, smoking emerged as a coping mechanism "when life's a drag" for white working-class women, particularly those with childcare responsibilities living in disadvantaged circumstances (Graham 1993). Similarly, for Australian and Canadian women the cigarette was an aid to coping with gender-specific feelings and experiences. Another explanation offered for the increased number of female smokers has been the tobacco industry's deftness in targeting women.

Targeting Women

Cigarettes were usually targeted at a mass market until the 1960s, although there were some experiments with specifically feminine cigarettes in Europe and North America before this. Rose Tips and Milo Violets were among the brands sold to North American women before World War I. Similarly in Britain there were a range of ladies' cigarettes including ones that were perfumed to order with tips covered in silver paper. In the interwar years these cigarettes were eclipsed by mass-produced brands. Specifically female cigarettes were tried unsuccessfully, for example, Fems in the United States. Producers also experimented with feminized versions of mainstream brands such as red-tipped cigarettes. As an advertisement for Minors Red Tips in a 1937 issue of *Woman* magazine explained, red tips are "the brilliant notion that prevents lipstick from showing on a cigarette, and helps men to preserve their beautiful illusions . . . Red tips for Red lips" (Tinkler 2001a).

It was not until 1968 that a specifically feminine cigarette was successful. Virginia Slims, launched by Philip Morris, was the eleventh best-selling cigarette in the U.S. in 1983 and quickly became established internationally. Other "female brands" followed include Kim, Capri, and, in Brazil, Charm cigarettes, which were Brazil's twelfth best-selling brand in 1983. Since the 1970s other brands have also been targeted primarily at women, and filter tips, along with low-**tar** and low-nicotine cigarettes, have been marketed as having feminine appeal.

Gender-specific marketing strategies have been central to targeting women as cigarette smokers. As early as 1929, debutantes brandishing lit cigarettes—their "torches of freedom"—were paraded through New York in an effort by one tobacco company to shift the taboos against women smoking in the streets and to thereby increase the number of cigarettes women could smoke. Women's magazines have also been enlisted to promote smoking to women through advertising.

tar a residue of tobacco smoke, composed of many chemical substances that are collectively known by this term.

WOMEN

Gender themes have a high profile in cigarette advertising. Smoking was aligned with weight loss in the promotion of Lucky Strike cigarettes in the United States in 1928, "Reach for a Lucky instead of a sweet." In the 1960s, with the launch of Virginia Slims, this angle was again brought to prominence with the declaration, "Fat smoke is history" (Greaves 1996). More generally, cigarettes have been associated with heterosexual attractiveness, romance, and glamour.

The association of women's smoking with emancipation and modernity is a particular common theme and one that has been surprisingly resilient and adaptable to a wide range of cultural contexts since the early twentieth century. Following its first appearance in interwar advertising, it reemerged in the context of the Women's Liberation Movement of the 1960s and 1970s. Olivier advertisements, for example, declared to British women that "in a man's world she smokes Olivier" (Tinkler 2001a). In the 1980s and 1990s, cigarettes have similarly been promoted as a sign of emancipation and also Western-style modernization in the former Eastern Germany and USSR, Japan, Hong Kong, China, and parts of Africa.

See Also Advertising; Age; Class.

▮ PENNY TINKLER

BIBLIOGRAPHY

Amos, Amanda, and Margaretha Haglund. "From Social Taboo to 'Torch of Freedom': The Marketing of Cigarettes to Women." *Tobacco Control* 9 (2000): 3–8.

Apperson, George L. *The Social History of Smoking.* London: Martin Secker, 1914.

ASH Factsheet no.1. January 2004. Available: <http://www.ash.uk/html/factsheets>.

Brandt, Allan M. "Recruiting Women Smokers: The Engineering of Consent." *Journal of the American Women's Association* 51, nos.1–2 (1996): 63–66.

Dale, Edgar. *The Content of Motion Pictures.* New York: Macmillan, 1935.

Ernster, Virginia L. "Mixed Messages for Women: A Social History of Cigarette Smoking and Advertising." *New York State Journal of Medicine* (July 1985): 335–340.

Goodman, Jordan. *Tobacco in History. The Cultures of Dependence.* London: Routledge, 1994.

Graham, Hilary. *When Life's a Drag: Women, Smoking and Disadvantage.* London: HMSO, 1993.

Greaves, Lorraine. *Smoke Screen: Women's Smoking and Social Control.* London: Scarlet Press, 1996.

Hilton, Matthew. *Smoking in British Popular Culture 1800–2000: Perfect Pleasures.* Manchester, U.K.: Manchester University Press, 2000.

Jacobson, Bobbie. *Beating the Ladykillers: Woman and Smoking.* London: Pluto, 1986.

Jacobson, Bobbie, and Amanda Amos. *When Smoke Gets in Your Eyes.* London: British Medical Association Professional Division, Health Education Council, 1985.

Koetzle, Michael. *Feu d'Amour: Seductive Smoke.* Koln: Benedikt Taschen, 1994.

Mitchell, Dolores. "Images of Exotic Women in Turn-of-the-Century Tobacco Art." *Feminist Studies* 18, no. 2 (1992): 327–350.

Schivelbusch, Wolfgang. *Tastes of Paradise: A Social History of Spices, Stimulants, and Intoxicants.* Translated by David Jacobson. New York: Pantheon Books, 1992.

Schudson, Michael. *Advertising: The Uneasy Persuasion: Its Dubious Impact on American Society.* New York: Basic Books, 1985.

Shafey, Omar et al., eds. "Tobacco Control Country Profiles 2003." Online. 2003. Available: <http://www.globalink.org.tccp>.

Tate, Cassandra. *Cigarette Wars: The Triumph of "the Little White Slaver."* New York: Oxford University Press, 1999.

Tinkler, Penny. "'Red Tips for Hot Lips': Advertising Cigarettes for Young Women in Britain, 1920–1970." *Women's History Review* 10 (2001a): 249–272.

———. "Rebellion, Modernity and Romance: Smoking As a Gendered Practice in Popular Young Women's Magazines, Britain 1918–1939." *Women's Studies International Forum* 24 (2001): 1–12.

———. "Refinement and Respectable Consumption: The Acceptable Face of Women's Smoking in Britain, 1918–1970." *Gender & History* 15, no. 2 (2003): 342–360.

Tyrrell, Ian. *Deadly Enemies. Tobacco and Its Opponents in Australia.* Sydney; University of New South Wales Press, 1999.

Wald, Nicholas et al., eds. *UK Smoking Statistics.* New York: Oxford University Press, 1988.

Warsh, Cheryl K. "Smoke and Mirrors: Gender Representation in North American Tobacco and Alcohol Advertisements before 1950." *Social History* XXXI, no. 62 (1998): 183–222.

 # Youth Marketing

Publicly the tobacco companies have always maintained that their advertising does not target youth. However, beneath the layers of public relations rhetoric, one cannot escape the essential fact that cigarette makers are in business to make a profit, and their profits depend on their ability to recruit new smokers to replace those who quit or die annually. Surveys show that in the United States 60 percent of smokers start by age sixteen, and fully 90 percent begin their smoking careers by age twenty. Few smokers begin smoking after age twenty-five. Brand loyalties are usually established during the first few years of smoking, with relatively few smokers switching brands annually.

Get 'em While They're Young

Internal industry documents reveal a clear understanding that the financial success of companies and given cigarette brands historically have depended in large measure on the percentage of new, primarily teenage, smokers that could be captured annually. A 1984 report from R.J. Reynolds stated the importance of young smokers as follows:

> *Young adult smokers have been the critical factor in the growth and decline of every major brand and company over the last 50 years. They will continue to be important to brands/companies in the future for two simple reasons: 1) the renewal of the market stems almost entirely from 18-year-old smokers, no more than 5% of smokers start after age 24; and 2) the brand loyalty of the 18-year-old smokers far outweigh any tendency to switch with age*
>
> (BURROWS).

Industry documents show that each company carefully researched and knowingly implemented marketing strategies to appeal to a teenager's normal desires to experiment, relieve stress, enhance self-confidence, rebel against authority, and be accepted among peers. According to a 1969 draft report to the Philip Morris Board of Directors, "a cigarette for the beginner is a symbolic act. I am no longer my mother's child, I'm tough, I am an adventurer, I'm not square" (Wakeham).

Saturate the Airways

Cigarette companies have always been among the earliest to pick up on new trends and exploit them through the mass media. In the 1930s, they were among the major advertisers on the radio. The American Tobacco Company promoted its Lucky Strike brand through the sponsorship of many musical and comedy shows, including the *Jack Benny Show* and the *Lucky Strike Hit Parade*. R.J. Reynolds sponsored the *Camel Pleasure Hour, Camel Caravan* and the *All-Star Radio Review*, all shows that had a large following among young people. From 1930 to 1950, it was common for public figures to endorse cigarette brands. For example, a 1934 advertisement for Camel cigarettes proclaimed that twenty-one out of twenty-three of the World Series Champion St. Louis Cardinals smoked Camels.

In the 1950s, cigarette companies began to move their advertising dollars from radio to television. In 1950 the industry sponsored over seven hours of television per week, which increased to 125 hours by 1963. In 1963 cigarette companies sponsored fifty-five television programs. Many of these shows had a large percentage of young viewers, as evident from the percentages of viewers less than twenty-one years of age:

TV Show	Sponsor	Percent
McHale's Navy	R.J. Reynolds	40
The Beverly Hillbillies	R.J. Reynolds	38
The Jackie Gleason Show	Philip Morris	38
The Red Skelton Show	Philip Morris	37
Wide World of Sports	Lorillard	38
The Dick Van Dyke Show	Lorillard	33
The Outer Limits	Liggett & Myers	46
The Price Is Right	Liggett & Myers	32
Combat	American Tobacco	46
The Twilight Zone	American Tobacco	30

(POLLAY).

Industry sponsorship of televised sports was also common. In 1963 R.J. Reynolds sponsored eight and American Tobacco sponsored six major league baseball teams. Philip Morris sponsored NFL games, and Brown & Williamson sponsored college football bowl games. In 1964

YOUTH MARKETING

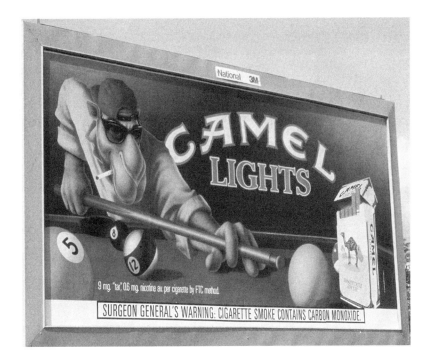

Camel Cigarette advertisement on a billboard, showing Joe Camel playing pool, a pack of Camel Lights prominently displayed in the foreground of the pool table. Launched in 1988 by R.J. Reynolds, the Joe Camel campaign met with immediate success, and controversy, in increasing sales in the 18–24 age group. When the campaign finally was phased out, sales fell dramatically. © JOEL W. ROGERS/CORBIS

Lorillard sponsored the Olympics and was a sponsor of the popular *Wide World of Sports* show on ABC. Overall, the switch to television helped transform cigarette advertising from words into pictures and information into images. Part of the tremendous success of Marlboro cigarettes has been attributed to the use of the cowboy image, which when it was first used in the 1960s, captured the youth market's desire for a symbol of independence and rebellion.

Sports, Movies, and MTV

A 1971 federal law prohibiting broadcast advertising of cigarettes merely caused cigarette makers to redirect their advertising to other media, including newspapers, magazines, billboards, and motion pictures, along with direct sponsorship of sporting and musical events.◆ Cigarette makers quickly learned that sponsorship of sporting and cultural events could be more cost-effective than any thirty-second television spot had ever been, especially since the association with sporting events conveyed a healthy, youthful image. For example, a 1987 Philip Morris report discussed the value of sponsoring auto racing:

◆ See "Sponsorship" for photographs of sporting and cultural events sponsored by tobacco companies.

> *Marlboro 500 at Michigan International Speedway was highly successful in creating brand awareness and generating positive publicity. The PM Sales Force did an exceptional job in placing banners and P.O.S. [point of sale] material in the surrounding area, as well as conducting sampling activities at the track itself. The race was broadcast live on ABC-TV, and Marlboro signage was visible throughout the 4 hour telecast*
>
> (PHILIP MORRIS COMPANIES).

Industry documents also reveal a strategic interest in placing youth-oriented cigarette brands, promotions, and advertising in locations where young people congregate, like the movies. A 1979 Philip

Most popular cigarette brands with U.S. teenagers, 1953–2002

Year	Brand	Market share among teenage smokers*	Cigarette company
1953	Lucky Strike	33%	American Tobacco
	Camel	17%	R.J. Reynolds
	Pall Mall	12%	American Tobacco
1959	Winston	12%	R.J. Reynolds
	Kent	11%	Lorillard
	Pall Mall	10%	American Tobacco
1964	Winston	32%	R.J. Reynolds
	Marlboro	13%	Philip Morris
	Pall Mall	12%	American Tobacco
1974	Marlboro	39%	Philip Morris
	Kool	24%	Brown & Williamson
	Salem	10%	R.J. Reynolds
1983	Marlboro	58%	Philip Morris
	Newport	13%	Lorillard
	Salem	7%	R.J. Reynolds
1993	Marlboro	61%	Philip Morris
	Camel	13%	R.J. Reynolds
	Newport	13%	Lorillard
2002	Marlboro	50%	Philip Morris
	Newport	25%	Lorillard
	Camel	10%	R.J. Reynolds

*Data come from industry tracking studies and government reports

Morris document discusses the placement of the Marlboro brand name in the children's movie *Superman II*, and a 1983 agreement between Associated Film Promotions and the actor Sylvester Stallone guarantees the usage of Brown & Williamson products in no less than five movies for a payment of $500,000 (Ripslinger). Another Brown & Williamson memorandum from 1983 discusses the "Kool record continuity promotional concept," suggesting that the program be tied-in with the RCA record club, which at the time offered records at a 50 percent discount. The memorandum explains that "smokers can order from a special Kool catalog and imagines that this will reach the Kool target audience, with the benefit of possible tie-ins such as MTV" (Hendricks).

A Philip Morris report from 1984 reviewing possible locations for the "Marlboro Spring Vacation Program" identifies places where young people congregate to offer free samples and brand promotions. The report describes ski resorts and ninety-eight different beach locations (Philip Morris Companies). In the 1990s, internal documents reveal that R.J. Reynolds sales representatives discussed ways to market its Joe Camel cigarette brand by identifying high-volume cigarette outlets in close proximity to colleges and high schools (McMahon). R.J. Reynolds launched Joe Camel in 1988. Between 1987 and 1992 Camel sales in the 18–24 age group rose more than 400% from 2.5% to 10.1% after being flat for the previous five years (R.J. Reynolds). In 1993, R.J. Reynolds commissioned a study to assess awareness of the Joe Camel character and slogan among children ages 10 to 17. The results of the survey revealed that, even though Joe Camel was not on TV, his total score of 86% awareness was not far behind the virtual 100%

awareness of the Energizer Bunny, Ronald McDonald, Tony the Tiger, and the Trix Rabbit (Roper Starch). Despite this alarming information, R.J.R. did not begin to phase out the Joe Camel advertising and promotional campaign until 1997. Almost immediately sales began to plummet.

The result of marketing to beginning smokers has been that they tend to smoke the cigarette brands that are among the most heavily advertised. Like other fads, the cigarette brands popular with teenagers have changed over time, and with these changes have ridden the fortunes of the cigarette makers (see table). Competition for the teenage smoker market remains keen even today as the lessons of the past clearly indicate that profitability and survival depend upon the concept "get' em while they're young."

See Also Advertising; Ethnicity; Marketing; Women; Youth Tobacco Use.

▌K. MICHAEL CUMMINGS
▌ANTHONY BROWN
▌CRAIG STEGER

BIBLIOGRAPHY

Burrows, D. "Younger Adult Smoker: Strategies and Opportunities." 29 February 1984. R.J. Reynolds Tobacco Company. Bates Number: 501928462–501929550.

Cummings, K. M., et al. "Marketing to America's Youth: Evidence from Corporate Documents." *Tobacco Control* 11 (1 March 2002): 5–17.

Cummings, K. M., and A. Brown. "History of Youth Brands." Roswell Park Cancer Institute. Available: <http://roswell.tobaccodocuments.org/media/youth_brand_chart_1950s–2000.pdf>.

Hendricks, J. L. "KOOL Record Continuity." 7 September 1983. Brown & Williamson Tobacco Corporation. Bates Number: 676026171–676026172.

Lynch, Barbara S., and Richard J. Bonnie, eds. *Growing Up Tobacco Free: Preventing Nicotine Addiction in Children and Youths*. Washington, D.C.: National Academy Press, 1994.

Philip Morris Companies, Inc. "840000 Marlboro Spring Resort Field Marketing Opportunities." 1984. Philip Morris Companies, Inc. Bates Number: 2044390059–2044390073.

Pollay, Richard W. "Promises, Promises: Self-Regulation of U.S. Cigarette Broadcast Advertising in the 1960s." *Tobacco Control* 3, no. 2 (June 1994): 134–144.

Pollay, Richard W., et al. "The Last Straw? Cigarette Advertising and Realized Market Shares among Youths and Adults, 1979–1993." *Journal of Marketing* 60 (April 1996): 1–16.

Ripslinger, J. F. Associated Film Promotions. [Re: Stallone Tobacco Use Agreement.] 14 June 1983. Brown & Williamson Tobacco Corporation. Bates Number: 685083120.

R.J. Reynolds. Camel: Crisis-Vision. 1993. R.J. Reynolds Company. Bates Number: 521896268–521896289.

Roper Starch. Advertising Character and Slogan Survey. November 1973. R.J. Reynolds Company. Bates Number: 517146060–517145108.

Wakeham, H. "Smoker Psychology Research." 26 November 1969. Philip Morris. Bates Number: 1000273741–1000273770.

Zollo, Peter. *Wise Up to Teens: Insights into Marketing and Advertising to Teenagers*. Ithaca, N.Y.: New Strategist Publications, 1995.

Youth Tobacco Use

Research published in 2001 indicates that most smokers worldwide start young, before the age of eighteen. In the United States, for example, some 3,000 children and young people begin to smoke each day. Thus research and public health efforts since the 1970s have focused on why young people start to smoke and how smoking initiation can be prevented. While there have been healthy changes in the attitudes of youth and adults toward tobacco since the early twentieth century, some of the same concerns about youth smoking voiced at that time continue to persist into the twenty-first century.

Smoking and Empire

Smoking among children and young people only began to be seen as a social issue with the appearance of cheap, attractively packaged cigarettes in the 1880s. In the early 1900s, smoking among children and young people was interpreted against a background of fears about national fitness, physical deterioration, and urbanization. In Britain, observers had been concerned during the Boer War (1899–1902), between the British Army and the Boer farmers in South Africa, about the high numbers of military recruits rejected on medical grounds, and by the poor health of soldiers in South Africa. The issue of juvenile smoking therefore had a wide appeal: It was attractive to observers concerned about health and physical deterioration, about hooliganism, and about working-class youth. It therefore was an issue with both health and moral implications.

In Britain, the main attempt to legislate against smoking came in the 1908 Children's Act, which prohibited the sale of tobacco to children under age sixteen and gave police power to seize tobacco from children caught smoking in public. However despite the 1908 Children's Act, smoking continued to form part of the vocabulary and activity of social reformers. Robert Baden-Powell, for example, wrote in *Scouting for Boys* (1908) that smoking harmed the sight and smell of boys who might become Scouts, and made them "shaky and nervous." Smoking was added to a list of moral failings that included getting up late, masturbation, drinking, and betting. In *Scouting for Boys*, the smoker was a "loafer," while the nonsmoker "has the ball at his feet" (Welshman 1996). Other contemporary observers drew smoking into their depiction of working-class youth and commented on the alleged dangers of juvenile smoking among girls. Smoking was also used as a badge of identity for working-class youth. While the 1908 Children's Act prohibited the sale of tobacco to children, middle-class youth movements continued to use smoking to define working-class children and adolescents up to World War I.

In the period between 1880 and 1914, therefore, the issue of smoking and young people had more to do with wider debates about urbanization and physical deterioration. The connection between smoking and lung cancer was yet to be made at the time. Rather than being linked to specific diseases, smoking was viewed as having general effects that compromised physical performance, especially in the context of work or

Youth tobacco use continues to be a major public health concern in the twenty-first century. According to some studies, smoking is part of a process of identity building and seems to be socially learned.
AP/WIDE WORLD PHOTOS

YOUTH TOBACCO USE

Boys smoking cigars and pipes, 1906. Smoking among children and young people only began to be seen as a social issue with the appearance of cheap, attractively packaged cigarettes in the 1880s. © CORBIS

military service. For example, it was alleged that the sense of smell of boys who smoked would be impaired. Similarly the potential effects of smoking on eyesight, and by extension on the accuracy of rifle shooting, was highlighted as an issue for potential soldiers. In the period up to 1914, therefore, those concerned about smoking and young people were interested more in morality than in health. Even after substantial publicity about the health hazards of smoking in the early 1950s, smoking among young people remained widespread. A survey in Britain in 1957 found that in a secondary modern school, 30 percent of the boys and 15 percent of the girls were regular smokers, despite the legal prohibition on the sale of tobacco to this age group.

Smoking in the Early Twenty-First Century

Youth tobacco use continues to be a major public health concern in the twenty-first century. However, researchers have proposed differing explanations for why young people start smoking and how tobacco use can be prevented among youth. Studies by psychologists on risk perception show that young people believe they can escape with a reduced amount of smoking before that risk becomes a reality. Young smokers are not informed about, and underestimate, the difficulties in stopping smoking. Moreover, young people do not process information and access results logically, and in fact calculate risks in ways that have more to do with emotion, deeply held beliefs, the views of people around

> "The remaining source of amusement is the street. The boy spends a large part of his spare time in loafing about the streets with his pals, playing games, singing, exchanging witticisms, and generally making himself obnoxious to the police and the public... Most boys seem to keep clear of the public-house for the first few years after leaving school; almost all of them smoke Woodbines or 'Coffin-Nails' as they facetiously term them" (Freeman 1914).

> "'Programmed' learning and visual aid, too, came to the child in the shape of cigarette card series. The value of the 'fag' card, from 1900 onwards, as a conveyor of up-to-date information was enormous... Before 1914 it would have been hard indeed to have found a boy in the working class without at least a few dog-eared cards about his person, dreaming of making up, by swap and gambling games, that complete set of fifty" (Roberts 1973).

> "I think I was about twelve and my granny used to smoke. I had one of my really good friends from school over and we stole one of her cigarettes and went to the bike shed and I hated every minute of it. I hated it. I tried to tell her how you do it, showed her... I hated it, really did not like it" (Hughes 2003).

them, and an inability to calculate simple odds. On the other hand, the economist Kip Viscusi has argued that young people overestimate the risks posed by smoking. He has portrayed young people as weighing up the benefits against the risks before deciding to light up. His arguments support the conventional defense of the tobacco companies in lawsuits brought by smokers, which is that smokers knew the risks involved and made an informed decision to smoke.

In addition to the pharmacological effects of nicotine, psychological studies suggest that social and personal attitudes toward smoking also affect the way people experience tobacco. Jason Hughes, for example, has studied why people smoke by conducting detailed interviews with individual smokers, nonsmokers, and ex-smokers. His interviews suggest that smoking is part of a broader process of identity building, and that smoking seems to be socially learned. Hughes argues that what might be called a "nicotine self-replacement model of smoking" is too static and that the scientific tendency to reduce the effects of tobacco to biological processes overlooks variations in how tobacco has been experienced in different socio-cultural environments (Hughes).

Additionally, he argues that, historically, tobacco use emerged in relation to broader trends that included medicalization, mass-consumerism, individualization, and informalization. The implications of all this for public policy are that policy makers should address motivations for starting smoking, and giving up may be as much social as biological conditioning. The development of tobacco use in individuals seems to follow a similar pattern to its broader development in the West: an initial period of civilization; a period of increasing opposition; and a third stage of habituation. Hughes argues from this that people need not just to stop smoking, but to "make the much broader transition involved in becoming a nonsmoker" (Hughes).

Over the course of the twentieth century there have been changes in the way that tobacco use among young people has been viewed. In the early 1900s, the issue of smoking and young people can be located within wider debates about urbanization and physical deterioration, rather than concern about specific health effects. In contrast, in the early twenty-first century the discourses around physical deterioration are much less apparent. Research is focused on identifying and understanding the factors that lead young people to start smoking. While a pharmacological explanation (that is, people continue to smoke because they are addicted to nicotine) came to dominate in the late twentieth century, it was not a complete explanation. Understanding why young people start smoking involves a complex web of social and environmental influences, including peer pressure, advertising, and popular culture.

See Also Advertising; Ethnicity; Marketing; Women; Youth Marketing.

▮ JOHN WELSHMAN

BIBLIOGRAPHY

Baden-Powell, Sir Robert. *Scouting for Boys: A Handbook for Instruction in Good Citizenship*, 2nd ed. London: Arthur Pearson, 1909.

Freeman, Arnold J. *Boy Life and Labour: The Manufacture of Inefficiency*. London: P. S. King and Son, 1914. Reprint, New York: Garland, 1980.

Hilton, Matthew. "'Tabs,' 'Fags' and the 'Boy Labour Problem' in late Victorian and Edwardian England." *Journal of Social History* 28, no. 3 (1995): 586–607.

Hughes, Jason. *Learning to Smoke: Tobacco Use in the West.* Chicago: University of Chicago Press, 2003.

Roberts, Robert. *The Classic Slum: Salford Life in the First Quarter of the Century.* Manchester, England: University of Manchester Press, 1971. Reprint, Harmondsworth, England: Penguin, 1973.

Slovic, Paul, ed. *Smoking: Risk, Perception, & Policy.* London: Sage, 2001.

Webster, Charles. "Tobacco Smoking Addiction: A Challenge to the National Health Service." *British Journal of Addiction* 79 (1984): 7–16.

Welshman, John. "Images of Youth: The Issue of Juvenile Smoking 1880–1914." *Addiction* 91, no. 9 (1996): 1,379–1,386.

Zimbabwe

The production and sale of tobacco has been a dynamic feature of Zimbabwe's colonial and post-colonial history just as Zimbabwe has been an important contributor to the global tobacco market in the last half of the twentieth century. But as events at the start of the twenty-first century demonstrate, the durability of tobacco as a key feature of Zimbabwe's political economy and the role of Zimbabwe in global markets are not guaranteed.

Colonial Period

Portuguese explorers and settlers brought tobacco to what is now Zimbabwe in the sixteenth century. This variety of *nicotiana rustica*, or *inyoka* tobacco (as it was called in the colonial period), was **air-cured,** mixed with water and other plants, and dried as loaves or cones. It was consumed as **snuff,** traded, or used as tribute payments to political or spiritual leaders.

In 1890 the British South African Company (BSAC) took control of the land that was to become Zimbabwe for the British Crown. White settlers swarmed into the newly named colony of Southern Rhodesia. By the early 1900s, the BSAC saw agriculture as the primary means for economic growth and for providing returns to its investors. Colonial governments allowed Europeans to settle on the land with the best soils and rainfall and displaced indigenous Africans into native reserves.

For the first three decades of colonial rule, however, the predominantly African farmers of the *inyoka* tobacco found a growing market as more and more Africans began working for European settlers in the new towns, mines, and farms. Government figures show that in the area that grew the most *inyoka* tobacco, African farmers sold over 25,000 pounds of their tobacco in 1906 and 300,000 pounds by 1923, the year when BSAC rule ended and self-rule for the white settlers began.

air-cured tobacco leaf tobacco that has been dried naturally without artificial heat.

snuff a form of powdered tobacco, usually flavored, either sniffed into the nose or "dipped," packed between cheek and gum. Snuff was popular in the eighteenth century but had faded to obscurity by the twentieth century.

ZIMBABWE

This farm worker stands with reaped tobacco in a field. Tobacco is reaped by hand in Zimbabwe, using a variety of tools to cut and carry the leaves. PHOTOGRAPH BY JASON LAURE

flue-cured tobacco also called Bright Leaf, a variety of leaf tobacco dried (or cured) in air-tight barns using artificial heat. Heat is distributed through a network of pipes, or flues, near the barn floor.

By the early 1900s, and with BSAC encouragement, Europeans started to grow different varieties of *nicotiana tabacum* such as Turkish, Burley, and, especially, Virginia. Heavy marketing helped to make European-style cigarettes more popular than *inyoka* tobacco in urban areas by the mid-1920s. The majority of this tobacco was consumed in South Africa and, increasingly, in the United Kingdom.

Flue-cured tobacco production fluctuated along with demand and labor supply as different producer organizations and marketing arrangements materialized and declined until the 1940s when Virginia tobacco emerged as the primary export for the colony. This was due to extensive government support for European agriculture at large and general lack of support for African farmers, who were discouraged from growing Virginia tobacco in the colonial period. The state provided land and loans to returning European soldiers, leading to several thousand new tobacco farmers. The government also ensured a cheap supply of African workers, often from neighboring colonies, who were legally treated as "servants" and not as "workers" by the European farmers. Moreover, the government facilitated the emergence of a strong tobacco farmer association (the Rhodesia Tobacco Association), which in turn effectively lobbied the government on many issues. One great accomplishment of this lobbying was the creation of a tobacco research board in 1950, with the farmers providing three times more of its funding than the government. The research and its dissemination to the farmers improved the yield and quality of tobacco in the region. Starting in 1948, long-term preferential purchase agreements with British tobacco manufacturers also helped considerably, as is shown in the following table. The table, based on Weinmann and Zimbabwe Tobacco Association, shows flue-cured acreage planted and the yield (in pounds per acre) of tobacco sold from 1924 to 2002.

Year	Acreage planted	Yield (lbs/acre) of tobacco sold
1924	7,001	489
1930	9,681	568

1940	61,283	572
1950	154,511	691
1960	212,239	1,029
1970	101,798	1,116
1980	158,910	1,701
1990	146,780	2,006
2000	206,146	2,526
2002	170,430	2,139

The situation changed dramatically after the Rhodesian government unilaterally declared independence from the United Kingdom in 1965. Many governments imposed sanctions that led to covert sales of Rhodesian tobacco and increased governmental controls over the industry. Intensifying fighting between African nationalist groups and Rhodesian forces throughout the 1970s further destabilized farming operations.

Post-Colonial Period

The start of post-colonial rule of Zimbabwe in 1980 brought greater stability for the industry. Tobacco again became the nation's chief export, and Zimbabwe resumed its place as one of the top three exporters of flue-cured tobacco in the world. Additionally, the white tobacco farmers actively sought to increase production of Burley and Virginia tobacco among African smallholder farmers as a way to entrench further the industry in the new post-colonial nation.

Although the worldwide antismoking movement influenced Zimbabwean tobacco farmers to diversify into other crops throughout the 1990s, a much greater threat to the industry has emerged in the highly politicized large-scale land acquisition movement unleashed by the government in 2000. The massive land redistribution from white farmers to black farmers increased the number of small-scale tobacco farmers. Yet there has been a dramatic decrease in tobacco production as many white tobacco producers lost their land, and the resulting economic chaos has led to a scarcity of many tobacco inputs, such as fertilizer, herbicides and pesticides. There was a 30-percent drop in flue-cured tobacco sold in the year 2002 compared to the year 2000, making Zimbabwe's future place in the global market uncertain.

See Also Africa; Antismoking Movement From 1950; British Empire.

▪ BLAIR RUTHERFORD

BIBLIOGRAPHY

Clements, Frank, and Edward Harben. *Leaf of Gold: The Story of Rhodesian Tobacco.* London: Methuen, 1962.

Kosmin, Barry. "The Inyoka Tobacco Industry of the Shangwe People: The Displacement of a Pre-Colonial Economy in Southern Rhodesia, 1898–1938." In *The Roots of Rural Poverty in Central and Southern Africa.* Edited by Robin Palmer and Neil Parsons. Berkeley: University of California Press, 1977.

Rubert, Steven C. *A Most Promising Weed: A History of Tobacco Farming and Labor in Colonial Zimbabwe, 1890–1945*. Athens: Ohio University Center for International Studies, 1998.

Weinmann, H. *Agricultural Research and Development in Southern Rhodesia, 1924–1950*. Salisbury: University of Rhodesia, 1975.

Zimbabwe Tobacco Association. "Zimbabwe Tobacco Statistics." 12 April 2004. Available: <http://www.zta.co.zw/otherimages/ZIM%20STATS%206a.htm.>

Tobacco in History and Culture
AN ENCYCLOPEDIA

Contributors List

VIRGINIA BARTOW
New York Public Library
New York, New York
   Arents Collection

MICHIEL BAUD
CEDLA
Amsterdam, The Netherlands
   Brazil
   South and Central America

CARA BAYLUS
University of Chicago
Chicago, Illinois
   Intellectuals

CAROL BENEDICT
Georgetown University
Washington, D.C.
   Therapeutic Uses

LISA BERO
University of California–San Francisco
San Francisco, California
   Tobacco Industry Science

VIRGINIA BERRIDGE
London School of Hygiene and Tropical Medicine
London, England
   Tobacco Control in the United Kingdom

FRANCISCO BETHENCOURT
Universidade Nova de Lisboa
Lisbon, Portugal
   Portuguese Empire

ALAN BLUM
University of Alabama
Tuscaloosa, Alabama
   Air Travel
   Antismoking Movement From 1950
   Philip Morris

ANTHONY BROWN
Roswell Park Cancer Institute
Buffalo, New York
   Documents
   Youth Marketing

IAN W. BROWN
University of Alabama
Tuscaloosa, Alabama
   Calumets

JOHN C. BURNHAM
Ohio State University
Columbus, Ohio
   Bad Habits in America
   Doctors

DAVID M. BURNS
University of California–San Diego School of Medicine
San Diego, California
   "Light" and Filtered Cigarettes

TRACY CAMPBELL
University of Kentucky
Lexington, Kentucky
   Architecture
   Black Patch War
   Kentucky

DAVID CANTOR
National Cancer Institute
Bethesda, Maryland
   Lung Cancer
   Nazi Germany

SIMON CHAPMAN
University of Sydney
Sydney, Australia
   Tobacco Control in Australia

JAMES COLGROVE
Columbia University
New York, New York
   Advertising Restrictions
   Smoking Restrictions

JEFF COLLIN
London School of Hygiene and Tropical Medicine
London, England
   Developing Countries

FRANCISCO COMÍN
Universidad de Alcalá
Madrid, Spain
   State Tobacco Monopolies

B. LEE COOPER
Newman University
Wichita, Kansas
   Music, Popular

ANDREW M. COURTWRIGHT
University of North Carolina, Chapel Hill
Chapel Hill, North Carolina
   Alcohol, Tobacco, and Other Drugs

DAVID T. COURTWRIGHT
University of North Florida
Jacksonville, Florida
   Alcohol, Tobacco, and Other Drugs

CONTRIBUTORS LIST

HOWARD COX
South Bank University Business School
London, England
   American Tobacco Company
   British American Tobacco
   Globalization

K. MICHAEL CUMMINGS
Roswell Park Cancer Institute
Buffalo, New York
   Documents
   Youth Marketing

KRISTEN DALEY
Northeastern University School of Law
Boston, Massachusetts
   Litigation

RICHARD DAYNARD
Northeastern University School of Law
Boston, Massachusetts
   Litigation

SUSAN DEANS-SMITH
University of Texas at Austin
Austin, Texas
   Mexico

TIMOTHY DEWHIRST
University of Saskatchewan
Saskatoon, Saskatchewan, Canada
   Public Relations
   Sponsorship

FRANK DIKÖTTER
School of Oriental and African Studies
London, England
   China
   Opium

ELLIOT N. DORFF
University of Judaism
Bel Air, California
   Judaism

JOSHUA DUNSBY
University of California-San Francisco
San Francisco, California
   Genetic Modification
   Secondhand Smoke

RUTH DUPRÉ
HEC Montréal
Montreal, Quebec, Canada
   Prohibitions

ROSEMARY ELLIOT
University of Glasgow
Glasgow, Scotland
   Appetite
   Body
   Product Design
   Soldiers

KARL FAGERSTRÖM
Smokers Information Centre
Helsingborg, Sweden
   Quitting Medications

PETER T. FURST
University of Pennsylvania Museum of Archaeology and Anthropology
Philadelphia, Pennsylvania
   Shamanism

ANNA B. GILMORE
London School of Hygiene and Tropical Medicine
London, England
   Developing Countries

TAMIKA GILREATH
Pennsylvania State University
University Park, Pennsylvania
   Ethnicity

STANTON A. GLANTZ
University of California-San Francisco
San Francisco, California
   Politics

MASAYUKI HANDA
Tobacco and Salt Museum
Tokyo, Japan
   Japan

MARK HANUSZ
Equinox Publishing
Jakarta, Indonesia
   Kretek

TERENCE E. HAYS
Rhode Island College
Providence, Rhode Island
   Missionaries
   Oceania

JACK E. HENNINGFIELD
Johns Hopkins University School of Medicine
Baltimore, Maryland
   Addiction
   Nicotine

MATTHEW HILTON
University of Birmingham
Birmingham, England
   Age
   Class
   Smoking Clubs and Rooms

MAI-ANH HOANG
Johns Hopkins Bloomberg School of Public Health
Baltimore, Maryland
   Disease and Mortality

JASON HUGHES
University of Leicester
Leicester, England
   Snuff
   Social and Cultural Uses

JENNIFER K. IBRAHIM
University of California-San Francisco
San Francisco, California
   Politics

LORI JACOBI
University of Alabama
Tuscaloosa, Alabama
   Philip Morris

JON D. KASSEL
University of Illinois at Chicago
Chicago, Illinois
   Psychology and Smoking Behavior

JORDAN KELLMAN
University of Louisiana at Lafayette
Lafayette, Louisiana
   French Empire

JEFFREY R. KERR-RITCHIE
University of North Carolina at Greensboro
Greensboro, North Carolina
   Slavery and Slave Trade

GARY KING
Pennsylvania State University
University Park, Pennsylvania
   Ethnicity

ROBERT KLINE
Northeastern University School of Law
Boston, Massachusetts
   Litigation

WIM KLOOSTER
Clark University
Worcester, Massachusetts
   Dutch Empire

CONTRIBUTORS LIST

KELLEY LEE
London School of Hygiene and Tropical Medicine
London, England
    Lobbying

CARLOS FRANCO LIBERATO
Federal University of Sergipe
Aracaju, Sergipe, Brazil
    Plantations

MICHAEL J. LICARI
University of Northern Iowa
Cedar Falls, Iowa
    Taxation

JOYCE LORIMER
Wilfrid Laurier University
Waterloo, Ontario, Canada
    Smuggling and Contraband

EVA MANTZOURANI
Canterbury Christ Church University College
Canterbury, England
    Music, Classical

DAWN MARLAN
University of Illinois at Chicago
Chicago, Illinois
    Film

ROBERT L. MARTENSEN
Tulane Health Sciences Center
New Orleans, Louisiana
    Therapeutic Uses

RUDI MATTHEE
University of Delaware
Newark, Delaware
    Iranian Tobacco Protest Movement
    Islam

J. ROSSER MATTHEWS
Virginia Tech
Blacksburg, Virginia
    Insurance
    Medical Evidence (Cause and Effect)

RUSSELL R. MENARD
University of Minnesota, Twin Cities
Minneapolis, Minnesota
    British Empire

ROBERT H. MILLER
United States Department of Agriculture, retired
Alexandria, Virginia
    Trade

DOLORES MITCHELL
California State University-Chico
Chico, California
    Visual Arts

HENEAGE MITCHELL
Bangkok, Thailand
    South East Asia

STEPHANIE MOLLER
Pennsylvania State University
University Park, Pennsylvania
    Ethnicity

GREGORY J. MORGAN
Johns Hopkins University
Baltimore, Maryland
    Tobacco Mosaic Virus

LUCY MUNRO
Keele University
Keele, England
    English Renaissance Tobacco

LAURA NÁTER
San Juan, Puerto Rico
    Caribbean
    Cuba

MICHAEL NEE
New York Botanical Garden
Bronx, New York
    Origin and Diffusion

MIMI NICHTER
University of Arizona
Tucson, Arizona
    South Asia

MARCY NORTON
George Washington University
Washington, D.C.
    Christianity
    Sailors

DIDIER NOURRISSON
Institut Universitaire de Formation des Maîtres de Lyon
Saint-Galmier, France
    Antismoking Movement in France
    Gitanes/Gauloises

JOHN PARASCANDOLA
Silver Spring, Maryland
    Warning Labels

MARK PARASCANDOLA
National Institutes of Health
Bethesda, Maryland
    "Safer" Cigarettes

JOSÉ; PARDO-TOMÁS
Institució "Milà i Fontanals" CSIC
Barcelona, Spain
    Botany (History)

JOSEPH PARKER
Broomes Island, Maryland
    Camel
    Chewing Tobacco
    Lucky Strike
    Marlboro
    Virginia Slims

CHRISTINA A. S. PERRON
Portland, Oregon
    Tobacco as an Ornamental Plant

ADRIENNE PETTY
Columbia University
New York, New York
    Labor
    Sharecroppers

JOHN PIERCE
University of California-San Diego Cancer Center
San Diego, California
    Consumption (Demographics)

TANYA POLLARD
Montclair State University
Montclair, New Jersey
    English Renaissance Tobacco

ELDRED E. PRINCE JR.
Coastal Carolina University
Conway, South Carolina
    Processing
    United States Agriculture

SEAN M. RAFFERTY
University of Albany
Albany, New York
    Archaeology
    Hallucinogens

BEN RAPAPORT
Reston, Virginia
    Connoisseurship
    Pipes

KURT RIBISL
University of North Carolina at Chapel Hill

CONTRIBUTORS LIST

Chapel Hill, North Carolina
  Retailing

FRANCIS ROBICSEK
Carolinas Heart Institute
Charlotte, North Carolina
  Mayas

JOSE MANUEL RODRÍGUEZ GORDILLO
University of Seville
Seville, Spain
  Spanish Empire

STEVEN C. RUBERT
Oregon State University
Corvallis, Oregon
  Africa

THEO RUDMAN
Good Living Publishing
Helderberg, South Africa
  Cigars

JARRETT RUDY
McGill University
Montreal, Quebec, Canada
  Cigarettes
  Industrialization and Technology

BLAIR RUTHERFORD
Carleton University
Ottawa, Ontario, Canada
  Zimbabwe

JONATHAN M. SAMET
Johns Hopkins Bloomberg School of Public Health
Baltimore, Maryland
  Disease and Mortality

PATRICIA B. SANTORA
Johns Hopkins University School of Medicine
Baltimore, Maryland
  Addiction
  Nicotine

STEVE SARSON
University of Wales Swansea
Swansea, Wales
  Chesapeake Region

RELLI SHECHTER
Ben-Gurion University
Beer Sheva, Israel
  Middle East

SAUL SHIFFMAN
University of Pittsburgh
Pittsburgh, Pennsylvania
  Psychology and Smoking Behavior

SIMON SMITH
University of York
Heslington, York, England
  Trade

ERIC SOLBERG
Anderson Cancer Center
Houston, Texas
  Antismoking Movement From 1950

GEORGE BRYAN SOUZA
University of Texas San Antonio
San Antonio, Texas
  Philippines

CRAIG STEGER
Roswell Park Cancer Institute
Buffalo, New York
  Documents
  Youth Marketing

COLIN TALLEY
Columbia University
New York, New York
  Quitting

CASSANDRA TATE
www.historylink.org
Seattle, Washington
  Antismoking Movement Before 1950
  Ostracism

CARSTEN TIMMERMANN
University of Manchester
Manchester, England
  Lung Cancer

PENNY TINKLER
University of Manchester
Manchester, England
  Advertising
  Women

KEITH VOLANTO
Collin County Community College
Dallas, Texas
  New Deal

CHERYL KRASNICK WARSH
Malaspina University-College
Nanaimo, British Columbia, Canada
  Marketing

GEOFFREY FERRIS WAYNE
Massachusetts Department of Public Health
Boston, Massachusetts
  Additives
  Chemistry of Tobacco and Tobacco Smoke
  Fire Safety
  Menthol Cigarettes
  Product Design
  Toxins

JOHN WELSHMAN
Lancaster University
Lancaster, England
  Sports
  Youth Tobacco Use

JOSEPH WINTER
Native American Plant Cooperative
Cuba, New Mexico
  Native Americans

MITCHELL ZELLER
Pinney Associates
Bethesda, Maryland
  Regulation of Tobacco Products in the United States

Tobacco in History and Culture
AN ENCYCLOPEDIA

Index

Page references for major articles are in **boldface**. *Illustrations are indicated by page references in* italics. *Tables are indicated by page references followed by* t.

A

Á bout de souffle/Breathless (film), 1: 235
AAA (Agricultural Adjustment Administration), 2: 385–387
Absinthe, tobacco use with, 1: 37
Accessories
  cigar accessories, 1: *159*, 159–160
  cigarette accessories, 1: *159*, 160; 2: 568
  pipe accessories, 1: 156–159, *157*, *158*, 165
  references to, in popular music, 1: 370–371
  snuff accessories, 1: 160–163, *161*; 2: 549–550, *560*, 667
  used by gallants, 2: 559, *560*
Accord (cigarette brand), 2: 493
Acetaldehyde, in Marlboro cigarettes, 1: 10
Acquisition, definition of, 1: 43
ACS (American Cancer Society), 1: 53; 2: 484–485
"Action against Access" campaign, by tobacco industry, 2: 477, 478
Action on Smoking and Health (ASH), 2: 624
Addiction to nicotine, **1: 1–6**
  additives and nicotine effects, 1: 8–9
  alcohol and drug use associated with, 1: 35–41; 2: 468
  cigarettes as most addictive form of, 1: *1*, 3–5
  criteria for, 2: 468
  FDA evidence on, 2: 493
  history of nicotine science and, 1: 1–3; 2: 388
  nicotine research and concepts of, 1: 2–3; 2: 467–468
  physiologic and behavior effects of, 2: 389–390
  popular music references to, 1: 370

  prevention of, 2: 391
  smoking behavior driven by, 2: 390–391
  tobacco "chippers" and, 2: 469
  tobacco industry documents on, 1: 4; 2: 494, 627
  treatment of, 1: 5–6
  withdrawal symptoms related to, 1: 3; 2: 391, 468, 486
  See also Chemistry of tobacco and tobacco smoke
  See also Nicotine
  See also Psychology and smoking behavior
  See also Quitting
Additives, **1: 6–10**
  ammonia and acetaldehyde as, 1: 10
  cigarette paper coated with, 2: 455
  health risks and regulation of, 1: 9–10
  physiological and behavioral effects, 1: 8–9
  role in product design, 1: 7–8; 2: 454–455
  in snuff, 2: 549–550
  See also Menthol
Adolescents. *See* Youth marketing; Youth tobacco use
Advance (tobacco company), 2: 493
Advertising, **1: 11–18**
  of alcohol, with tobacco, 1: 36
  antismoking advertising, 1: 19–20, *21*, 53–54, *83*, 345; 2: 430
  brand loyalty linked to age, 2: 689
  branding in, 1: 11; 2: 564
  Camel cigarettes, 1: *106*, 106–107, 172; 2: 670, 690
  in China, 1: 15, 132–133, *133*
  cigarettes as appetite suppressants, 1: 62–63, *64*, 172, 336

  classical music in, 1: 368
  dominance of cigarette use due to, 1: *171*, 171–173
  early tobacco promotions, 1: 11–12
  exoticism in, 2: 669–670, 681
  globalization of cigarette market, 1: 265
  impact of, 1: 17–18
  Joe Camel character in, 1: 22, 107, 173, 335; 2: 692–693
  Kool cigarettes, 1: 332, 334–335; 2: 692
  Lucky Strike cigarettes, 1: 62–63, *64*, 172, *252*, 318–320, *319*, 335–336; 2: 686, 690
  Marlboro cigarettes, 1: 337–339, *338*; 2: 407, 671, 691
  mass advertising of cigarettes, 1: 246
  outdoor advertising displays, 1: 328
  package inserts/coupons, 1: 327–328
  Philip Morris, 2: 407; *2 408*
  product placement in film and television, 1: 13, 232; 2: 477–478, 691–692
  public relations vs., 2: 474
  in retail outlets, 2: 499–500, 501
  shift from broadcast advertising to sponsorship, 2: 594–595
  sports and tobacco advertising, 2: 597–599, *598*, 671–672
  targets of, 1: 14–15, *15*, 228–229, 334–336; 2: 685–686, 691–692
  themes of, 1: 15–16
  in twentieth century, 1: *12*, 12–14, *14*
  unregulated health claims for tobacco products, 2: 493
  Virginia Slims, 1: 16, 63, 336; 2: 663, 664
  during World War I and II, 2: 670
  See also Advertising restrictions

707

INDEX

Advertising CONTINUED
  See also Cigarette cards
  See also Marketing
  See also Sponsorship
  See also Women, advertising targeted to
  See also Women, in advertisements
  See also Youth marketing
Advertising restrictions, **1: 19–23**
  advertising effects on youth and, 1: 22
  antismoking advertising and, 1: 19–20, *21*, 53–54; 2: 430
  Australia, 2: 620
  banning of tobacco advertising, 1: 19–20; 2: 463, 502, 594, 672
  economic and political constraints on, 1: 20
  European community, 2: 438
  France, 1: 60
  freedom of speech issues, 1: 21–22
  Internet tobacco advertising, 2: 502
  Japan, 1: 20
  limits of, 1: 22–23
  Nazi Germany, 1: 322; 2: 383
  during 1960s and 1970s, 1: 19–20
  politics of, 2: 430
  in retail outlets, 2: 501
Africa, **1: 23–30**
  Angolan tobacco jar from, 1: *157*
  domestic cultivation of tobacco in, 1: 26–29
  early propagation of tobacco in, 1: 24–25, *25*
  French colonies in, 1: 244
  importance of tobacco trade in, 1: 25–26
  late twentieth-century effects of tobacco in, 1: 29–30
  Malawi tobacco production, 1: 28–29; 2: 647
  pipes in, 1: *28*, 68; 2: 421–422
  Portuguese tobacco trade with, 1: 23, 26; 2: 442–443
  Zambian tobacco production, 1: 29
  See also Zimbabwe
African Americans
  menthol cigarettes preferred by, 1: 228, 349
  as post–Civil War tobacco workers, 1: 297
  smoking prevalence among, 1: 226–227
  as target of tobacco marketing, 1: *228*, 228

women, in nineteenth-century advertisements, 2: 670
women union leaders, 1: 297
See also Slavery and slave trade
Age and tobacco use, **1: 31–33**
  early twentieth century, 1: 31–33
  initiation of smoking related to age, 2: 689–690
  later twentieth century, 1: *32*, 33
  See also Children
  See also Youth tobacco use
Agricultural Adjustment Act of 1933, 1: 287; 2: 385, 524, 660
Agricultural Adjustment Act of 1938, 2: 387
Agricultural Adjustment Administration (AAA), 2: 385–387
Agriculture. *See* Processing; Production of tobacco; Sharecroppers and sharecropping; Slavery and slave trade; Tenant farmers and farming; United States agriculture
AHA (American Heart Association), 2: 484–485
Air-cured tobacco
  in Africa, 1: 27
  chemistry of, 1: 113
  definition of, 1: 27
  in modern cigarettes, 2: 454
  process of curing, 1: 27; 2: *448*, 448–449, 655
Air travel, **1: 33–35**
  airline flight attendants class action lawsuit, 1: 34, 312
  banning of smoking during, 1: 34; 2: 537, 539
al-Din Shah, Nasir, 1: 272, *273*, 273
ALA (American Lung Association), 2: 484–485
Alaska Natives
  diffusion of tobacco to, 2: 522
  Inuit ash and quid boxes of, 1: *69*
  smoking and tobacco use of, 1: 229
Alcohol
  depressant effect of, 1: 40
  tobacco use combined with, 1: 35–36, *36*, *37*, 38–39
Alice in Wonderland (Carroll), 1: *304*
Alkaloids
  in Chinese tobacco, 1: 130
  definition of, 1: 114
  formation of toxic compounds from, 2: 632
  in hallucinogens, 1: 259; 2: 518
  in *Nicotiana rustica*, 2: 519
Allen & Ginter (tobacco company), 1: 251

Altadis (French/Spanish tobacco company), 1: 244
Altria (Philip Morris), 2: 409, 438
AMA (American Medical Association), 1: 53
American Baptist churches, antitobacco resolution of, 1: 142–143
American blend cigarette
  definition of, 1: 105
  types of tobacco in, 2: 454, 644–645
American Cancer Society (ACS), 1: 53; 2: 484–485
American Heart Association (AHA), 2: 484–485
American Lung Association (ALA), 2: 484–485
American Machine and Foundry Company, 1: 266
American Medical Association (AMA), 1: 53
American Stop Smoking Intervention Study (ASSIST), 1: 57; 2: 435–436
American Tobacco Company, **1: 42–44**
  acquisition by British American Tobacco, 1: 44, 95
  acquisition of Middle Eastern manufacturers, 1: 358
  advertising costs of, 1: 265
  blending of tobaccos in cigarettes, 1: 358
  diversification of, 1: 44
  foreign expansion of, 1: 43, 253, 265
  growth and dissolution of, 1: 43–44, 146–147
  origins of, 1: 42, 252–253
  See also Lucky Strike cigarettes
Amerindians. *See* Native Americans
Ammoniation
  definition of, 1: 114
  of Marlboro cigarettes, 1: 10, 114
Amphetamine, for weight control, 2: 484
Anadenanthera peregrina, 1: 259
Angola
  tobacco jar from, 1: *157*
  tobacco production in, 2: 443
Anti-Cigarette League of America
  founding and activities of, 1: 48–49
  popular support of, 1: 32
Antismoking legislation
  airline travel ban on smoking, 1: 34; 2: 537, 539
  Australian tobacco regulation, 2: 619–621

INDEX

banning of broadcast advertising, 1: 19–20; 2: 463, 594, 672
California Smoke-Free Workplace Act, 2: 465
Children's Act of 1908, 1: 334; 2: 694
Cigarette Labeling and Advertising Act, 2: 429–430, 672, 673, 675
clean indoor air legislation, 2: 431–432, *432*, 465, 516, 537, 621
in early twentieth-century United States, 2: 461
Framework Convention on Tobacco Control provisions, 1: 23; 2: 438–439, *439*, 495, 677–678
French legislation, 1: 59–60; 2: 539
Master Settlement Agreement provisions, 1: 22, 57–58, 311; 2: 436–437
outlawing of cigarettes in U.S., 1: 32
preemption of local tobacco control activity, 2: 429–430, 431, 433, 675
Public Health Cigarette Smoking Act, 2: 594, 675–676
in South East Asia, 2: 584
state cigarette prohibition laws and, 1: 45–46
for warning labels, 2: 675–678
youth access laws, 2: 431–432, 477, 478
See also Advertising restrictions
See also Lobbying
See also Prohibitions
See also Smoking restrictions
See also Tobacco control
Antismoking movement before 1950, **1: 45–51**
cigarettes as gateway drug and, 1: 46–47
cigarettes as target of, 1: 45–46, 47, 147–148
early critics of tobacco use and, 1: 83–84
in France, **1: 59–60**
health risks of smoking and, 1: 50–51
Iranian Tobacco Protest movement as, 1: 272–273
moral basis of, 1: 48, 84, 140–141; 2: 459, 462
in Nazi Germany, 1: 322; 2: 382
in pre-eighteenth-century literature, 1: 303
prohibitions in United States, 2: 461–463
"Smokes for Soldiers" campaigns and, 1: 49–50

social Darwinism and, 1: 47–48
state cigarette prohibition laws, 1: 45–46
youth smoking and, 1: 31–32, 334
See also Prohibitions
Antismoking movement from 1950, **1: 51–58**
antismoking advertising by, 1: 19, 53–54, *55*
in Australia, **2: 619–621**
call for government action by, 1: 53–54
Christian churches and, 1: 142–143
emerging independent organizations in, 1: 54
in France, **1: 59–60**
India, 2: *578*
litigation as strategy of, 1: 56–58
moral and scientific basis of, 1: 52–53
ostracism of smokers and, 2: 404, *404*
successes and failures of, 1: 57–58
tobacco industry as target of, 1: 54–57
See also Prohibitions
Anxiety
smoking associated with onset of, 2: 470
smoking for relief of, 2: 469
Aortic aneurysm, 1: 201t
Apocalypse Now (film), 2: 477
Appetite suppression, **1: 61–65**
Chinese use of tobacco for, 1: 131
historical context of, 1: 61–63
scientific context of, 1: 63–65, 84
shamanistic tobacco use and, 2: 521
tobacco advertising and, 1: 62–63, 64, 172, 336; 2: 686
Apprentice cigar makers, 1: 266
Archaeology of tobacco, **1: 66–70**
direct and indirect evidence in, 1: 66–67, 69
in Europe and Africa, 1: 68
in Mayan culture, 1: *340*, 341–342, *342*
Native American tobacco use and, 2: 376–377
in North and South America, 1: 68
pipes and, 1: 68–70
tobacco-related accessories and, 1: 67, *67*, 69
Architecture, **1: 71–73**
Arents Collection, **1: 73–76**, *74*
Arents, George, Jr., 1: 73, 305

Argillite
definition of, 1: 104
pipes made from, 1: 104; 2: 415
Queen Charlotte Islands, pipe of, 1: 157
ARIC (*Asociación Rural de Interés Colectivo*), 1: 354
Ariva (noncombusting cigarette), 2: 493
Arizona, smoking restriction legislation in, 2: 537
Aromatic hydrocarbons, polynuclear (PAH), 2: 632, 633
Art. *See* Film; Music, classical; Music, popular; Visual arts
Art Deco
in cigarette cases, 2: 667
definition of, 1: 163
in French cigarette packaging, 1: 248
Art Nouveau
definition of, 2: 667
Job cigarette poster as, 2: 668, *668*
Arts events, tobacco industry sponsorship of, 1: 331–332, *333*
ASH (Action on Smoking and Health), 2: 624
Ash stand, 1: 163
Asia. *See* China; Japan; South Asia; South East Asia
Asian Americans
smoking prevalence among, 1: 227
as target of tobacco marketing, 1: 228
Asociación Rural de Interés Colectivo (ARIC), 1: 354
ASSIST (American Stop Smoking Intervention Study), 1: 57; 2: 435–436
Atelier, definition of, 2: 419
Atherosclerosis, 1: 201t
Australia
Aboriginal smokers in, 2: 619
advertising of tobacco in, 1: 13, 14, 15
advertising restrictions in, 1: 20
origin of tobacco in, 2: 398
tobacco consumption in, 2: 619
tobacco control by, **2: 619–621**, *620*
Authoritarian, definition of, 1: 80
Auto racing
Philip Morris sponsorship of, 2: 691
tobacco industry sponsorship of, 1: 333–334; 2: 594, *594*, 691
youth tobacco marketing and, 1: 335; 2: 691
Aversion therapy, for quitting, 2: 485–486

INDEX

Ayahuasca (Banisteriopsis caapi),
    1: 259; 2: 518
Azores, tobacco production in,
    2: 443

B

Bad habits in America, **1: 77–79**
    consumer culture and, 1: 79
    relationship of tobacco use to
        vice, 1: 78–79
    stereotypical behavior associated
        with, 1: 77–78
    See also Psychology and smoking
        behavior
Baden-Powell, Robert, 2: 597, 694
Bahian tobacco, 1: 26, 89–91
Bal Tiga Nitisemito (*kretek*
    manufacturer), 1: 288
Ballet performances, tobacco
    consumption during, 1: 365
Banisteriopsis caapi, 1: 259; 2: 518
Banning of smoking. *See*
    Advertising restrictions;
    Antismoking legislation;
    Prohibitions; Tobacco control
Banzhaf, John, 1: 19, 53–54
Bara, Theda, 1: 233
Barns, 1: 71–72, *72*; 2: 448
Barrie, James M., 1: 304; 2: 422, 534
Barrymore, Maurice, 1: *162*
Bars
    banning of smoking in, 2: 465,
        540, *623*
    cigar bars, 2: 535
    One Double Oh Seven Club and
        Smoking Parlor, 2: 465
Basic Instinct (film), 1: 235–236;
    2: 566
BAT. *See* British American Tobacco
    (BAT)
Beans, cultivation by Native
    Americans, 2: 379, *380*
Behavior. *See* Bad habits in America;
    Psychology and smoking
    behavior
Belmondo, Jean-Paul, 1: 235
Benzo(a)pyrene, 2: 633
Bermuda, contraband trade in,
    2: 544
Bernays, Edward, 1: 335; 2: 474,
    475
Betel quid
    pan, in India, 2: 577–578
    in Philippines, 2: 411–412
    tobacco combined with, 1: 36–37
Bidi cigarettes
    additives in, 1: 8
    competitive price of, 1: 169
    popularity of, in South Asia,
        1: 195; 2: 576, 577

production of, in India,
    2: 576–577
toxic substances in, 2: 577
women smokers of, 2: 685
The Big Heat (film), 1: 231
Biotechnology
    definition of, 1: 245
    See also Genetic modification
Birth weight, smoking effects on,
    1: 209, 210*t*
Bizet, Georges, 1: 365; 2: 670
Black market for tobacco products
    definition of, 2: 492
    prohibition as cause of, 2: 492
    See also Smuggling and
        contraband
Black Patch War, **1: 80–81**; 1: 285
Black tobacco
    Black Patch area of Kentucky,
        1: 80, 285
    cultivation in South America,
        2: 572
    in Gitanes and Gauloises
        cigarettes, 1: 248, 249
Blacks. *See* African Americans;
    Slavery and slave trade
Blade Runner (film), 1: 236
Blending of tobaccos in cigarettes,
    1: 358; 2: 454, 644–645
Blonde Venus (film), 1: 233–234,
    236
Blum, Alan, 1: 55
Body, **1: 81–85**
    advertising themes related to,
        1: 16
    body weight set-point, 1: 65
    health effects of tobacco, 1: *82*,
        84
    methods of tobacco consumption
        and, 1: 81–83
    See also Appetite suppression
    See also Disease
    See also Disease risk
Bogart, Humphrey, 1: 235, 236,
    330
Bonsack, James A., cigarette
    machine of, 1: 145–146, 171,
    252, 264; 2: 452
Books about tobacco. *See* Arents
    Collection; Literature of tobacco
Botany (history), **1: 85–88**
    early European classifications of
        tobacco, 1: 85–87, *86*
    Linnaeus's classification of
        tobacco, 1: 87–88
    origination of *Nicotiana*, 1: 85
    tobacco as an ornamental plant,
        2: 616–618; 2: *617*, *618*
Boycott, definition of, 1: 94
Branding
    advertising value of, 1: 11; 2: 564

brand loyalty linked to age,
    2: 689
brand-stretching, 1: 14
    as objective of sponsorship,
        2: 595
    purpose of, 1: 11
Brando, Marlon, 1: 231
Brazil, **1: 88–92**
    Bahian tobacco, 1: 26, 89–91
    British American Tobacco
        expansion in, 1: 254
    chewing tobacco use by Brazilian
        warrior, 2: *572*
    contraband trade in, 2: 543
    plantations in, 2: 427–428
    sharecropping in, 1: 294
    slave labor in, 1: 292
    southern states of, 1: 91–92
    tobacco consumption in,
        2: 441–442
    tobacco exports of, 2: 646, 647
    tobacco production in, 1: 91–92;
        2: 442–443
    warning labels on cigarette packs,
        1: *90*
    See also Portuguese Empire
Breast cancer, 1: 202*t*
Breathless/À bout de souffle (film),
    1: 235
Briar pipes, 1: 157, 167; 2: 418
Bright leaf tobacco
    in cigarettes, 1: 149; 2: 659
    flue-cured, 1: 28, 149, 251;
        2: 449, 659
    in Navy plug, 1: 127
    New Deal support for, 2: 524
    processing of, 2: 659
British American Tobacco (BAT),
    1: 92–95
    acquisition of American Tobacco
        Company, 1: 44, 95
    Brazilian tobacco interests of,
        1: 91
    Chinese market of, 1: 93–94,
        132–133, 254
    Cunliffe-Owen's leadership of,
        1: 94
    formation of, 1: 43, 93–94, 147,
        254
    growth and international
        expansion of, 1: 93–94, 254
    as multinational corporation,
        2: 438
    postwar difficulties and
        diversification of, 1: 94–95
British Anti-Tobacco Society, 1: 334
British Empire, **1: 96–101**
    American colonial settlements of,
        1: 96–97
    Colonial Debts Act of 1732,
        1: 98–99

INDEX

expansion of tobacco market in, 2: 641–642
Navigation Acts of, 1: 96
planters' debts to tobacco merchants, 1: 98–99, 98t
price of tobacco and importation of, 1: 98t, 99–101; 2: 638, *639*, 640, 641
productivity of tobacco growers and, 1: 99–101
re-export of tobacco to European countries, 2: 642, 642t
taxes on Chesapeake tobacco, 1: 97; 2: 641
Zimbabwe tobacco production and, 2: 699–701, 700t–701t
See also Chesapeake region
See also United Kingdom
British South African Company, 2: 699
Broadcast advertising
banning of broadcast advertising, 1: 19–20; 2: 463, 594, 672
shift from broadcast advertising to sponsorship, 2: 594–595, 691–693
youth tobacco marketing on, 2: 690–691
See also Television
Broin v. Philip Morris Companies, 1: 312
Brooks, Louise, 1: 231
Brown & Williamson Tobacco Corporation
advertising of Kool cigarettes, 1: 332
confidential documents of, 1: 216, 310
genetically modified Y-1 tobacco and, 1: 247
product placement by, 2: 477–478, 692
protection of documents by lawyers, 2: 626
settlement of lung cancer lawsuit against, 1: 218
Brown, Chuck, 1: 333
Brugmansia (Datura arborea), 1: 259, 260; 2: 518
Die Büchse der Pandora (film), 1: 231
Buckingham cigarettes, 1: 331
Bull Durham Tobacco, 1: 11, 13; 2: 497, 670
Burley barns, 1: 71
Burley tobacco
air-curing of, 2: 448, 449
expansion of, in United States, 2: 660
Latin American production of, 2: 574
in Navy plug, 1: 127

White Burley variety of, 1: 246
Burma (Myanmar), 2: *580*, *581*, 582
Burney, Leroy, 1: 53

C

Cabaret image, 1: 79
Caffeine
ban on coffee, in Ottoman Empire, 1: 355
nicotine use related to, 1: 40
smoking and coffee drinking, 1: 35–36
Cage, Nicholas, 1: 235
Califano, Joseph, 1: 54–55
California
clean indoor air legislation in, 2: 431, 537
large-scale state tobacco control program of, 2: 434–435
Smoke-Free Workplace Act, 2: 465
warning label requirements, 2: *678*
California Environmental Protection Agency (CalEPA), 2: 514
"Call for Philip Morris" slogan, 2: 407, *408*
Calumets, **1: 103–105**
for cementing alliances, 1: 105; 2: 554
ceremonial use of, 1: 103–104; 2: *553*
definition of, 1: 105
examples of, 1: *104*, *169*
prehistoric pipes associated with, 1: 69–70
short-stemmed bowl of, 1: 105
Calvert, George, 1: 120–121
Camel cigarettes, **1: 105–107**
advertising and marketing of, 1: *106*, 106–107, 172; 2: 670, 690
"Camel Scoreboard" promotion, 1: 332
health concerns and marketing of, 1: 331
Middle Eastern images in packaging of, 1: 359
youth marketing of, 2: 690
See also Joe Camel
Campaign for Tobacco-Free Kids, 1: 56
Canada
banning of tobacco sponsorships in, 1: 335
freedom of speech and advertising restrictions in, 1: 21
warning label requirements, 2: 677

Canadian tobacco industry
arts events sponsored by, 1: 331
sports events sponsored by, 1: 332, 334
Cancer
additives and, 1: 9
case control studies of, 1: 205–206
early research on risk of, 1: 267–268
insurance rates based on risk of, 1: 268–269
mortality rates related to, 1: 205–207, *206*
relative risk of, 1: 200
secondhand smoke as cause of, 1: 346; 2: 209–210, 210t, 513, 514, 538, 627
smoking cessation for risk reduction, 1: 5–6
tobacco combined with drugs and, 1: 41
tobacco industry's response to evidence of, 1: 206–207, 298–299, 307–308, 323
U.S. Surgeon General reports/statements on, 1: 53, 201, 201t–203t, 207, 208, 323; 2: 506
See also Carcinogens
See also Disease risk
See also Lung cancer
"Cancer by the Carton" (Norr), 1: 208, 337; 2: 429
Cannabis
African consumption of, 1: 24
cancer risk related to, 1: 41
definition of, 1: 24
tobacco smoking with, 1: 37, 41
Cape Verde islands, tobacco production in, 2: 443
Capital investment, definition of, 2: 530
Caporal tobacco, 1: 248
Caracas, tobacco trade of, 1: 110
Carbon monoxide
from bidi cigarettes, 2: 632
in tobacco smoke, 2: 632
Carcinogens
additives as, 1: 9
in chemically changed tobacco products, 1: 116
definition of, 1: 9
polynuclear aromatic hydrocarbons (PAHs) as, 2: 632, 633
in secondhand smoke, 1: 209; 2: 512, 514, 538
in smokeless tobacco, 2: 632
tar as, 2: 506, 631, *631*
in tobacco smoke, 2: 632

INDEX

Carcinogens CONTINUED
  tobacco-specific nitrosamines
    (TSNAs) as, 1: 113; 2: 632
Cardiovascular diseases
  links between smoking and,
    1: 207–209
  mortality with, 2: 513
  secondhand smoke and, 1: 210;
    2: 513
  Surgeon General's findings on,
    1: 202t
Caribbean, **1: 107–111**
  Caracas and Puerto Rican trade,
    1: 110
  contraband trade in, 2: 542–544
  control of tobacco production by
    Spain, 1: 108–111; 2: 586
  crisis of late eighteenth-century
    trade and, 1: 111
  French tobacco colonies in,
    1: 240–241
  non-Hispanic islands in, 1: 108
  plantations in, 2: 427–428
  Santo Domingo tobacco trade,
    1: 109–110
  See also Cuba
Carmen (Bizet), 1: 365; 2: 670, 681
Carotte (carrot)-shaped tobacco, for
  nasal snuff, 1: 261–262; 2: 549
Cartel
  American Tobacco Company as,
    1: 42–43
  definition of, 1: 42
  Imperial Tobacco Company as,
    1: 253–254
Carter, Helena Bonham,
  2: 672–673, 680
Casablanca (film), 1: 236, 330
Casals, Pablo, 1: 367
Castano group, class action
  litigation by, 1: 311–312
Catherine de Medici, 2: 388, 611
Catholic church
  banning of smoking in church,
    1: 138, 168, 361; 2: 458
  communion and tobacco use,
    1: 137–138
  condemnation of smoking,
    2: 403
  conversion of Amerindians to
    Christianity, 1: 136–137
  financial support from tobacco,
    1: 141–142, 187
  prohibition of smoking in Brazil,
    2: 441
  tobacco use by clergy, 1: 137–138
  See also Missionaries
Catlin, George, catlinite named for,
  1: 104
Cause and effect. *See* Medical
  evidence (cause and effect)

CDC (Centers for Disease Control
  and Prevention), 2: 495
Celebrities
  on cigarette cards, 1: 14
  endorsement of tobacco products
    by, 2: 665, 690
  health-oriented testimonials by,
    2: 671
  influence on smoking initiation,
    1: 32
  lung cancer in, 2: 672
  See also Film
  See also Visual arts
Center for Indoor Air Research
  (CIAR), 2: 516, 539, 626
Centers for Disease Control and
  Prevention (CDC), functions of,
  2: 495
Central America. *See* South and
  Central America
Cerebrovascular disease, 1: 202t
Cervical cancer, 1: 202t
Chairs, for smokers, 1: 163
*The Changing Cigarette: A Report of
  the Surgeon General*, 2: 507
Chattel, definition of, 1: 124
Chemistry of tobacco and tobacco
  smoke, **1: 111–116**
  in bidi cigarettes, 2: 577
  chemical analysis and product
    design, 1: 114–115;
    2: 632–633
  chemical compounds in smoke,
    1: 112, *112*; 2: 390, 632
  curing method effect on, 1: 113
  health effects of product changes,
    1: 115–116
  manufacturing and processing
    effects on, 1: 113–114
  See also Additives
  See also Toxins
Cheroots, 2: 579, *580*, *581*
Chesapeake region, **1: 117–124**
  absence of large towns in, 1: 123;
    2: 640
  British taxes on tobacco in, 1: 197
  Civil War and end of slavery in,
    1: 124; 2: 531
  contraband trade in, 2: 543–544,
    544
  creole population in, 2: 638
  difficulties faced by colonizers of,
    1: 118–119; 2: 637–638
  diversification of crops in, 2: 529,
    640
  Dutch tobacco imports from,
    1: 221
  expansion and contraction of
    production in, 2: 529–531
  export of tobacco from, 1: 121;
    2: 525, 635, *636*

  foundations of success in,
    1: 119–120
  human cost of tobacco trade in,
    2: 642–643
  indentured servants in, 1: 99, 123,
    291–292; 2: 424, 637–638
  Maryland colony, 1: 120–121
  Navigation Acts and, 1: 96, 121
  overdependency on tobacco,
    2: 640
  plantation model in, 2: 424, 427
  planters' debts to British
    merchants, 1: 98, 98–99
  political institutions in, 1: 119
  population of, 1: 122
  price of tobacco in, 1: 98t,
    99–101; 2: 2: 640, 638, *639*,
    641
  private property (headrights) in,
    1: 119
  productivity of tobacco growers
    in, 1: 99–101; 2: 638–639
  rivers in, for transporting
    tobacco, 2: 426
  settlement patterns in,
    1: 122–123
  slavery in, 1: 99, 123–124,
    292–293; 2: 526, 526–527
  tobacco cultivation in,
    1: 121–122
  tobacco trade as basis for
    settlement of, 1: 96–97,
    122–124; 2: 637–638
  Virginia colony, 1: 117–119, *118*
Chewing tobacco, **1: 125–128**
  baseball players and, 1: 126, *126*,
    173
  Brazilian warrior with, 2: 572
  carcinogens in, 2: 632
  current brands of, 1: *128*
  in developing countries, 1: 195
  expectoration associated with,
    2: 655–666
  fine-cut and long-cut tobacco for,
    1: 128
  forms of, in India, 2: 577–578
  graduated product marketing by
    U.S. Tobacco, 1: 115–116
  Indian women's use of, 1: 195
  plug form of, 1: 126, 127
  popularity in United States,
    1: 168; 2: 655
  references to, in popular music,
    1: 372
  scrap tobacco in, 1: 127–128
  twist form of, 1: 127
  warning labels required for,
    2: 676
  working-class use of, 1: 153–154
  youth marketing of, 1: 334
  See also Smokeless tobacco

INDEX

Chibouk (Middle Eastern pipe),
1: 355; 2: 420
Children
advertising targeted to, 1: 15, *15*,
334–335
Campaign for Tobacco-Free Kids,
1: 56
secondhand smoke effects on,
1: 209, 210*t*; 2: 513
smoking by Indonesian Dani
child, 1: *198*
as tobacco workers, 1: 263, 354;
2: 658
See also Youth marketing
See also Youth tobacco use
Children's Act of 1908, 1: 334;
2: 694
China
advertising of tobacco in, 1: 15
British American Tobacco market
in, 1: 93–94, 132–133, 254
cigarette smoking in, 1: 132–135,
133
Communist Party and tobacco
production in, 1: 134
Jesuit missionaries in, 1: 362
map of, 1: *130*
mortality due to tobacco in,
1: 205, 211
names for tobacco, 1: 129
opium use and anti-opium
movement in, 1: 132–135
pipes used in, 1: 131–132;
2: 420–421
snuff usage in, 1: 131–132;
2: 550, 613
tobacco epidemic in, 1: 211
tobacco imports for blending,
2: 645
See also Chinese medicine
Chinese medicine
debates over therapeutic use of
tobacco, 2: 613–614
materia medica of eighteenth and
nineteenth century, 2: 613
perceived benefits of tobacco in,
1: 130–131
seventeenth century, 2: 612–613
twentieth century, 2: 614–615
yin and yang cosmology in,
2: 612
"Chippers," and resistance to
tobacco addiction, 2: 469
Christianity, **1: 135–143**
changing views on tobacco,
1: 360–361
condemnation of smoking, 2: 403
contemporary antitobacco efforts,
1: 142–143
conversion of Amerindians to,
1: 136–137

financial benefits of tobacco for,
1: 141–142, 187
Protestant stance on tobacco use,
1: 139–141
tobacco as diabolical idolatry,
1: 136–137
See also Catholic church
See also Missionaries
Chronic obstructive lung disease
(COPD), 1: 202*t*
Chutta
in India, 2: 579
reverse-*chutta* smoking, 2: 583,
684
CIAR (Center for Indoor Air
Research), 2: 516, 539, 626
Cigamod *(Cigarrera La Moderna)*,
1: 353
Cigar. *See* Cigars
Cigar bars, definition of, 2: 535
Cigar boxes
advertising on, 1: 11; 2: 669
Good Shot cigars, 1: 78
Cigar Makers International Union,
1: 266
Cigar-store figures, 1: 163–164,
164; 2: 497
Cigarette cards
as advertising medium, 1: 11–12
of Allen and Ginter firm, 1: 73
athletes on, 2: 597
classical music celebrities on,
1: 368
as collectibles, 1: 327; 2: 696
definition of, 1: 11
Duke, Buck, advertising use by,
1: 79, 327
scantily clad women on,
1: 11–12, 32, 79; 2: 681
women theatre stars on, 1: *14*
Cigarette Labeling and Advertising
Act
ban on radio and television
advertising, 2: 672
bypassing of, by tobacco
industry, 2: 673
preemption of other regulations
by, 2: 429–430, 675
warning label required by, 2: 429,
672, 675
Cigarette warning labels. *See*
Warning labels
Cigarettes, **1: 144–149**
accessories for, 1: *159*, 160;
2: 568
addiction due to, 1: 3–5
additives in, 1: 6–10
advantages (convenience) of,
1: 168–169
Bright leaf tobacco in, 1: 149;
2: 659

China, 1: 132–135, *133*
class differences among smokers,
1: 154
cost of, consumption related to,
1: 184, 184*t*
disclosure of tar and nicotine
levels in, 2: 506, 619
as dominant mode of
consumption, 1: *171*, 171–173
as drug delivery system, 1: 3–5,
82
effeminacy associated with,
2: 452, 562–563
feminized versions of, 2: 452,
453, 685
fire-safe cigarettes, **1: 238–239;**
2: 493
as gateway to alcohol/drug use,
1: *39*, *40*, 46–47
Gitanes/Gauloises, 1: *241*,
248–250, *249*, *250*
global marketing and sales of,
1: 146–147, 251–256, 265
hand-rolled, 1: 144–145, 169;
2: 452
immigrants associated with,
2: 452
Indian production of, 2: 577
kretek, 1: 195, 287–289; 2: 581
mass production and
consumption of, 1: 145–146,
251–253, 264; 2: 452–454,
453, 564, 659
mental functions enhanced by,
2: 562, 669
menthol cigarettes, 1: 63, 228,
348–349; 2: 584
Middle Eastern consumption of,
1: 358–359
mildness as selling point for,
2: 453
modern cigarettes, 2: 454–455
Native American types of, 2: 379
noncombusting types of, 2: 493
opposition to, by antismoking
movement, 1: 45–47, *47*,
147–148
origins of, 1: 144–145, 168
ostracism of smokers, 2: 404, *404*
purchase of, by pack vs. carton,
2: 497, 499
smoking by soldiers during world
wars, 1: 32, 49–50, 148, 172;
2: 568–569, *569*
as supplementary tool, 2: 562
See also Bidi cigarettes
See also Camel cigarettes
See also "Light" and filtered
cigarettes
See also Lucky Strike cigarettes
See also Marlboro cigarettes

INDEX

Cigarettes CONTINUED
  See also Product design
  See also Prohibitions
  See also "Safer" cigarettes
  See also Smoking restrictions
  See also Virginia Slims
Cigarrera La Moderna (Cigamod),
  1: 353
Cigarrera La Tabacalera Mexican
  (Cigatam), 1: 353
Cigars, **1: 150–153**
  accessories for, 1: *159*, 159–160
  American production of,
    1: 151–152
  apprenticeships for
    manufacturing of, 1: 266
  British production of, 1: 151
  cigar-store figures, 1: 163–164,
    164
  companion set for, 1: 160
  Connecticut Valley leaf for,
    2: 449, 655
  current consumption of,
    1: 152–153
  dispensers for, 1: 160
  European import of South
    American tobacco for, 2: 573
  humidors for, 1: 163
  Indonesian cigar wrappers, 2: 580
  Kipling, Rudyard, comparison of
    cigars to harem, 2: 669,
    680–681
  lighters (lamps) for, 1: *159*,
    159–160
  machine-production of, 1: 266
  manufacturing of, 1: 265–266
  in Mayan artifacts, 1: *340*, 341
  McCall, June, as "Miss Perfecto,"
    2: *682*
  origins of, 1: 150–151, 167–168
  in popular music, 1: 372
  popularity after Napoleonic wars,
    2: 644
  resurgent popularity of, in
    England, 2: 561
  sexual images associated with,
    1: 78; 2: 669, 680–681, *682*
  sik'ar as origin of "cigar," 1: 150,
    167
  smoking of lit end of (reverse-
    chutta), 2: 583, 684
  South and Central American cigar
    makers, 2: 573
  warning labels requirements,
    2: 676, *678*
Cigatam *(Cigarrera La Tabacalera
  Mexican)*, 1: 353
Ciliatoxic agents
  definition of, 1: 114
  in tobacco smoke, 2: 632
Cipollone, Rose, 1: 310

Cipollone v. Liggett Group, Inc.,
  1: 309–310
Citrate, definition of, 1: 238
Civil disobedience, against tobacco
  industry, Australia, 2: 620
Civil War
  emancipated slaves as soldiers in,
    2: 531
  pipes and tobacco accessories
    from, 1: 67
Class, **1: 153–155**
  aristocratic snuff use, 1: 153;
    2: *548*, *549*, 559–560
  British smoking habits and,
    2: 622
  cheap cigarettes and, 1: 154, *154*
  cigarette availability to lower
    social classes, 1: 78
  individualism and smoking,
    1: 153
  Japanese tobacco consumption,
    1: 279
  sailors as lower-class citizens,
    2: *510*, 510–511
  of women smokers, 2: 680, 683
  working-class tobacco use,
    1: 153–154; 2: 694
Class action litigation, 1: 311–312,
  312
Classical music. *See* Music, classical
Clay pipes
  cases for, 1: 158–159; 2: *417*
  colonial American, 2: 416–417
  developing countries, 1: 195
  European, 2: *416*, 417
Clean indoor air legislation
  in Australia, 2: 621
  California's Smoke-Free
    Workplace Act, 2: *465*
  international politics and, 2: 438
  municipal ordinances as,
    2: 431–432, *432*
  states with indoor air legislation,
    2: 516, 537
Clothing, for smokers, 1: 151, *162*,
  163
Clove cigarettes. *See* Kretek
Coalition on Smoking OR Health,
  1: 56
Cocaine, 1: 41
Cocoa, as tobacco additive, 1: 7
Coffee
  ban on, in Ottoman Empire,
    1: 355
  dissemination of, tobacco
    diffusion vs., 2: 635
  relationship between nicotine and,
    1: 40
  smoking and coffee drinking,
    1: 35–36
"Coffin nails," 1: 46; 2: 462, 696

Colbert, Jean-Baptiste, 1: 241–242
Colonial Debts Act of 1732,
  1: 98–99
Colonialism and imperialism. *See*
  British Empire; Caribbean;
  Chesapeake region; Cuba; Dutch
  Empire; French Empire;
  Globalization; Portuguese
  Empire; Spanish Empire; Trade
Columbus, Christopher, 2: 585,
  586
Comedy, in English Renaissance
  literature, 1: 225
Compañia Arrendataria de Tabacos
  (CAT), Spain, 2: 589
Comprehensive Smokeless Tobacco
  Health Education Act of 1986,
  2: 676
Concerts, smoking at, 1: 366,
  366–368; 2: 534
Concessions, definition of, 1: 272
Conestoga cigar (stogie), 1: 151
Conglomerate, definition of, 1: 256
Connecticut, tobacco control
  legislation in, 2: 433
Connecticut Valley leaf, 2: 449, 655
Connoisseurship, **1: 155–166**
  Arents Collection as, 1: 73–76
  cigar accessories, 1: *159*, 159–160
  cigar-store figures, 1: 163–164,
    164
  cigarette accessories, 1: *159*, 160;
    2: 568
  Civil War accessories, 1: 67
  clothing, 1: *162*, 163
  definition of, 1: 155
  furniture, 1: 163
  historical significance of tobacco
    use, 1: 164
  Inuit ash and quid boxes, 1: 69
  lighters, 1: 164
  match safes, 1: 164
  pipes and pipe accessories,
    1: 156–159, *157*, *158*, *165*
  snuff accessories, 1: 160–163,
    161; 2: 549–550, *560*, *667*
  types of tobacciana, 1: 156
  See also Accessories
Consolidation, definition of, 1: 85
Consumption (demographics),
  1: 167–184
  advertising and cigarette
    consumption, 1: *171*, 171–173
  Africa, 1: 29
  age level of smokers and,
    1: 31–33
  Australia, 2: 619
  changes in per capita
    consumption (1990-2000),
    1: 183–184, 183t, 184t
  cigar consumption, 1: 152–153

colonial Brazil, 2: 441–442
cost of cigarettes and, 1: 184, 184t
demographic of quitting, 1: *176*, 176–177
developing countries, 1: 194–195
early twentieth century, 1: 31–33; 2: 462
England and Wales, in seventeenth century, 2: 525–526
health considerations in, 1: 173–174, *174*
Japan, 1: 277–280
later twentieth century, 1: 33; 2: 471, 684
male and female differences in smoking initiation, 1: *175*, 175–176
male-female smoking rates, 1: 177t, 178–181, 178t, 180t; 2: 684
methods of consuming tobacco, 1: 167–169
Middle East, 1: 358–359
Nazi Germany, 2: 384
Oceania, 2: 395
origins of tobacco consumption, 1: 170
peak per capita consumption, 1: 179t, 180t, 182
per capita consumption during 2000, 1: 181t, 182–183
per capita consumption in U.S. (1880-1950), 1: *171*, 172
Portuguese colonies and Portugal, 2: 441–442
prohibition effects on, 2: 462, 465
rising proportion of women smokers, 2: 684–685
South Asia, 2: 577–579, *578*
starting and quitting patterns in U.S., 1: *175*, 175–177, *176*, *177*
tobacco control events and per capita consumption, 2: 430, *430*
trends in consumption (1970-2000), 1: 179t, 180t, 181–182, 181t
in World War I, 1: 32, 172; 2: 568–570
worldwide patterns of smoking, 1: 177–181
worldwide spread of tobacco consumption, 1: 170–171
Contraband trade
 definition of, 1: 110
 See also Smuggling and contraband
Control of tobacco use. *See* Antismoking legislation; Regulation of tobacco products in United States; Smoking restrictions; Tobacco control
Convenience stores
 tobacco sales in, 2: 498
 See also Retailing
Cooperative(s)
 African tobacco growers, 1: 27
 definition of, 1: 27
COPD (chronic obstructive lung disease), 1: 202t
Cope's Tobacco Plant, A Monthly Periodical, Interesting to the Manufacturer, the Dealer, and the Smoker, 1: 304
Coppola, Sophia, 1: 237
Corn, cultivation by Native Americans, 2: 379, *380*
Coronary heart disease, 1: 202t, 210
Cosmology, of Native Americans, 2: 554–555, *555*
Cotton, Charles, 1: 225
Cotton Club cigarette girls, 1: *330*
Cotton farming, in United States, 2: 656–657
Council for Tobacco Research (CTR), 1: 217; 2: 429, 476
A Counterblaste to Tobacco (King James I of England), 1: 75, 83, 139, 303; 2: 459
Coupons, on cigarettes packages, 1: 327–328
Covington and Burling (law firm), 2: 627
Cowboy, in Marlboro cigarettes advertising, 1: 338, *338*; 2: 407, 671, 691
Crack cocaine, cancer risk related to, 1: 41
Creation stories, of Native Americans, 2: 375
Creoles
 in Chesapeake region, 2: 638
 definition of, 1: 137
Crimean War, 2: 667
Cross Cuts (cigarettes), 1: 332
Crow Tobacco Planters, 2: 554
Cruella De Vil (film character), 1: *234*
CTR (Council for Tobacco Research), 1: 217
Cuba, **1: 186–190**
 cigar production in, 1: 150–151, *152*, 188
 decline of tobacco industry during eighteenth century, 1: 111
 financing of Spanish purchases from, 1: 190
 land conflicts and tobacco production in, 1: 189–190
 protection of tobacco trade by Spain, 1: 109, 150–151, 187
 quality of tobacco from, 1: 186
 sharecropping in, 1: 294
 slave labor in, 1: 292
 Spanish monopoly of tobacco trade in, 1: 187–189
 tobacco factory labor in, 1: 295–296
 See also Spanish Empire
Cullman, Joseph F., III, 1: *21*
Cultivation of tobacco. *See* Production of tobacco; United States agriculture
Cunliffe-Owen, Hugo, 1: 94
Curing. *See* Processing
Cytrel (tobacco substitute), 2: 507

D

Dandies
 posturing of, 1: *162*
 snuff use by, 2: 548
 See also Gallants
Dannemann (tobacco company), 1: 90, 92
Dannemann, Gerhard (Geraldo), 1: 89
DAP (diammonium phosphate), in Marlboro cigarettes, 1: 10
Dark tobacco. *See* Black tobacco
Darnel, Linda, 2: 671
Datura arborea (Brugmansia), 1: 259, 260; 2: 518
Davies, John, 1: 224, 225
Davis, Bette, 2: 564, 670–671
Davis, Sammy, Jr., 1: 332
de Jerez, Rodrigo, 2: 403
Dead Souls (Gogol), 2: 534
Dealer promotions, 1: 328
Dean, James, 1: 235, *235*
Death
 and resurrection, in shamanistic initiation rituals, 2: 519–520
 smoking as defiance of, 2: 564–566
 See also Mortality
Decroix, Émile, 1: 59
Deer tongue extract, 1: 9
Defiance
 smoking as act of, 2: 565–566
 See also Rebelliousness
Dekker, Thomas, 1: 224
Delaware, tobacco control legislation in, 2: 433
Demographics. *See* Consumption (demographics)
Denmark, snuff use in, 2: 550

INDEX

Depressant
  alcohol as, 1: 40
  definition of, 1: 40
Depression, smoking associated with, 2: 470
Developing countries, **1: 193–198**
  consumption patterns in, 1: 194–195
  cultivation of tobacco in, 1: 196–197
  forms of tobacco consumed in, 1: 195, *196, 198*
  health impacts of tobacco in, 1: 195–196
  mortality due to tobacco in, 1: 193, 205, 211, *212*
  street vendors as tobacco retailers, 2: *498*
  tobacco control in, 1: 197–198
  tobacco industry targeting of, 1: 254–255; 2: 438
  warning labels for cigarettes, 2: 676–677
Diammonium phosphate (DAP), in Marlboro cigarettes, 1: 10
Dietrich, Marlene, 1: 231, *232, 234*; 2: 564, 670
Diffusion of tobacco
  Africa, 1: 24–25, *25*
  to Europe, 2: 634–635
  by missionaries, 1: 362–363
  Native American tobacco, 2: 377, *378*
  to Oceania, 2: 394–395
  to Philippines, 2: 410–412
  by sailors, 2: 509–511, 634
  by shamans, 2: 518–519
  worldwide spread of tobacco consumption, 1: 170–71
  See also Consumption (demographics)
  See also Origin of tobacco
al-Din Shah, Nasir, 1: 272, *273*, 273
Disease, **1: 199–212**
  cardiovascular diseases, 1: 202t, 207–209, 210; 2: 513
  in Chesapeake settlements, 2: 637–638
  in developing countries, 1: 193, 195–196
  doctors' resistance to smoke-related disease evidence, 1: 216, 321–322
  global tobacco epidemic of future, 1: 210–212, *212*
  health effects of active smoking, 1: 201–203, 201t–203t
  health expenditures for, user fees for payment of, 2: 607–608
  insurance for, 1: 267–269

medical evidence (cause and effect) for, 1: 321, 343–347
psychopathology related to smoking, 2: 470
secondhand smoke as cause of, 1: 209–210, 210t; 2: 512–516, *515*, 537–538, 624, 627
smoking cessation for prevention of, 1: 5–6
Surgeon General's list of smoke-related diseases, 1: 201, 201t–203t
tobacco use by doctors and, 1: 214–215
See also Cancer
See also Disease risk
See also Lung cancer
See also Mortality
See also Toxins
Disease risk
  additives in, 1: 9–10
  of chemically changed tobacco products, 1: 115–116
  early recognition of, 1: 50–51
  early research on cancer risk, 1: 267–268
  emerging awareness of, in U.S., 1: 208
  global tobacco epidemic and, 1: 210–212, *212*
  insurance rates for nonsmokers based on, 1: 268–269
  Jewish prohibition of tobacco use due to, 1: 282–283
  main types of, 1: 84
  marketing of cigarettes and, 1: 331
  medical evidence (cause and effect) in, 1: 321, 343–347
  methods of smoking for risk reduction, 2: 506
  prohibition of tobacco use based on, 2: 464–465
  relative risk of cancer, 1: 200
  scientific verification of, 1: 52–53
  secondhand smoke in, 1: 209–210, 210t; 2: 346–347, 512–516, *515*, 537–538, 624, 627
  smoking cessation for cancer risk reduction, 1: 5–6
  tobacco combined with drug use in, 1: 41, 46
  tobacco industry's research on, 2: 625–628
  tobacco industry's response to evidence of, 1: 206–207, 298–299, 307–308, 323
  See also Disease
  See also Secondhand smoke
Dispensers, for cigars and cigarettes, 1: 160

Dispersal of tobacco. *See* Diffusion of tobacco
Diversification, definition of, 1: 256
Divinity, tobacco associated with, 1: 224
DOC (Doctors Ought to Care), 1: 55–56
Doctors, **1: 214–216**
  resistance to smoke-related disease evidence, 1: 216, 321–322
  tobacco use by, 1: 214–215
Doctors Ought to Care (DOC), 1: 55–56
Documents of tobacco industry, **1: 216–219**
  addictive nature of cigarettes in, 1: 4; 2: 494, 627
  availability on Internet, 1: 218; 2: 437
  Cipollone case and access to, 1: 309–310
  as courtroom evidence, 1: 217–218
  exposure of, 1: 218–219; 2: 625
  hiding and protection of, 2: 626–627
  on industry-sponsored Websites, 1: 218
  MSA requirement for release of, 1: 218; 2: 437
  See also Litigation
Doll, Richard, 1: 199–200, 206, 207, 344
Dominican Republic
  export of cigar tobacco from, 2: 573
  newspaper description of tobacco trade in, 2: 573
  Santo Domingan tobacco trade, 1: 109–110
  tobacco traders in, 2: 572
Dopamine
  definition of, 1: 40
  tobacco effects on, 1: 40; 2: 389
Douglas, Michael, 1: 235
Doyle, Arthur Conan, 1: 270; 2: 534
Drama, in English Renaissance literature, 1: 224
Drugs
  addiction to, 1: 2–3
  cannabis as, 1: 24, 37, 41
  crack cocaine as, 1: 41
  hallucinogens as, 1: 259–260, 341; 2: 518, 520, 555
  See also Drugs, tobacco combined with
  See also Opium

Drugs, tobacco combined with, **1: 35–41**
  alcohol use with, 1: 35–36, *36*, *37*, 38–39
  betel chewing with, 1: 36–37
  biosocial linkages in, 1: 38
  cannabis with, 1: 37
  difficulty of quitting tobacco vs. drugs, 2: 468
  gateway hypothesis in, 1: 39, *40*, 46–47
  health consequences of, 1: 41
  history of, 1: 35–37
  opium with, 1: 37
  scientific research on, 1: 39–40
Du Maurier (tobacco company), 1: 334, 336
Duke, James Buchanan
  of American Tobacco Company, 1: *42*, 42–45
  licensing of Bonsack Cigarette Machine by, 1: 146, 171; 2: 452
  "Plug Wars" and, 1: 127
  promotion of tobacco to retailers, 2: 497
  use of cigarette cards for advertising, 1: 79, 327
Dupin, Armandine Aurore Lucille (George Sand), 2: 668, 680, 682
Dutch Empire, **1: 219–222**
  African tobacco trade of, 1: 24
  Caracas and Puerto Rican tobacco exports to, 1: 110
  Caribbean contraband trade of, 2: 543
  Chesapeake tobacco trade of, 2: 544
  clay pipes made in, 2: *417*
  colonial tobacco cultivation by, 1: *220*, 220–221
  cultivation of tobacco in Holland, 2: 635
  decline of tobacco trade in, 1: 222
  re-export of tobacco from British ports, 2: 642, 642t
  taxation of tobacco by, 1: 222
  tobacco trade and industry in, 1: 221
Duty
  definition of, 2: 502
  duty-free cigarettes in World War I, 2: 569, *569*
  duty-free vendors, 2: 502
Dysphoria, definition of, 2: 562

E

Earhart, Amelia, 1: *319*
Eclipse (cigarette), 2: 493
Ecology, definition of, 1: 246
Econometrics, definition of, 1: 22
Edison, Thomas, 1: 46–47
Education, demographic differences in quitting smoking, 1: *176*, 177
The Effects of Intemperance (Steen), 2: 666
Die Ehe der Maria Braun/The Marriage of Maria Braun (film), 1: 234–235
Ejrup, Borje E.V., 2: 483–484
Elbow pipes, 1: 69
Emancipation
  as advertising theme, 1: 16
  manumission of slaves, 2: 529
  of slaves, 2: 531–532
Embryonic, definition of, 1: 254
Emphysema, 1: *203*
Endometrial cancer, 1: 202t
English Renaissance literature, **1: 223–226**
  male and female smokers depicted in, 1: 225
  medical use of tobacco in, 1: 223–224
  recreational use of tobacco in, 1: 224–225
Environmental Protection Agency
  report on secondhand smoke, 1: 346–347; 2: 514
  tobacco industry lawsuit against, 2: 516
Environmental tobacco smoke. *See* Secondhand smoke
Ephemeral, definition of, 1: 156
Epidemiological, definition of, 1: 51
Epidemiological studies
  case control studies in, 1: 205–206
  causality criteria for, 1: 345–346
  of "light" and filtered cigarettes, 1: 299, 300–301
  of lung cancer, 1: 200–201, 207, 323–324, 344; 2: 622–623
  of secondhand smoke, 2: 513–514
  statistical significance of, tobacco industry and, 2: 515
Epidemiology
  causation and mid-twentieth-century epidemiology, 1: 344–346
  definition of, 1: 200
  of lung cancer, 1: 200
"Epigram 82. Of Tobacco" (Davies), 1: 224, 225
Epinephrine
  definition of, 2: 389
  shamanistic vision and, 2: 521
  stimulation by tobacco, 2: 389
Eroticism. *See* Sexual symbolism of smoking
Esophageal cancer, 1: 202t

Estado da India (Portuguese colonies), 2: 441, 445, 446
Ethnic groups, **1: 226–229**
  smoking among, 1: 226–227
  as target of tobacco advertising, 1: *228*, 228–229
Ethnicity, definition of, 1: 226
Ethnohistory, definition of, 1: 67
Eugenics, and antismoking movement, 1: 48
Eugenol
  definition of, 1: 195
  released from *kreteks*, 1: 195
European colonialism. *See* British Empire; Caribbean; Chesapeake region; Cuba; Dutch Empire; French Empire; Globalization; Portuguese Empire; Spanish Empire
European Common Market, 2: 646
European Union
  Common Agricultural Policy (CAP) of, 2: 646
  warning label requirements, 2: 677
Evin law (France), 1: 60
Exalt (noncombusting cigarette), 2: 493
Expanded tobacco
  definition of, 1: 238
  in modern cigarettes, 2: 454

F

Factoría de Tabacos de La Habana, 1: 187, 189
Faerie Queene (Spenser), 1: 223, 303
Far East. *See* China; Japan
Farm labor. *See* Labor
Farm subsidies. *See* New Deal
Fassbinder, Rainer Werner, 1: 234
Fatwa
  definition of, 1: 273
  permission of tobacco use by, 1: 355
FDA. *See* United States Food and Drug Administration (FDA)
Federal Food, Drug, and Cosmetic Act, 2: 493
Federal Trade Commission (FTC)
  testing method for tar and nicotine levels, 1: 5
  warning labels proposed by, 2: 675
Females. *See* Women
Femmes fatales
  in films, 1: 233–235, *234*
  and respectability of women smokers, 2: 680
  Spanish women as, 2: 670
Fems (cigarettes), 2: 685
Fernö, Ove, 2: 488, 489

INDEX

Fight Club (film), 2: 672
Figurine pipes, 1: 69
Film, **1: 231–237**
   acceptability of smoking enhanced by, 1: 79
   femme fatale role in, 1: 233–235, *234*
   glamorization of smoking in, 2: 564, 670–671, *671*
   humanness portrayed in, 1: 236–237
   meaning of tobacco in, 1: 231–233
   product placement as advertising in, 1: 13, 232; 2: 477–478, 691–692
   rebel role in, 1: 235, *235*
   tobacco industry's influence on, 1: 232–233
   vamp as rebel in, 1: 235–236
   See also Visual arts
Film noir
   characteristics of, 1: 231
   definition of, 1: 231
   smoking in, 2: 671, *671*
Filtered cigarettes. *See* "Light" and filtered cigarettes
Fine arts events, tobacco industry sponsorship of, 1: 331–332, *333*
Finely cut tobacco
   for chewing tobacco, 1: 128
   kizami, in Japan, 1: 279
Fire-curing of tobacco, 2: 449, 655
Fires
   fire-safe cigarettes, **1: 238–239**, *239*
   noncombusting types of cigarettes, 2: 493
   tobacco-related, San Nicolas Convent, Philippines, 2: 412
Flavorants. *See* Additives
Flight attendants, class action lawsuit by, 1: 34, 312
Florida, clean indoor air legislation in, 2: 431, 433
Flue-cured tobacco
   black Kentucky tobacco, 1: 285
   Bright leaf tobacco, 1: 28, 149, 251; 2: 449, 659
   chemistry of, 1: 113
   definition of, 1: 28
   Latin American production of, 2: 574
   Malawi, Africa, 1: 28
   in modern cigarettes, 2: 454
   New Deal price supports for, 2: 385
   process of curing, 1: 28; 2: 449–450, 659
   production in foreign countries, 1: 255

Food
   sacramental, tobacco as, 2: 518, 521–522
   See also Appetite suppression
Food and Drug Administration (FDA). *See* United States Food and Drug Administration (FDA)
A Fool There Was (film), 1: 233
Framework Convention on Tobacco Control
   advertising restriction as requirement of, 1: 23
   international politics and, 2: 438–439, *439*
   provisions of, 2: 438, 495
   warning label requirements of, 2: 677–678
Framingham Heart Study, 1: 208
France
   antismoking movement in, **1: 59–60**
   cigarettes named by, 1: 144
   clay pipes made in, 2: 416
   Gitanes/Gauloises cigarettes, 1: 248–250, *249–250*
   "le smoking" (smoking jacket) in, 1: 151
   literature about tobacco, 1: 304–305
   monopoly of tobacco as alternative to prohibition, 2: 460
   re-export of tobacco from British ports, 2: 642, 642t
   SEITA for maximizing tobacco industry in, 1: 244
   seventeenth- and eighteenth-century music in, 1: 365
   size of tobacco factories in, 1: 263
   smoking restriction in, 2: 539
   smuggling in, 2: 545–546
   snuff use in, 2: 548
   tobacco cultivation and production in, 1: 243–244
   See also French Empire
Franciscan missionaries, 1: 361
Fredrickson, Donald T., 2: 484
Freebase nicotine, 1: 4, 113
Freedom of speech, advertising restrictions and, 1: 21
French Empire, **1: 240–244**
   African colonies of, 1: 244
   American revolution and tobacco policy of, 1: 243
   Caribbean tobacco colonies of, 1: 240–241
   local vs. colonial tobacco production, 1: 243–244
   North American tobacco colonies of, 1: 242–243
   royal monopoly of tobacco in, 1: 241–242

   tax revenues from tobacco, 1: 243
   tobacco imports from U.S., 1: 243
   See also France
Freud, Sigmund, 1: 270; 2: 667
FTC. *See* Federal Trade Commission (FTC)
Fumeuse (smoking chair), 1: 163
Furniture, for smoking, 1: 163

G

Gallants
   definition of, 1: 2
   in English Renaissance literature, 1: 224
   manner of smoking and tobacco accessories of, 2: 559, *560*
   snuff use by, 2: 550
   See also Dandies
Gandhi, Mahatma, 2: 575–576
Gant, Harry, 1: 334
Gardening, of ornamental tobacco, 2: *2: 618*; **2: 616–618**; *2: 617*
Gas phase of smoking, 1: 112
GASP (Group Against Smoking Pollution), 1: 54, 56; 2: 536
Gaston, Lucy Page, 1: 48; **1: 49**; 1: 141
Gauloises cigarettes, **1: 248–250**
   advertising for, 1: *241*
   Disque Bleu variety of, 1: 249, *249*
Gender. *See* Masculine identity; Men; Sexual symbolism of smoking; Women
Genetic modification, **1: 245–248**
   commercial efforts in, 1: 245–247, *246*
   for controlling nicotine level, 1: 247
   hybridization in, 1: 246
Genetic ("race") poison, tobacco as, 1: 47–48, 322; 2: 383
Germany
   Bahian tobacco monopoly, 1: 90
   literature about tobacco, 1: 304–305
   seventeenth- and eighteenth-century music in, 1: 365
   wood *gesteckpfeife*, 2: 415
   See also Nazi Germany
Gitanes cigarettes, **1: 248–250**; 1: *250*
Glamorization of smoking. *See* Celebrities; Film; Visual arts
Globalization, **1: 251–256**
   British American Tobacco and, 1: 254–255
   cartels and Tobacco War in, 1: 253–254

cigarette marketing and,
   1: 146–147, 265
foreign investments by tobacco
   companies, 1: 253
global tobacco oligopoly in,
   1: 255–256
international politics and,
   2: 438–439
marketing methods and, 1: 254
mass production of cigarettes
   and, 1: 251–253
modern global industry in, 1: 256
smoke-related diseases associated
   with, 1: 210–212, 212
spread of flue-cured leaf tobacco
   and, 1: 255
See also American Tobacco
   Company
See also British American Tobacco
   (BAT)
See also Industrialization and
   technology
See also Philip Morris
Glover, Danny, 2: 478
Goa, tobacco trade of, 2: 445–446,
   575
Godard, Jean-Luc, 1: 235
Gogol, Nikolay, 2: 534
Goh, Jessica, 2: 457
Golf, sponsorship by tobacco
   industry, 1: 334
Goodness, as theme of *Lost in
   Translation*, 1: 237
Gore, Al, 2: 457
Great Britain. *See* United Kingdom
Great Depression
   New Deal programs and,
      2: 385–387
   sharecropping system and, 2: 524
   tobacco leaf quota during, 2: 660
Green, John R., 2: 497
Greer, Jane, 2: 671
Group Against Smoking Pollution
   (GASP), 1: 54, 56; 2: 536
Guerilla warfare, definition of, 1: 80
Gum, nicotine, 2: 488–489
Gutkha, 2: 578, 578
Gutta-percha
   definition of, 1: 164
   pipes made from, 2: 415

H

Hallucinogens, **1: 259–260**
   in Mayan ceremonial tobacco use,
      1: 341
   Nicotiana rustica as, 2: 555
   ritual use of, 1: 259–260
   shamanistic use of tobacco as
      hallucinogen, 1: 259, 260, 341;
      2: 518, 520, 555

types of hallucinogenic plants,
   1: 259
Harem
   in cigarette advertising, 2: 669,
      670
   Kipling, Rudyard, comparison of
      cigars to, 2: 669, 680–681
   smoking practices in, 1: 356, 357,
      358
Hats, for smokers, 1: 163
Havana cigars, 1: 150, 151
Headrights, in Chesapeake colonies,
   1: 119
Health claims for tobacco,
   unregulated, 2: 492, 493
*The Health Consequences of
   Involuntary Smoking* (U.S.
   Surgeon General), 1: 34; 2: 514
Health insurance. *See* Insurance
Health risks of smoking. *See*
   Disease risk
Health use of tobacco. *See*
   Therapeutic use of tobacco
Heart disease. *See* Cardiovascular
   diseases
Heathen(s)
   definition of, 1: 136
   Native Americans as, 2: 559
   slaves considered as, 2: 527
   tobacco viewed as heathen habit,
      1: 361; 2: 441
Hegemony, definition of, 1: 44
Helmar Turkish cigarettes, 1: 145;
   2: 563
Hemminger, Graham Lee, 2: 672
Henbane *(hyoscyamus)*, tobacco vs.,
   1: 86–87
Herb of Nicot, 2: 388
   See also Nicot, Jean
Hill, Austin Bradford
   "Bradford Hill criteria" named for,
      1: 346
   criteria for causal relationships,
      1: 345–346
   epidemiologic studies of lung
      cancer, 1: 206, 207, 344
Hill, George Washington
   chairman of American Tobacco
      Company, 1: 44
   Lucky Strike advertising
      campaign and, 1: 318, 328,
      335–336
Hill, Percival S., 1: 44
Hirayama, Takeshi, 2: 513–514,
   515, 624, 627
Hispanic Americans
   smoking prevalence among,
      1: 227
   as target of tobacco marketing,
      1: 228–229
Hoffman, Frederick L., 1: 267–268

Hoffmann, Dietrich, 1: 114
Hogshead
   definition of, 1: 119
   storage and shipping of tobacco
      in, 2: 655
Holders, for cigarettes, 1: 160
Holland. *See* Dutch Empire
Holmes, Sherlock, 1: 270; 2: 534
Homes, smoking restrictions and,
   2: 540
Hookahs. *See* Water pipes (hookahs)
Hopkins, Miriam, 1: 234
Houses, of tenant farmers, 1: 71
Hubble-bubble (water pipes),
   1: 167, 195; 2: 420
Hughes, Lloyd "Spud," 1: 348
Humanness, portrayal in films,
   1: 236–237
Hume, Tobias, 1: 226
Humidors, for cigars, 1: 163
Humoral theory
   definition of, 1: 360
   Galen's four humors, 2: 611, 611
   Monardes' health claims based
      on, 2: 610–611
   See also Therapeutic use of
      tobacco
Hunger suppression. *See* Appetite
   suppression
Hybridization
   definition of, 1: 246
   of ornamental tobacco, 2: 617,
      618
Hydraulic, definition of, 1: 127
Hydraulic tobacco presses, 1: 263
Hydrocarbons, aromatic
   polynuclear (PAH), 2: 632, 633
Hyoscyamus (henbane), tobacco vs.,
   1: 86–87
Hyperbolic, definition of, 1: 231

I

I Love Lucy (television show),
   1: 332
Iconography, definition of, 1: 330
Identity
   identity building and youth
      tobacco initiation, 2: 694, 696
   See also Masculine identity
Immigrants, cigarette smoking
   associated with, 2: 452
Imperial Tobacco Company, 1: 43,
   93, 146, 253–254
Impotence, smoking as cause of,
   1: 55, 55
Impressionism, 2: 669
Indentured servants
   as chief labor force in Chesapeake
      region, 1: 291–292
   definition of, 2: 525

INDEX

Indentured servants CONTINUED
  hardships faced by, 2: 637–638
  immigration from England
    2: 526
  interaction with slaves, 2: 527
  protection from overwork, 1: 99
  replacement by slavery, 1: 99
  terms of service of, 1: 123; 2: 424
Independent Scientific Committee on Smoking and Health (ISCSH), 2: 624
India
  British American Tobacco expansion in, 1: 254
  chewing tobacco use by women in, 1: 195
  colonial introduction of tobacco in, 2: 575
  consumption of tobacco in, 2: 577–579, 578
  cost of cigarettes vs. bidis in, 2: 577
  forms of tobacco consumption in, 1: 172–173
  modern cultivation and production of tobacco in, 2: 576–577
  mortality due to tobacco in, 1: 205
  Mughal rulers of, 2: 575
  Portuguese colonies in, 2: 441, 445
  Portuguese tobacco trade with, 2: 446
  prohibition of tobacco use in, 2: 459
  See also Bidi cigarettes
Indian colonies of Portugal, tobacco trade in, 2: 441, 445–446
Individualism
  definition of, 1: 153
  of European middle-class smokers, 1: 153
  link between smoking and, 1: 270
  smoking as metaphor of, 2: 565
  snuff use as mark of, 2: 559–560
  tailoring of cigarette smoking and, 2: 562
Indonesia (Sumatra), 2: 580–582
  kretek manufacturing in, 1: 287–289; 2: 581
  tobacco exports of, 2: 645
  tobacco industry in, 2: 580–581
Indoor air. See Clean indoor air legislation; Secondhand smoke
Industrialization and technology, 1: 261–266
  cigar production and, 1: 265–266
  early industrial production, 1: 263
  globalization related to, 1: 265

mass production of cigarettes, 1: 145–146, 251–253, 264; 2: 452–454, 453, 561, 659
Mexican tobacco industry, 1: 352–353
preindustrial production methods, 1: 261–263, 262
product design and, 2: 452–454, 453
second industrial revolution and, 1: 263–264, 264
See also Labor
Inhaler, for nicotine delivery, 2: 490
Inipi sweat lodge ceremony, 2: 382
Initiation of smoking
  advertising and women's smoking initiation, 1: 17
  age-related factors in, 2: 470, 689–690
  factors in, 2: 470–471, 472
  identity building and youth tobacco initiation, 2: 694, 696
  male and female differences in smoking initiation, 1: 175, 175–176
  risk calculation by youth and, 2: 695–696
  starting and quitting patterns in U.S., 1: 175, 175–177, 176, 177
  Virginia Slims and adolescent girls, 1: 173
  See also Psychology and smoking behavior
  See also Youth marketing
Insurance, 1: 267–269
  basic principles of, 1: 267
  smoke-related cancer research and, 1: 267–269
  tobacco cessation treatment covered by, 1: 269
  for tobacco trading losses, 2: 640
Intellectuals, 1: 270–271; 1: 271; 2: 562, 669
Internet
  prohibition of tobacco advertising on, 2: 502
  sale of tobacco on, 2: 501–502
Inyoka tobacco, Zimbabwe, 2: 699
Ion exchanger complex, definition of, 2: 488
Iranian Constitutional Revolution of 1905-1911, 1: 272
Iranian Tobacco Protest movement, 1: 272–273
Ireland, banning smoking in workplace, 2: 540
Irony, definition of, 2: 664
Iroquois, smoking by, 1: 170; 2: 553
Irritation of tobacco smoke, additives for reducing, 1: 9

ISCSH (Independent Scientific Committee on Smoking and Health), 2: 624
Islam, 1: 274–276
  coffee drinking combined with smoking, 1: 35–36
  Iranian Tobacco Protest movement and, 1: 272–273
  medical use of tobacco and, 1: 274
  on Philippine island of Mindanao, 2: 412
  prohibition of tobacco use, 1: 275–276; 2: 459
  theological scrutiny of tobacco use in, 1: 274–275
  tobacco use in Qajar period, 1: 274, 275
Iwaya Co., Japan, 1: 280

J

Jaguar, identification of shamans with, 2: 518, 521
"Jaguar stones," 2: 518
James I, King of England
  A Counterblaste to Tobacco, 1: 75, 83, 139, 303; 2: 459
  taxation of tobacco by, 2: 603
  vilification of tobacco by, 1: 139
Japan, 1: 277–281
  advertising restrictions in, 1: 20
  consumption of tobacco in, 1: 277–280
  diffusion of tobacco into, 1: 277
  epidemiological studies of secondhand smoke in, 2: 513–514
  kiseru pipes in, 1: 277–279, 278; 2: 420–421
  production and manufacturing of tobacco in, 1: 280–281
  prohibition of tobacco use in, 1: 277, 280; 2: 459
  regulation of tobacco in, 1: 277, 280–281
  smoking restriction in, 2: 538, 539
  tobacco accessories in, 1: 278
  warning labels for cigarettes in, 2: 676
  World War II and smoking in, 1: 279, 279
Japan Tobacco, Inc., 1: 281; 2: 438
Jars, for tobacco, 1: 157–158, 157–158
Jarvik, Murray, 2: 488, 489
Jesuit missionaries, 1: 361, 362
Job cigarette poster, 2: 668, 668

Joe Camel
   advertising impact on youths, 1: 22, 173
   awareness of, among children and teens, 2: 692–693
   on billboard advertisement, 2: *691*
   as "Joe Chemo," 1: *200*
   rejuvenation of Camel cigarettes by, 1: 107
   in youth marketing campaigns, 1: 335
   See also Camel cigarettes
Johansson, Scarlett, 1: 237
Johnson, Samuel, 1: 303
Johnston, Frances Benjamin, 2: 669
Jonson, Ben, 1: 224, 225
José, Edward, 1: 233
Joseph of Cupertino, 1: 138
Judaism, **1: 281–283**
   health risks of tobacco and prohibition of, 1: 282–283
   ritual concerns with tobacco use, 1: 281–282, *282*
   tobacco manufacturing by Jews, 1: 282
Junta da Administraçào do Tabaco, 2: 443, 445, 446
Junta de Tabaco, 2: 588

K

Karuk people, 2: 556
Kent cigarettes, 2: 505–506
Kentucky, **1: 285–287**
   Black Patch War and, 1: 80–81, 285
   tobacco consumption in, 1: 287
   tobacco production in, 1: 285–286, *286*
Kerr-Smith Tobacco Act of 1934, 2: 386, 387
Khaini, 2: 578
Kidder, Margot, 1: 236
Kidney cancer, 1: 202t
King, Billie Jean, 1: 336; 2: 672
Kipling, Rudyard, 2: 669, 680–681
Kiseru pipes, 1: 277–278, *278*; 2: 420–421
Kizami tobacco, 1: 279
Koch, Robert, 1: 343
Kool cigarettes
   advertising of, 1: 332
   Willie the penguin mascot for, 1: 334–335
   youth marketing of, 2: 692
Koop, C. Everett, 1: 34; 2: 463, *464*
Kretek, **1: 287–289**
   as cultural signifier, 1: 289
   definition of, 1: 195
   eugenol released from, 1: 195
   manufacture of, in Indonesia, 1: *288*, 288–289; 2: 581

L

Labor, **1: 291–297**
   child tobacco workers, 1: 263, 354; 2: 658
   farm labor, 1: 291–295, *294*
   indentured servants in, 1: 99, 123, 291–292; 2: 424, 525, 526, 527, 637–638
   manufacturing labor, 1: 295–297
   migrant labor, 1: 295
   on plantations, 2: 424, *425*, 427
   racism and sexism in, 1: 297
   references to, in popular music, 1: 370–371
   sharecropping and, 1: 293–294; 2: 522–524, *523*, 532, 572, 657–658
   slaves in, 1: 99, 123–124, 292–293; 2: 424, *425*, 427, *526*, 526–527, 657
   smallholder farmers and, 1: 29, 293, 295
   tenant farming and, 1: 71, 294; 2: 524, 658–659
   tobacco cultivation requirements, 2: 425–426, 638
   wage work, 1: 293–295
   women workers, 1: 295, 296, 297
   Zimbabwe tobacco workers, 2: 700
   See also Industrialization and technology
   See also Processing
Labor unions
   Cigar Makers International Union, 1: 266
   Cuban cigar makers in United States, 1: 296
Lakotas, *Inipi* sweat lodge ceremony, 2: 382
Landscape of tobacco, 1: 71
Langley, John Newport, 2: 388
Laryngeal cancer, 1: 202t
Lasker, Albert, 1: 331
Lasker, Mary, 1: 331
Latin America. *See* South and Central America
Latter-Day Saints (Mormon Church), 1: 141
Law, John, 1: 242
Lawsuits. *See* Litigation
Lawyers
   direction of tobacco industry research by, 2: 625–628
   See also Litigation
Lee, Ivy, 2: 474
Legislation. *See* Antismoking legislation; Lobbying; Politics; Regulation of tobacco products in United States

Leisure, as advertising theme, 1: 16
Lesbians, smoking as rebellious act of, 2: 683
Leukemia, 1: 202t
Lewin, Louis, 1: 2
Liberty, prohibition of tobacco vs., 2: 456–457
License to Kill (film), 2: 673
Lichtneckert, Stefan, 2: 488
Licorice, as tobacco additive, 1: 7
Liggett Group
   Cipollone v. Liggett Group, Inc., 1: 309–310
   safer cigarette development by (XA Project), 2: 508
   settlement of lawsuit against, 1: 218
"Light" and filtered cigarettes, **1: 298–301**
   addiction and design of, 1: 5
   chemical analysis of smoke produced by, 1: 114–115
   cigarette design manipulation and, 1: 299–300
   compensation for reduced nicotine delivery with, 1: 5, 300; 2: 633
   epidemiologic studies and, 1: 299, 300–301
   genetic modification of tobacco and, 1: 247
   Japanese-made filter cigarette, 1: 279–280
   Kent cigarettes as, 2: 505–506
   lack of effectiveness of, 1: 201, 298; 2: 493
   machine measurement of, 1: 5, 300, 301
   menthol cigarettes positioned as, 1: 348–349
   public health recommendations and, 1: 299
   for reduction of toxin delivery, 2: 632–633
   Strickman, Robert L., filter of, 1: *115*
   tar and nicotine delivery by, 1: 5
   as tobacco industry's response to health risks, 1: 298–299
   toxin reduction as goal of, 2: 630
   types of filter materials in, 2: 455
   ventilation holes in, 1: 300; 2: 455
   See also "Safer" cigarettes
Lighters
   Chinese use of, 1: 134
   cigar lighters (lamps), 1: *159*, 159–160
   cigarette lighters, 1: *159*
   connoisseurship of, 1: 164
Linnaeus, Carl, taxonomic classification of, 1: 85, 87–88

INDEX

Literature of tobacco, **1: 302–306**
  Arents Collection, 1: 73–76
  early literature, 1: 302–303
  eighteenth century to present, 1: 303–305, *304*, *305*
  English Renaissance literature, 1: 223–226
  future of, 1: 3–6
  pipes in art and literature, 2: 422
  smoking rooms described in, 2: 533, 534
  writers and tobacco use, 2: 562
Litigation, **1: 307–313**
  airline flight attendants' class action lawsuit, 1: 34, 312
  American Tobacco Company dissolution by, 1: 44
  as antismoking movement strategy, 1: 56–57
  Cipollone, Rose, verdict in favor of, 1: 309–310
  class actions, 1: 311–312, *312*
  against Environmental Protection Agency, 2: 516
  expensive legal tactics by tobacco industry, 1: 307–308
  first verdict in favor of plaintiff, 1: 310
  first wave of (1954-1979), 1: 307–308
  against Food and Drug Administration regulation of tobacco, 2: 494
  freedom of speech and advertising restrictions, 1: 21
  by nonsmoking plaintiffs, 1: 312–313
  plaintiffs' responsibility for their disease and, 1: 309
  second wave of (1980s-1990s), 1: 308–31: 310
  state Medicaid reimbursement actions, 1: 311
  third wave of (1990s-present), 1: *309*, 310–313
  tobacco industry documents and, 1: 309–310; 2: 437
  See also Documents of tobacco industry
  See also Lobbying
  See also Master Settlement Agreement (MSA)
"Little Johnny" (Johnny Roventini), in Philip Morris advertising, 2: 407, *408*
Liver cancer, 1: 202t
Lobbying, **1: 314–317**
  definition of, 1: 314
  direct and indirect forms of, 1: 315
  expenditures for, 1: 315–316; 2: 432–433
  guidelines for, in U.S., 1: 317
  in other countries, 1: 317
  by U.S. tobacco industry, 1: 314–317, *316*
  See also Litigation
  See also Politics
Lobeline, as quitting medication, 2: 482, 483
Lost in Translation (film), 1: 237
Lozenges
  for nicotine delivery, 2: 490
  powdered tobacco in, 2: 508
Lucky Strike cigarettes, **1: 318–320**; 1: *319*
  advertised as appetite suppressant, 1: 62–63, *64*, 172; 2: 686
  declining market share of, 1: 319–320
  green packaging of, women's dislike of, 1: 318–319, 336; 2: 474–475
  health concerns and marketing of, 1: 331
  international advertising for, 1: *252*
  success in Germany, 1: *93*
  women as target of advertising for, 1: 318–319, *319*, 335–336
  youth marketing on radio and television, 2: 690
Lump tobacco, 1: 262
Lundgren, Claes, 2: 488
Lung cancer, **1: 320–325**
  calculation of risk of, 1: 201
  case control studies of, 1: 205–206
  causes of, uncertainty about, 1: 321–322
  in celebrity smokers, 2: 672
  denial of smoking link by tobacco industry, 1: 206–207, *217*, 323, 324
  epidemiologic studies of, 1: 200–201, 207, 323–324, 344; 2: 622–623
  lung diseases vs., 1: 322, *322*
  in Marlboro Man model, 2: 672
  mortality rates related to, 1: 205–206, *206*, 320
  Nazi identification of link to smoking, 1: 322, 383
  patient's responsibility for their disease, 1: 309, 325
  politicization of issues related to, 1: 324–325
  public knowledge about smoking and, 1: 174, *174*
  recognition of smoking as cause of, 1: 53, 174, 199–200, 203t, 323–325
  relative risk of, 1: 200
  Royal College of Physicians report on, 2: 623
  secondhand smoke as cause of, 1: 210; 2: 514
  U.S. Surgeon General reports/statements on, 1: 53, 203t, 323; 2: 506
  worldwide smoking patterns and, 1: 177–178
  See also Cancer
Lung diseases
  chronic obstructive pulmonary disease, 1: 202
  emphysema, 1: *203*
  lung cancer vs., 1: 322, *322*
Lynch, David, 1: 235

M

Madak, as blend of opium, 2: 396
Magazines, tobacco advertising in, 1: 13
Mainstream smoke, definition of, 2: 512
Malawi
  tobacco exports of, 2: 647
  tobacco growing and processing by, 1: 28–29
Malaysia, 2: 583
Males. *See* Men
Manumission of slaves, definition of, 2: 529
Marginal populations. *See* Class; Ethnic groups; Women, advertising targeted to; Youth marketing
Marginalization
  definition of, 1: 272
  ostracism of tobacco users as, 2: 403–405, *404*
  of smoking, 2: 565
  See also Smoking restrictions
Marijuana (cannabis), 1: 24, 37, 41
Market share
  definition of, 1: 44
  international investments by tobacco companies, 1: 253
  of Virginia Slims, 2: 664–665
Marketing, **1: 327–336**
  addressing health concerns, 1: 331
  Camel cigarettes, 1: *106*, 106–107; 2: 690
  dealer promotions, 1: 328
  ethnic groups as target of, 1: *228*, 228–229
  fine arts promotion for, 1: 331–332, *333*
  global marketing and sales of cigarettes, 1: 146–147, 251–256

graduated product marketing by U.S. Tobacco, 1: 115–116
outdoor displays for, 1: 328
package inserts/coupons for, 1: 327–328
in retail outlets, 2: 499–500, 501
sports promotion, 1: 332–334
of Virginia Slims, 2: 663–665, *664*
during World War I and II, 1: 329–331
See also Advertising
See also Retailing
See also Sponsorship
See also Youth marketing
Marketing quota, definition of, 2: 387
Marlboro cigarettes, **1: 337–339**
advertising of, 1: 337–339, *338*; 2: 407
ammonia and acetaldehyde in, 1: 10, 114
auto racing sponsorship by, 1: 334
branding and image of, 2: 595
change in image of, 1: 337, 338; 2: 564
cowboy figure in advertising of, 1: 338, *338*; 2: 407, 671, 691
international sales of, 1: 339
lung cancer in Marlboro Man model, 2: 672
"Marlboro Spring Vacation Program," 2: 692
as top-selling brand, 2: 408, 409
Marlboro Country Music Festival, 1: 332
Marquette, Jacques, 1: 103, *361*
The Marriage of Maria Braun/Die Ehe der Maria Braun (film), 1: 234–235
Martin, Dean, 1: 332
Maryland, 1: 120–121
Civil War and, 1: 124
contraband trade in, 2: 544–545
population of, 1: 122
settling of, 1: 120–121
See also Chesapeake region
Masculine identity
as advertising theme, 1: 15–16
Bogart, Humphrey, as ideal of, 1: 236
cigar and pipe smoking as symbol of, 2: 683
cigarettes as symbol of, 1: 148
European middle-class smokers and, 1: 253
in film, 1: 235, *235*
in Marlboro cigarettes advertising, 1: 338, *338*; 2: 407, 595
rebellious behavior of smokers and, 1: 77

tobacco linked to ownership of women's bodies, 2: 680–681
war and masculine image of smoking, 1: 148, 329; 2: 667
Mass production of cigarettes. *See* Industrialization and technology
Master Settlement Agreement (MSA)
advertising restrictions in, 1: 22, 311; 2: 501
cash flow to states from, 2: 437
individual state settlements and, 1: 311
as landmark development in antismoking movement, 1: 57–58
lobbying by tobacco industry for, 1: 315
Philip Morris and, 2: 409
provisions of, 1: 57, 311; 2: 436
release of industry documents as requirement of, 1: 218; 2: 437
state antismoking programs and, 1: 58; 2: 436–437
Match(es), Chinese use of, 1: 134
Match safes, 1: *67*, 164
Mathis-Swanson, Beverly, 2: 465
Matsigenka shamans of Peru, 2: 517–518
Mayas, **1: 339–342**
habit of tobacco smoking, 1: 340–341
importance of tobacco to, 1: 341–342
pagan saint Maximon of, 2: 571
sik'ar as origin of "cigar," 1: 150, 167
tobacco portrayal on artifacts of, 1: *340*, 341–342, *342*
McCall, June, as "Miss Perfecto," 2: *682*
McFarland, Wayne, 2: 482–483
McLaren, Wayne (Marlboro Man model), 2: 672
Mechanization
of tobacco cultivation, 2: 660
See also Industrialization and technology
Medicaid
definition of, 1: 287
Kentucky Medicaid expenditures, 1: 287
litigation for reimbursement of smoke-related expenses, 1: 311
Medical evidence (cause and effect), **1: 343–347**
criteria for causal relationships, 1: 345–346
early twentieth-century view on, 1: 343–344
experimental causation model and, 1: 321

lung cancer causation and, 1: 321
mid-twentieth-century epidemiology and, 1: 344–346
sampling bias and, 1: 345
secondhand smoking debate, 1: 346–347
Medicinal use of tobacco. *See* Therapeutic use of tobacco
Medicine societies, Native American, 2: 381–382
Meerschaum pipes, 2: 418–419, *420*
Men
advertising targeted to, 1: 14
lung cancer death rates for, 1: 320
male and female differences in uptake of smoking, 1: *175*, 175–176
male-female smoking rates, early twenty-first century, 1: 177*t*, 178–181, 178*t*, 180*t*
portrayal in English Renaissance literature, 1: 225
smoking clubs for, 2: 533, *534*, 535
tobacco consumption in developing countries, 1: 195
tobacco shops as social clubs for, 2: 497
See also Masculine identity
See also Sailors
See also Soldiers
Mental disorders, smoking associated with, 2: 470
Mental function
smoking for enhancement of, 2: 562, 669
See also Intellectuals
Menthol
as additive, 1: 7
physiological and respiratory effects of, 1: 349
tobacco chemistry affected by, 1: 113, 114
Menthol cigarettes, **1: 348–349**
African American use of, 1: 228, 349
early brands of, 1: 348
Philippine consumption of, 2: 584
taste of, and appetite suppression, 1: 63
Meprobamate, as quitting medication, 2: 484
Mercantilism. *See* British Empire; Dutch Empire; French Empire; Retailing; Spanish Empire; Trade
Mergers and acquisition, of tobacco companies, 1: 43, 44, 95, 358
Mérimée, Prosper, 2: 670
Mexico, **1: 350–354,** *351*
colonial period of, 1: 350–352
early cigarettes in, 2: 451

INDEX

Mexico CONTINUED
  pre-Columbian period of, 1: 350
  Spanish control of tobacco trade
    in, 1: 35–352
  tobacco factories in, 1: 352
  tobacco industry development in,
    1: 352–354
  See also Spanish Empire
Middle East, **1: 355–360**
  cigarette consumption in,
    1: 358–359
  cultivation and processing of
    tobacco in, 1: 356–358
  modern production and
    consumption in, 1: 359–360
  prohibition of tobacco use in,
    1: 275–276; 2: 355, 459
  socio-cultural significance of
    smoking in, 1: 355–356, 357
  tobacco trade in, 1: 356–358
  water pipes in, 1: 355; 2: 420,
    421
Migrant labor, 1: 295
Mill, John Stuart, 2: 456–457
Milo Violets (cigarettes), 2: 683,
  685
Minnesota
  clean air legislation in, 2: 431, 537
  litigation against tobacco
    industry by, 2: 436
  tobacco control program of, 2: 434
Minnesota Clean Indoor Air Act,
  2: 431
Minors Red Tips (cigarettes), 2: 685
Missionaries, **1: 360–363**, *361, 363*
  changing views of tobacco by
    church, 1: 360–361
  conflicting uses of tobacco by,
    1: 361–362
  promotion of tobacco by,
    1: 362–363
  See also Christianity
Mississippi, litigation against
  tobacco industry, 1: 56–57;
  2: 436
Mitchum, Robert, 2: *671*
Moll Catpurse, 1: 225
Monardes, Nicolás, 1: 86–87;
  2: 558, 610–611
Monopoly
  definition of, 1: 80
  See also Spanish tobacco
    monopoly
  See also State tobacco monopolies
Montez, Lola, 2: 680
Morgan, James J., 2: 477
Mormon Church, 1: 141
Mortality
  cardiovascular disease, 2: 513
  deaths attributable to tobacco,
    1: 204–205, *205*

developing countries, 1: 193, 205,
  211, *212*
Kentucky tobacco users, 1: 287
lung cancer rates, 1: 205–206,
  206, 320
Nazi Germany, 2: 384–385
reduced life span of smokers,
  1: 203–204, *204*
secondhand smoke as cause of,
  2: 513
smoking as defiance of,
  2: 564–566
See also Disease
Movies. See Film
Mozambique, tobacco production
  in, 2: 443
MSA. See Master Settlement
  Agreement (MSA)
Mucha, Alphonse, 2: 667–668, *668*
Mufti, definition of, 1: 355
Murad IV, Sultan, 1: 275, 355;
  2: 459
Murai Bros. and Co., Japan, 1: 280
Murray, Bill, 1: 237
Murrow, Edward R., 1: 332; 2: 672
Music, classical, **1: 364–368**
  early-seventeenth century
    England, 1: 364–365
  nineteenth- and twentieth-
    century compositions,
    1: 365–366
  seventeenth- and eighteenth-
    century Europe, 1: 365
  smoking concerts, 1: *366*,
    366–368; 2: 534
  in tobacco advertising, 1: 368
Music, popular, **1: 369–373**
  chewing tobacco and cigar
    references in, 1: 372–373
  chronological list of songs,
    1: 371t
  Native American influence in,
    1: 372
  slang and metaphor in, 1: 372
  smoking themes reflected in,
    1: 369–372
Muslims. See Islam
My Lady Nicotine: A Study in Smoke
  (Barrie), 1: 304; 2: 422, 534
Myanmar (Burma), 2: *580, 581,*
  582

N

Narghile (water pipe), 1: 355;
  2: 420
Nasal spray, nicotine, 2: 489
NASCAR racing, tobacco industry
  sponsorship of, 1: 333
Nasir al-Din Shah, 1: 272, *273,*
  273

National Academy of Science report
  on environmental tobacco
  smoke, 2: 514
National Cancer Institute (NCI),
  2: 495, 507
National politics. See Lobbying;
  Politics; Regulation of tobacco
  products in United States
Nationalism, definition of, 1: 147
Native Americans, **2: 375–382**
  ancient farming by, 2: 377–380
  calumet use by, 1: 69–70,
    103–105, *169*; 2: *553*, 554
  cosmology of, 2: 554–555, *555*
  creation stories of, 2: 375
  Crow Tobacco Planters, initiation
    and role of, 2: 554
  early tobacco use by, 2: 376–377,
    378
  hallucinogen and tobacco use by,
    1: 259–260, 341; 2: 518, *520,*
    555
  Inipi sweat lodge ceremony of,
    2: 382
  Internet sales of tobacco from
    reservations of, 2: 502
  Iroquois, smoking by, 1: 170;
    2: *553*
  knowledge of tobacco cultivation,
    2: 424
  medicine societies of, 2: 381–382
  methods of tobacco ingestion by,
    2: *379*, 379–380
  origins of tobacco and,
    2: 398–399, 518–519, 552
  postcontact tobacco use of, 2: 556
  pre-Columbian tobacco use of,
    1: 350; 2: 552–556
  recreational tobacco use of,
    2: 555–556
  references to, in popular music,
    1: 372
  religious and spiritual significance
    of tobacco to, 2: 376, 380–382,
    381, 553–554
  sailors' tobacco trade with, 2: 511
  smoking prevalence among,
    1: 227–228
  tobacco as sign of alliances
    between humans, 1: 105;
    2: 554
  women in cigarette advertising,
    2: 669
  See also Shamanism
Navigation Acts
  Chesapeake region tobacco trade
    and, 1: 121
  Dutch tobacco trade and, 1: 221
  purpose of, 1: 96
Navy plug tobacco, 1: 127
Nayarit, Mexico, 1: 353, 354

Nazi Germany, **2: 382–385**
  bans on smoking, 1: 322; 2: 383
  identification of link between smoking and cancer, 1: 322; 2: 383
  pre-war tobacco imports, 2: 645
  smoking by soldiers in, 2: 570
  tobacco as genetic (race) poison, 1: 48, 322; 2: 383
  tobacco consumption in, 2: 384, *384*
  youth smoking and sports participation, 2: 597
NCI (National Cancer Institute), 2: 495, 507
Near East, pipes used in, 2: 420, 421
Nervine, definition of, 2: 562
Nervous system, nicotine in study of, 2: 388
New Amsterdam (New York), 2: 544–545
New Deal, **2: 385–387**
  See also Agricultural Adjustment Act of 1933
New Guinea, 2: 393, 394
New Jersey, tobacco tax rate of, 2: 434
New smoking material (NSM), as tobacco substitute, 2: 507
Newspapers, tobacco advertising in, 1: 13
Ni Zhumo, 2: 612
Nicot, Jean
  introduction of tobacco in Europe, 2: 388, *388*
  nicotine named after, 1: 59, 87–88
Nicotian therapy
  definition of, 2: 612
  in Europe, 2: 611
Nicotiana
  chromosome pairs in, 2: 399
  early European classifications of, 1: 85–87
  Linnaeus's classifications of, 1: 87–88
  number of species of, 2: 397
  origination of, 1: 85; 2: 397–402
Nicotiana alata, 2: 616, 617
Nicotiana attenuata
  early use by Native Americans, 2: 376
  evolutionary geography of, 2: *400*
  sequence of development of, 2: 377, *378*
Nicotiana augustifolia, 2: 616
Nicotiana axillaries, 2: 616
Nicotiana clevelandii, 2: 376
Nicotiana forgetiana, 2: 616–617

Nicotiana fruticosa, 2: 616
Nicotiana glauca, 2: 376, 377, *378*
Nicotiana glutinosa, 2: 616, 617
Nicotiana noctiflora, 2: 617
Nicotiana otophora, 2: *400*, 401
Nicotiana paniculata, 2: 399, *400*, 616
Nicotiana plumbaginifolia, 2: 616
Nicotiana pusilla, 2: 616
Nicotiana quadrivalvis
  cultivation by Native Americans, 2: 398
  early use by Native Americans, 2: 376
  evolutionary geography of, 2: *400*
  multvalvis variety of, 2: 379, 381
  quadrivalvis variety of, 2: 381
  sequence of development of, 2: 377, *378*
Nicotiana rotunifolia, 2: 616
Nicotiana rugosa, 2: 616
Nicotiana rustica
  classification of, 1: 85–88
  early use by Native Americans, 2: 376
  evolutionary geography of, 2: *400*, 401
  inyoka tobacco of Zimbabwe as variety of, 2: 699
  origination of, 2: 399–401, *400*
  sequence of development of, 2: 377, *378*
  shamanistic use of, 2: 518–519, 555
  smoked in calumets, 1: 104
Nicotiana silvestris, 2: *400*, 401, 616
Nicotiana tabacum
  classification of, 1: 85–88
  cultivation of, in Zimbabwe, 2: 700
  early use by Native Americans, 2: 376
  evolutionary geography of, 2: *400*
  origination of, 2: 401–402
  as ornamental plant, 2: 617
  in Philippines, 2: 410
  sequence of development of, 2: 377, *378*
Nicotiana tomentosa, 2: 616, 617
Nicotiana tomentosiformis, 2: 398, *400*, 401–402
Nicotiana trigonophylla
  early use by Native Americans, 2: 376
  evolutionary geography of, 2: *400*
  sequence of development of, 2: 377, *378*

Nicotiana tristis, 2: 616
Nicotiana undulata, 2: 399, *400*, 616
Nicotiana x sanderae, 2: 617
Nicotine, **2: 387–391**
  addiction to, 1: 1–6; 2: 389–391, 467–468
  additive effects on, 1: 8–9
  appetite suppression due to, 1: 65
  disclosure of levels in cigarettes, 2: 506, 619
  drug and alcohol use combined with, 1: 35–41; 2: 468
  effect on body, 2: 389–390
  freebase (non-ionized), 1: 4, 113
  genetic modification for control of, 1: 247
  hallucinogenic effect of, 1: 259
  history of scientific study of, 1: 1–3; 2: 388
  in "light" and filtered cigarettes, 1: 298, 299
  product design and delivery of, 2: 390
  smoking behavior driven by, 2: 390–391
  toxicity of, 2: 521
  See also Hallucinogens
  See also Product design
  See also Psychology and smoking behavior
Nicotine Anonymous, 1: 143
Nicotine cholinergic receptors, 2: 389
Nicotine-delivery medications. *See* Quitting medications
Nicotine gum, 2: 488–489
Nicotine intoxication
  of shamans, 2: 519
  See also Shamanism
Nicotine nasal spray, 2: 489
Nicotine patch, 2: 489, *489*
Nicotine withdrawal
  symptoms/syndrome, 1: 3, 391; 2: 468, 486
Nicotinic receptors, 2: 388, 389
Nitrosamines, tobacco-specific (TSNAs), 1: 113; 2: 632, 633
Nonsmokers' rights
  clean indoor air legislation and, 2: 431–432, *432*, 537
  Non-smokers' Bill of Rights, 1: 34, 54
  political action of tobacco industry and, 2: 431–432, *432*
  smoking restrictions and, 2: 536–537
Norr, Isroy (Roy) M., 1: 52, 337
Norr Newsletter about Smoking and Health, 1: 52
Norway, snuff use in, 2: 550

INDEX

NSM (new smoking material, as tobacco substitute), 2: 507
Nurses' Health Study, 1: 208

O

Oceania, **2: 393–395**, *394*
 demographics of consumption in, 2: 395
 diffusion and trade of tobacco in, 2: 394–395
 early history of tobacco use in, 2: 393–394
 smoking and tobacco use of, 1: 229
 See also Philippines
"Of Tobacco" (Davies), 1: 224, 225
Office of the Surgeon General
 functions of, 2: 495
 See also United States Surgeon General reports
Oligopoly, definition of, 1: 256
Omni cigarettes, 2: 493
"On Tobacco" (Cotton), 1: 225
101 Dalmatians (film), 1: *234*
Opera
 Carmen (Bizet), 1: 365; 2: 670, 681
 smoking linked with sexuality in, 2: 681
 tobacco consumption during, 1: 365, *366*
Opium, **2: 396–397**
 cigarettes as replacement for, 1: 132–133
 definition of, 1: 37
 madak blend of, 2: 396
 morphine in, 2: 389
 pipes for, 1: 132
 pure opium smoking, 2: 396
 as snuff additive, 2: 549
 tobacco smoking with, 1: 37, 47; 2: 396
Oral cancer
 in India, due to *gutka*, 2: *578*
 Surgeon General's findings on, 1: 203t
Oriental leaf tobacco, 2: 451, 454
Origin of tobacco, **2: 397–402**
 in Australia, 2: 398
 Nicotiana rustica and, 2: 399–401, *400*, 518–519
 Nicotiana tabacum and, 2: 401–402
 in North and South America, 2: 398–399, *400*, 552
 shamanism and, 2: 518–519
 See also Diffusion of tobacco
Orinoco colony, Venezuela
 contraband trade in, 2: 542–543
 Warao shamans of, 2: 518, 519–520

Ornamental tobacco plants, **2: 616–618**, *617*, *618*
Ostracism of tobacco users, **2: 403–405**, *404*
Ottoman Empire
 banning of tobacco use in, 1: 275, 355
 cigarette consumption in, 1: 358–359
 cultivation and processing of tobacco in, 1: 356–358
 modern production and consumption in, 1: 359–360
 socio-cultural significance of smoking in, 1: 355–356, *357*
 tobacco trade in, 1: 356–358
 See also Middle East
Ouida (English novelist), 2: 533, 668, 681
Out of the Past (film), 2: 671
Outdoor advertising displays, 1: 328
Outdoor areas, smoking restrictions in, 2: 540
Ovarian cancer, 1: 203t
Oviedo, Gonzalo Fernández de, 1: 136, 302

P

Pacific Ocean islands. *See* Oceania
Package inserts and coupons, 1: 327–328
Packaging
 in advertising, 1: 11–12
 use by soldiers in World War I, 2: 568
PAH (polynuclear aromatic hydrocarbons), 2: 632, 633
Palatinate, definition of, 1: 120
Pall Mall cigarettes, 1: 331
Pan (betel quid), in India, 2: 577–578
Pancreatic cancer, 1: 203t
Panegyric, definition of, 1: 118
Papalettes, 1: 144; 2: 451, 644
Paper cigarettes wrappers
 porosity of, 2: 455
 ventilation holes in, 1: 300; 2: 455
Papua New Guinea, 2: 394, 395
Park, Roswell, 1: *324*
Particulate phase of smoking, 1: 112; 2: 632
Passive smoking. *See* Secondhand smoke
Patch, for nicotine delivery, 2: 489, *489*
Paternalistic, definition of, 1: 124
Patriotism, cigarette smoking linked to, 1: 148, 330; 2: 463, 569

Peace pipes. *See* Calumets
Pearl, Raymond, 1: 203–204, *204*, 268
Pearson, Karl, 1: 343
Peptic ulcer disease, 1: 203t
Personal liberty, prohibition vs., 2: 456–457
Pesticide, tobacco as, 2: 614
Petrified Forest (film), 1: 236
pH
 processing effects on, 1: 113
 "smoke pH," 1: 7
Pharmacotherapy, definition of, 1: 269
Philip Morris, **2: 407–409**
 advertising by, 2: 407; 2 *408*
 appetite satient product developed by, 1: 63
 cigarette girls and sales of, 1: *330*
 diversification of, 2: 408–409
 fire-safe paper technology of, 1: 239
 Master Settlement Agreement effects on, 2: 409
 product placement in films, 2: 477
 promotion of tobacco to retailers, 2: 497
 renamed as Altria, 2: 409
 sponsorship of arts events by, 1: 332
 television program sponsorship by, 1: 332; 2: 408
 See also Marlboro cigarettes
 See also Virginia Slims
Philip V, King of Spain, 2: 588, *589*
Philippines, **2: 410–413**
 diffusion and consumption of tobacco in, 2: 411–412
 flue-cured tobacco production in, 2: 584
 introduction of tobacco in, 2: 410–411
 menthol cigarette smoking in, 2: 584
 production and trade of tobacco in, 2: *411*, 412
 revenue from and governance by Spain, 2: 412–413
 smoking of lit end of cigar, by women, 2: 583, 684
 Spanish tobacco monopoly in, 2: 591
 tobacco production in, 2: 584
Physicians. *See* Doctors
Physiology, definition of, 1: 3
The Picture of Dorian Gray (Wilde), 2: 669
Pipes, **2: 414–423**
 accessories for, 1: 156–159, *157*, *158*, *165*
 African, 1: 28, 68; 2: 421–422

argillite pipes, 1: 104, 157; 2: 415
in art and literature, 2: 422–423
bore diameter of, 1: 70
briar pipes, 1: 157, 167, 214
cases for, 1: 158–159
chibouk as, 1: 214, 355
Chinese, 1: 131–132; 2: 420–421
Civil War pipes and tobacco, 1: 67
clay pipes, 1: 158, 195; 2: 416–417, *417*
definition of, 2: 414
demographics of smokers, 2: 422
European, 1: 68, 70
of Far East, 2: 420–421
German wood *gesteckpfeife*, 2: *415*
as high art, auctioning of, 1: 156, 157
historical importance of, 2: 414
as indirect evidence of tobacco use, 1: 67
Japanese *kiseru* pipes, 1: 277–278, *278*; 2: 420–421
Kuba tribe, Africa, 1: *28*
materials used for, 2: 414–416
meerschaum pipes, 2: 418–419, *420*
Middle Eastern, 1: 355–356, 358; 2: 420, *421*
Native American pipes, 2: *379*, 379–380
Near Eastern, 2: 420, *421*
Oceania, 2: *394*
opium pipes, 1: 132
porcelain pipes, 2: 417–418, *418*
pottery pipes, 2: 418
prehistorical, 1: 68–70
tampers (stoppers) for, 1: 157
tobacco jars for, 1: 157–158, *157–158*
types of, 1: 167
water pipes (hookahs), 1: 131, 167, 195, *304*, 355; 2: 420, *421*, 575
of western world, 2: 416–419
wood pipes, 2: *415*, 418, *419*
See also Calumets
Pipestone, 1: 104
See also Argillite
Pitt, Brad, 2: 672
Plantations, **2: 423–428**
in Caribbean islands and Brazil, 2: 427–428
in Chesapeake region, 2: 424, 427
culture and cultivation of tobacco on, 2: 424–427
definition of, 1: 26
in India, 2: 575
size of slaveholdings of, 1: 293
slave labor on, 2: 424, *425*, 427, 657
See also Architecture

Planters' Protective Association (PPA), 1: 80
Platform pipes, 1: 69
Plug tobacco
definition of, 1: 43
as form of chewing tobacco, 1: 126, 127
hydraulic presses for making, 1: 263
"Lucky Strike" as name for, 1: 318
"Plug Wars," 1: 127
wooden tobacco presses for making, 1: 262
Pocahontas, marriage to John Rolfe, 1: 122
Politics, **2: 428–440**
Chesapeake region, 1: 119
clean indoor air ordinances and, 2: 431–432, *432*, 537
early battles on smoking and health, 2: 429–430, *430*
international politics, 2: 438–439, *439*
lawsuits against tobacco industry and, 2: 436–437
national politics, 2: 432–433
New Deal policies and, 2: 385–387
nonsmokers' rights and, 2: 431–432
preemption of local tobacco control activity and, 2: 429–430, 431, 433
state politics, 2: 433–434
state tobacco control programs and, 2: 429–430, *430*, 434–436
See also Lobbying
Pollen, archaeological evidence of, 2: 377
Polynuclear aromatic hydrocarbons (PAH), 2: 632, 633
Popular music. *See* Music, popular
Porcelain pipes, 2: 417–418, *418*
Portuguese Empire, **2: 440–446**
consumption of tobacco in, 2: 441–442
control of tobacco production, 2: 442–443
end of, by military coup d'état, 2: 445
extent of tobacco commerce in, 2: 445–446
Indian colonies of, 2: 441, 445–446
introduction of tobacco to India, 2: 575
royal tobacco monopoly and finances of, 2: 444–445
smuggling in Portugal, 2: 546
spread of tobacco consumption by sailors, 1: 170

tobacco trade in Africa, 1: 23, 26; 2: 442–443
See also Brazil
Pottery pipes, 2: 418
PPA (Planters' Protective Association), 1: 80
Preemption of tobacco control programs
along with Cigarette Labeling and Advertising Act, 2: 429–430, 675
overturn of preemption in, 2: 433
by tobacco industry, 2: 429–430, 431
Pregnancy, smoking effects on, 1: 209
Premier, as safer cigarette, 2: 508
Priesthoods, Native American, 2: 381
Priests, tobacco use by, 1: 137–138
Private property (headrights), Chesapeake colonies, 1: 119
Processing, **2: 447–450**; 2: *448*
air-curing in, 1: 27; 2: *448*, 448–449, 655
chemistry of tobacco affected by, 1: 113–114
definition of, 2: 447
fire-curing in, 2: 449, 655
flue-curing in, 1: 28; 2: 449–450, 659
grading of leaves in, 2: 655, *666*
prizing of leaves in, 2: 426
by slaves, 1: 293; 2: 530
stripping in, 1: 71, 261; 2: 655
sun-curing of tobacco in, 1: 113; 2: 454
See also Air-cured tobacco
See also Flue-cured tobacco
See also Production of tobacco
Product design, **2: 450–455**
addiction as goal of cigarette design, 1: 4
additives in, 1: 7–8; 2: 454–455
air-cured vs. flue-cured tobacco in, 2: 454
chemical analysis in, 1: 114–115
chemistry of tobacco affected by, 1: 113
cigarettes for women, 2: 452
colored cigarette tips for women, 2: 453
cork-tips for cigarettes, 2: 453
of early cigarettes, 2: 451–452
elements of, 2: 450
fire-safe cigarettes, 1: 238–239
"light" and filtered cigarettes, 1: 5, 299–300; 2: 455, 632–633

INDEX

Product design CONTINUED
  mechanized production of cigarettes and, 2: 452, 454, 453
  of modern cigarettes, 2: 454–455
  nicotine delivery and, 2: 390
  paper characteristics in, 2: 455
  reconstituted tobacco sheets (RTS) in, 2: 454
  safer cigarettes, 2: 507–508
  types of tobacco in, 2: 454
  unregulated health claims for newer products, 2: 493
Product placement in film and television
  advertising potential of, 1: 13
  goals of tobacco industry in, 1: 232
  public relations value of, 2: 477–478
  youth as target of, 2: 691–692
Production of tobacco
  Africa, 1: 26–29
  Brazil, 1: 91; 2: 442–443
  Caribbean region, 1: 108
  Chesapeake region, 1: 99–100, 121–122; 2: 529–531, 638–639
  Cuba, 1: 187, 189
  cultivation requirements for, 2: 425–426, 638
  culture and methods of cultivation, 2: 424–427, 654
  developing countries, 1: 196–197
  Dutch Empire, 1: 220, 220–221; 2: 635
  France, 1: 243–244
  India, 2: 576–577
  Japan, 1: 280–281
  Malawi, 1: 28–29; 2: 647
  Middle East, 1: 356–358
  Native Americans, 2: 398, 424
  Philippines, 2: 411, 412, 584
  on plantations, 2: 424–427, 442
  Portuguese colonies, 2: 442–443
  slavery in, 1: 99, 123–124, 292–293; 2: 526, 526–527
  South and Central America, 2: 570–574
  South Asia, 2: 576–577
  stemming in, 1: 296, 297
  stripping in, 1: 71, 261
  suckering in, 2: 426, 654–655
  Zambia, 1: 29
  Zimbabwe, 2: 699–701, 700
  See also Labor
  See also Processing
  See also United States agriculture
Prohibitions, **2: 456–466**
  in Australia, 2: 619–621
  black market for tobacco resulting from, 2: 492
  in Brazil, 2: 442–443

  by Catholic church, 1: 138, 168, 361; 2: 441, 458
  consumption decline related to, 2: 465
  health as focus of, 2: 464–465
  in Holy Roman Empire, 2: 459
  ineffectiveness of, 2: 456
  in Islamic countries, 1: 275–276, 355
  in Japan, 1: 277, 280; 2: 459
  by Jewish rabbis and law, 1: 283
  in Middle East, 1: 275–276; 2: 355, 459
  monopoly and taxation as alternative to, 2: 460–461
  in Nazi Germany, 1: 322; 2: 383
  from 1970 to present, 2: 463–465
  ostracism of tobacco use and, 2: 403–405, 404
  personal liberty vs., 2: 456–457
  philosophical views of, 2: 456–457, 458
  by Protestant churches, 1: 139–141
  of retail sale of tobacco to youth, 2: 500–501
  in seventeenth century, 2: 458–461
  of smoking in public places, 1: 280; 2: 463
  sumptuary regulations and, 2: 456
  in twentieth-century United States, 2: 461–463
  in United Kingdom, 2: 624
  of youth access to tobacco, 2: 461–462, 461t, 464, 500–501
  See also Advertising restrictions
  See also Antismoking legislation
  See also Antismoking movement before 1950
  See also Antismoking movement from 1950
  See also Regulation of tobacco products in United States
  See also Smoking restrictions
  See also Tobacco control
Project ASSIST (American Stop Smoking Intervention Study), 1: 57; 2: 435–436
Protestant churches
  antitobacco position of, 1: 139–141
  financial support from tobacco, 1: 141–142
  tobacco as salary for ministers, 1: 142
Psilocybin, 1: 259
Psychiatric disorders, smoking associated with, 2: 470

Psychoactive drugs
  definition of, 1: 24
  prehistoric use of tobacco as, 1: 69; 2: 398
  as snuff additive, 2: 549
  See also Drugs
  See also Drugs, tobacco combined with
  See also Hallucinogens
Psychology and smoking behavior, **2: 467–473**
  addiction to nicotine and, 1: 3–6, 8–9; 2: 467–468
  bad habits in America, 1: 77–79, 78
  biological and physiological factors in, 2: 467
  consumer culture and, 1: 79
  disreputable behavior associated with tobacco, 1: 77–79
  identity building and youth tobacco initiation, 2: 694, 696
  initiation of smoking and, 2: 470–471, 472, 689–690, 694, 696
  psychopathology associated with smoking, 2: 470
  quitting smoking and, 2: 468, 484, 485–486
  reasons for smoking, 2: 468–469
  relaxation associated with smoking, 2: 469
  sensorimotor aspects of cigarette handling, 2: 469
  shift of attitude toward smoking, 2: 471–472
  stereotypical behavior of tobacco users, 1: 77–78
  tobacco "chippers" and resistance to, 2: 469
  youth tobacco use and, 2: 464, 696
  See also Rebelliousness
  See also Social and cultural use of tobacco
Public Health Cigarette Smoking Act, 2: 594, 675–676
Public places, restriction of smoking in. *See* Prohibitions; Secondhand smoke; Smoking restrictions
Public relations, **2: 473–479**
  advertising vs., 2: 474
  definition of, 2: 474
  early efforts of tobacco industry at, 2: 474–475, 475
  image enhancement as purpose of, 2: 476
  key tools for, 2: 476–479
  new product publicity and, 2: 476–477

product placement for,
2: 477–478
secondhand smoke controversy
and, 1: 347; 2: 476, 514–516,
539
sponsorship of issues and events
as, 2: 478
Tobacco Industry Research
Committee (TIRC) and,
1: 206–207; 2: 429, 475–476
tobacco industry's response to
disease risk and, 1: 206–207,
298–299, 307–308, 323
Tobacco Institute and, 2: 476
youth "prevention" campaigns in,
2: 478
See also Advertising
See also Marketing
See also Sponsorship
Public transportation, smoking
restrictions on, 1: 34; 2: 537, 539
Puerto Rico, tobacco trade of,
1: 110
Puritans, prohibition of tobacco use
by, 1: 140
Puzzle (pottery) pipes, 2: 418

Q

Quitting, **2: 481–486**
adaptations of models for, 2: 484
aversion therapy in, 2: 485–486
cancer risk reduction by, 1: 5–6
Christian teachings in, 1: 143
current therapies, from 1990
onward, 2: 486
demographic differences in,
1: *176*, 176–177
difficulty of, 2: 468
Ejrup, Borje E.V., treatment for,
2: 483–484
experimental studies of,
2: 485–486
Fredrickson, Donald T., treatment
for, 2: 484
insurance coverage for treatment
programs, 1: 269
McFarland, Wayne, Five Day Plan
of, 2: 482–483
nicotian therapy, 2: 611, 612
popular music references to,
1: 370
programs in the 1960s and
1970s, 2: 484–486
public health service assistance
for, 2: 391
silver nitrate as early "cure," 1: 49
starting and quitting patterns in
U.S., 1: *175*, 175–177, *176*, *177*
twentieth-century models for,
2: 481–484

unsuccessfulness of, 2: 468
weight gain after, 1: 61, 65
willpower and, 2: 481, *482*
withdrawal symptoms with,
2: 391, 468, 486
Quitting medications, **2: 488–490**
as alternative for tobacco, 1: 5;
2: 490
clinical practice guidelines and,
1: 269
early developments of
replacement therapy in, 2: 488
future of, 2: 490
initial marketing of nicotine gum
and, 2: 488–489
lobeline as, 2: 482, 483
lozenge for, 2: 490
meprobamate as, 2: 484
nasal spray for nicotine delivery,
2: 489
nicotine patch as, 2: 489, *489*
oral inhaler for, 2: 490
reduction of cancer risk with, 1: 5

R

"Race (genetic) poison," tobacco as,
1: 47–48, 322; 2: 383
Radio advertising. *See* Broadcast
advertising
Radiocarbon dating, definition of,
1: 68
Raleigh, Walter, 2: 403
Rat Pack, smoking by, 1: 332
Reader's Digest, 1: 208, 337; 2: 429
Reagan, Ronald, 1: *329*
Rebel Without a Cause (film), 1: 235,
235
Rebelliousness
of *kabukimono* gangs, in Japan,
1: 277
lesbian smoking as symbol of,
2: 683
rebel image in films, 1: 235, *235*
smoking as act of defiance,
2: 565–566
smoking by women as visual
statement of, 2: 682
youth smoking as symbol of,
1: *32*; 2: 464, 566
Receptors
discovery of, 2: 388
nicotine cholinergic, 2: 389
nicotinic, 2: 388, 389
Reconstituted tobacco sheets (RTS),
2: 454, 507
Reducing Tobacco Use (U.S. Surgeon
General's report), 2: 677
Reeve, Christopher, 1: 236
Referendum, definition of, 2: 385
The Regimental (pipe), 2: 418

Regulation of tobacco products in
United States, **2: 491–495**
of additives, 1: 9–10
exemption of tobacco from
consumer product statutes,
2: 491
FDA's attempt at regulation,
2: 493–494, 500
federal agencies with role in, 2: 495
future of, 2: 495
large-scale state tobacco control
programs, 2: 434–436
lobbying against tobacco control
legislation, 1: 315
model for regulation, 2: 492
outlawing of cigarettes in, 1: 32
preemption of legislation by
tobacco industry and,
2: 429–430, 431, 433
reasons for regulating tobacco
products, 2: 491–492
of retail tobacco sales, 2: 500–501
tobacco control effects on per
capita consumption, 2: 430,
430
unregulated health claims for
cigarettes and, 2: 492, 493
See also Antismoking legislation
See also Politics
See also Prohibitions
See also Tobacco control
Religion
influence on visual arts, 2: 666
Judaism and, 1:281–283
missionaries of, 1: 360–363
tobacco described as divine,
1: 224
tobacco in Native American
religions, 2: 376, 380–382,
381, 553–554
See also Catholic church
See also Christianity
See also Islam
See also Shamanism
Renoir, Pierre Auguste, 2: 669
Reproductive disorders, links
between smoking and, 1: 209
Republican Attorneys General
Association (RAGA), 2: 437
Republicanism, definition of, 2: 530
Restaurants, banning of smoking
in, 2: 540
Retailing, **2: 496–502**
advertising and marketing in,
2: 499–500, 501
current methods of, 2: 498
history of, 2: 496–497
importance of outlets for,
2: 498–500, *499*
Internet tobacco sales and,
2: 501–502

INDEX

Retailing CONTINUED
   preventing youth access to tobacco, 2: 500–501
   pricing of tobacco and, 2: 501
   regulation of tobacco sales and, 2: 500–501
   See also Marketing
Revel (noncombusting cigarette), 2: 493
Reverse-*chutta*, 2: 583, 684
Reynolds Tobacco Company. *See* R.J. Reynolds Tobacco Company
Rhodesia
   tobacco exports from, 2: 646
   unilateral declaration of independence by, 2: 646, 701
   See also Zimbabwe
Rhodesia Tobacco Association, 2: 700
Ribisi, Giovanni, 1: 237
Ribonucleic acid (RNA), definition of, 2: 629
Richelieu, Armand Jean du Plessis, Cardinal de, 1: 240
Riggs, Bobby, 2: 672
Right to smoke. *See* Nonsmokers' rights; Smokers' rights
Rivers, for transporting tobacco, 2: 426
R.J. Reynolds Tobacco Company
   as leading cigarette manufacturer, 1: 44
   Premier brand, as safer cigarette, 2: 508
   R.J. Reynolds Co. v. Engle, 1: 312
   See also Camel Cigarettes
RNA (ribonucleic acid), definition of, 2: 629
The Roaring Girl (Dekker), 1: 224, 225
Rococo, definition of, 1: 168
Rolfe, John, 1: 119, *120*, **122**
Roosevelt, Franklin, 2: 385, 386
Rose, Jed, 2: 489
Rose Tips (cigarettes), 2: 685
Roventini, Johnny ("Little Johnny"), in Philip Morris advertising, 2: 407, *408*
Royal College of Physicians report, *Smoking and Health*, 2: 623–624
RTS (reconstituted tobacco sheets), 2: 454, 507
Ruby Queen cigarettes, 1: 254
Rush, Benjamin, 1: 38, 140; 2: 461
Russell, Michael, 2: 488, 489, 490
Russia
   Iranian tobacco monopoly and, 1: 272
   prohibition of tobacco use in, 2: 459
Russian Orthodox Church, 1: 138–139

S

Saccharine, in chewing tobacco, 1: 127
Sacramental food, tobacco as, 2: 518, 521–522
"Safer" cigarettes, **2: 505–508**
   disclosure of tar and nicotine levels in, 2: 506
   filter and tar issues related to, 2: 505–506
   fire-safe cigarettes as, 1: 238–239
   genetic modification of tobacco and, 1: 247
   Kent cigarettes as, 2: 505–506
   methods of smoking for risk reduction, 2: 506
   novel cigarette design for, 2: 507–508
   scientific research on, 2: 506–507
   unregulated health claims for newer products, 2: 493
   See also "Light" and filtered cigarettes
Sailors, **2: 509–511**
   chewing tobacco usage of, 1: 125
   dispersal of tobacco by, 2: 509–511
   Navy plug made for, 1: 127
   Portuguese sailors, and spread of tobacco, 1: 170
   snuff usage of, 2: 509
Salvation Army, 1: 48
Sand, George (Armandine Aurore Lucille Dupin), 2: 668, 680, 682
Santo Domingo, tobacco trade of, 1: 109–110
Sartre, Jean Paul, 1: 270, *271*
Schneider, Nina, 2: 488
Schygulla, Hannah, 1: 234
Scientific studies. *See* Epidemiological studies; Medical evidence (cause and effect); Tobacco industry science
Scottish merchants, in colonial tobacco trade, 2: 639, 641
Scouting for Boys (Baden-Powell), 2: 597, 694
Scrap tobacco, in chewing tobacco, 1: 127–128
Secondhand smoke, **2: 512–516**
   activism against, after 1960, 1: 54
   airline passengers and flight attendants affected by, 1: 34, 312
   British tobacco control policy and, 2: *623*, 624
   as carcinogen, 1: 209; 2: 512, 514, 538
   clean indoor air legislation and, 2: 431–432, *432*, 438, 465, 516, 537, 621
   composition of, 2: 512
   controversy over, created by tobacco industry, 2: 476
   Environmental Protection Agency report on, 1: 346–347; 2: 514, 538
   epidemiologic studies of, 2: 513–514
   health effects of, 1: 209–210, 210t; 2: 512–513, *515*, 537–538
   Hirayama, Takeshi, study of nonsmoking wives, 2: 513–514, 515, 624, 627
   international network to prevent legislation on, 2: 438
   Japanese spousal study sponsored by tobacco industry, 2: 627
   lung cancer due to, 1: 210; 2: 514
   "mainstream smoke" as, 2: 512
   names for, 2: 512
   Non-smokers' Bill of Rights and, 1: 34, 54
   public awareness of risks of, 2: 513–514
   "sidestream smoke" as, 1: 50; 2: 512
   smoking restrictions based on, 2: 537–540
   tobacco industry's efforts to prevent restrictions on, 1: 347; 2: 476, 514–516, 539
   trends in public knowledge about smoking and, 1: *174*
   U.S. Surgeon General's report on, 1: 34; 2: 514
Seedbeds, in tobacco cultivation, 2: 425, 654
SEIT (Société d'Exploitation Industrielle des Tabacs), 1: 244
SEITA (Société d'Exploitation Industrielle des Tabacs et Allumettes), 1: 244, 248
Seventh-Day Adventist Church, 1: 143
Sexual symbolism of smoking
   attitude toward women smokers and, 2: 679–681
   bad behavior associated with smoking and, 1: 77–78, *78*
   in *Carmen*, 2: 670, 681
   cigars associated with, 1: 78; 2: 669, 680–681, *682*
   effeminacy of cigarettes and, 2: 452, 562–563
   exotic women in advertising, 2: 669–670, 681
   femme fatale role in films, 1: 233–235, *234*; 2: 670, 680
   glamorization of smoking in films, 1: 79; 2: 564, 670–671, *671*

INDEX

harem women and, 1: 1: 358, 356, *357*; 2: 669, 670, 680–681
male control of women's body and, 2: 680–681
McCall, June, as "Miss Perfecto," 2: *682*
Mucha, Alphonse, Job cigarette poster of, 2: 667–668, *668*
scantily clad women in advertising, 1: 11–12, 32, 78, *78*, 79; 2: 681, *682*, 711
vamp role in films, 1: 233–236
Virginia Slims marketing and, 1: 16, 336; 2: 663, *664*
women's liberation and, 1: 16, 148, 335; 2: 563, 667–669, 686
See also Masculine identity
Shag tobacco, 1: 261, 263
Shamanism, **2: 517–522**
amount of tobacco consumption by, 2: 519–521, 555
in Brazil, 2: 441
characteristics of shamans, 2: 521
Crow Tobacco Planters, initiation and role of, 2: 554
death and resurrection concepts in, 2: 519–520, 555
definition of, 1: 259
diffusion of tobacco and, 2: 518–519
hallucinogen and tobacco use in, 1: 259–260, 341; 2: 518, *520*, 555
healing role of shamans, 2: 519, 554
Matsigenka shamans of Peru, 2: 517–518
Mayan shamans, 1: 340
methods of tobacco intake, 2: 522
Nicotiana rustica use by, 2: 518–519, 555
therapeutic use of tobacco by, 2: 519, 554, *555*
tobacco as sacramental food of, 2: 518, 521–522
tobacco shamanism, 2: 381, 518
transformation of sight and voice in, 2: 520–521
Warao shamans of Orinoco Delta, 2: 518, 519–520
Sharecroppers and sharecropping, **2: 522–524**
after abolition of slavery, 1: 294; 2: 522–523, *523*
after Great Depression, 2: 524
in Brazil and Cuba, 1: 294
definition of, 1: 295
former slaves as, 2: 532, 657–658
in South and Central America, 2: 572
system of sharecropping, 2: 523–524
Sheen, Martin, 2: 477
Sherlock Holmes, 1: 270; 2: 534
Sherman Antitrust Act of 1890, 1: 43
Shi'ism, 1: 275
Sidestream smoke
definition of, 1: 50
mainstream smoke vs., 2: 512
SIDS (Sudden Infant Death Syndrome), 1: 209, 210t
Sight, paranormal, of shamans, 2: 520–521
Signifier, definition of, 1: 289
Sinatra, Frank, 1: 332
Situados de tabaco, 1: 190
Skoal Bandit race car, 1: 333, 334
Slang, in popular music, 1: 372
Slavery and slave trade, **2: 525–532**
in Chesapeake tobacco production, 1: 99, 123–124, 292–293; 2: 526, *526*–527
Civil War and end of, 1: 124
culture and community of slaves, 2: 528–529, 530–531
emancipation of slaves, 2: 531–532
English slave trade, 2: 527
expansion and contraction of tobacco production and, 2: 529–531
exportation to southern states, 2: 530–531
gender of imported slaves, 2: 527–528
Igbo and Biafran slaves, 2: 527
interaction with indentured servants, 2: 527
internal trading of slaves, 2: 529
laws for codifying slavery, 2: 527
manumission of slaves, 2: 530
movement to Western territories, 2: 657
natural increase in slave population, 1: 292–293; 2: 528
on plantations, 2: 424, *425*, 657
population of slaves in Chesapeake, 2: 527
punishment of slaves, 2: 528
quasi-freedom of slaves, 2: 530
runaway slaves, 2: 528
sharecropping by former slaves, 2: 532, 657–658
slaves as chattel, 1: 124
task system in, 2: 657
tobacco processing by, 1: 293; 2: 530
tobacco revolution and, 2: 527–529
women and children in, 2: 528
Slimness. *See* Appetite suppression; Virginia Slims; Weight control
Smallholders
definition of, 1: 29
tobacco production by, 1: 293, 295
Zambia, Africa, 1: 29
Smoke. *See* Chemistry of tobacco and tobacco smoke; Secondhand smoke
Smoke-free public places. *See* Secondhand smoke
"Smoke pH," 1: 7
Smokeless tobacco
banning of, in Australia, 2: 621
carcinogens in, 2: 632
forms of, in India, 2: 577–578
reduction of toxins in, 2: 633
Snus, produced in Sweden, 2: 633
warning labels required for, 2: 676
See also Chewing tobacco
See also Snuff
Smokers' rights
ostracism of smokers and, 2: 404–405
tobacco industry support of, 1: 347; 2: 539
See also Nonsmokers' rights
Smoking and Health (Royal College of Physicians), 2: 623–624
Smoking and Health: A Report of the Surgeon General's Advisory Committee (1964), 1: 53, 208; 2: 675
Smoking cessation. *See* Quitting
Smoking clubs and rooms, **2: 533–535**
Smoking concerts, 1: *366*, 366–368; 2: 534
Smoking jackets, 1: 151, *162*, 163
Smoking restrictions, **2: 535–541**
in air travel, 1: 34; 2: 537–539
in bars, 2: 465, 540, *623*
international restrictions, 2: *538*, *539*, *540*
nonsmokers' rights and, 2: 431–432, *432*, 536–537
ostracism of tobacco use and, 2: 403–405, *404*
public acceptance of smoking and, 2: 536
in public places, 2: 463, 540–541
secondhand smoke and, 1: 347; 2: 476, 514–516, 537–540
in South and Central America, 2: 574
in twenty-first century, 2: 540–541

INDEX

Smoking restrictions CONTINUED
  in workplace, 1: 34; 2: 463, 465, 537, 510, *623*
  See also Antismoking legislation
  See also Prohibitions
  See also Regulation of tobacco products in United States
  See also Tobacco control
Smuggling and contraband, **2: 542–546**
  associated with state tobacco monopolies, 2: 601
  in Bermuda, 2: 543
  in Brazil, 2: 543
  in Caribbean region, 2: 542–543
  in Chesapeake region, 1: 97; 2: 543–544, *544*
  contraband trade at new world plantations, 2: 542–544
  in Cuban tobacco trade with Spain, 1: 187
  in Dominican tobacco trade, 1: 110
  French tobacco trade, 2: 545–546
  in French tobacco trade, 1: 240–241
  Great Britain, 2: 545, 636
  Maryland, 2: 543–544
  Mexican tobacco trade, 1: 351
  Portugal, 2: 557
  South American colonies, 2: 571
  South East Asia, 2: 582
  Spain, 2: 557
  Spanish tobacco monopoly, 2: 588, 589, 591
  taxation as stimulus for, 2: 608
  Thailand, 2: 582
  Virginia, 2: 543–544
Sneeshing (snuff) mull, 1: 162
Snuff, **2: 547–551**
  art (method) of using, 1: 81–82
  carcinogens in, 2: 632
  carotte (carrot)-shaped tobacco for, 1: 261–262; 2: 549
  Chinese use of, 1: 131–132; 2: 613
  current brands of, 1: *128*
  definition of, 1: 3
  Dutch production of, 1: 221
  etiquette for use of, 1: 168; 2: 548
  European aristocrats' use of, 1: 153, 168; 2: 548, *549*, 559
  flavorings for, 2: 549–550
  in India, 2: 579
  individualization of use of, 2: 559–560
  medical use of, 1: 132
  moist dip form of, 1: 168
  nasal (dry) form of, 2: 547
  oral, 1: 128; 2: *548*, 550

preindustrial production of, 1: 261–262
production in Seville, Spain, 1: 187
shipboard use by sailors, 2: 509
social class of users of, 2: 550, 560
toxic and psychotropic substances in, 2: 550
women's use of, 2: *549*, 679
youth marketing of, 1: 334
See also Smokeless tobacco
Snuff accessories, 1: 160–163; 2: 549–550
  handkerchiefs, 1: 163
  snuff bottles, 1: 161–162; 2: 550
  snuff grater, 1: 161
  snuff mull, 1: 162
  snuffboxes, 1: 161, *161*; 2: *549*, 560, 667
Social and cultural use of tobacco, **2: 551–566**
  association of tobacco use with vice, 1: 78–79
  bad habits in America, 1: 77–79
  calumets and alliances, 1: 103–105, *104*; 2: 554
  in China, 1: 134
  defiance symbolized by smoking, 2: 564–566
  early European uses, 2: 558–559
  feminine connotation of cigarette smoking, 2: 562–564
  glamour associated with smoking, 2: 564, 670–671, *671*
  limits on growth of tobacco trade related to, 2: 642
  long-term trends in, 2: 552
  for losing vs. gaining self-control, 2: 552
  mass consumption and branding of cigarettes, 2: 564
  in Middle East, 1: 355–356, *357*
  modern understanding and uses, 2: 556–561
  popular music and, 1: 369–372
  pre-Columbian (Amerindian) understanding and uses, 2: 552–556
  return to smoking by Europeans, 2: 561
  by sailors, 2: *510*, 511
  smoking as supplementary tool, 2: 551–562
  in smoking clubs and rooms, 2: 533–535, *534*, 557
  smoking gallants and, 2: 559
  snuff use, 1: 153, 168; 2: *548*, 549, 550, 559–561
  tobacco shops as social club, 2: 497

transfer and acceptance of tobacco in Europe, 2: *557*, 557 558
youth tobacco use related to, 2: 696
See also Class
See also Native Americans
See also Psychology and smoking behavior
See also Women tobacco users
Social class. See Class
Social Darwinism, and antismoking movement before 1950, 1: 47–48
Société d'Exploitation Industrielle des Tabacs (SEIT), 1: 244
Société d'Exploitation Industrielle des Tabacs et Allumettes (SEITA), 1: 244, 248
Soil Conservation and Domestic Allotment Act (SCDAA), 2: 386
Soil conservation, definition of, 2: 386
Solanaceae family of plants, 1: 86, 259
Soldiers, **2: 568–570**
  cigarettes as support for soldiers, 2: 568–569
  cigarettes included as rations for, 2: 569
  endorsement of smoking by doctors, 1: 215, *215*
  Nazi soldiers, tobacco consumption by, 2: 384, *384*
  smoking of cigarettes during World War I, 1: 32, 49–50, 148, 172; 2: 568–569
  smoking of cigarettes during World War II, 1: 172; 2: 569
Songs. See Music, classical; Music, popular
South and Central America, **2: 570–574**
  archaeology of tobacco in, 1: 68
  black tobacco cultivation in, 2: 572
  blond tobacco production in, 2: 574
  colonial beginnings of, 2: 570–572
  consumption of tobacco in, 2: 572
  contraband trade in, 2: 571
  European markets for tobacco from, 2: 571
  modern tobacco use and restriction in, 2: 574
  origin of tobacco in, 2: 398–399
  rural tobacco production in, 2: 572–573
  shamanism in, 2: 517–522
  tobacco industry in, 2: 573

See also Native Americans
See also Spanish Empire
See also Caribbean
South Asia, **2: 575–579**
  advertising of tobacco in, 1: 15
  colonial introduction of tobacco in, 2: 575
  consumption of tobacco in, 2: 577–579, *578*
  modern cultivation and production of tobacco in, 2: 576–577
  See also India
South East Asia, **2: 579–584**
  advertising of tobacco in, 1: 15
  antitobacco legislation in, 2: 584
  bidi popularity in, 1: 195
  Indonesian *kretek*, **1: 287–289**
  Malaysia, 2: 583
  Myanmar, 2: 582
  South Korea, 2: 584
  Sumatra, 2: 580–582
  Thailand, 2: 580–582: 582
  Vietnam, 2: 580–582: 582
  See also Philippines
South Korea, 2: 584
Southern Rhodesia. *See* Rhodesia; Zimbabwe
Souza Cruz (tobacco company), 1: 91, 254
Spanish Empire, **2: 585–592**
  acceptance and use of tobacco in, 2: 585–586
  Bourbon dynasty in, 2: 588, *589*, 590
  Caribbean tobacco trade of, 1: 107–111; 2: 542–543, 586
  cigar popularity in, 1: 150–151; 2: 644
  cigarette smoking as symbol of decline of, 1: 48
  conversion of Amerindians to Christianity by, 1: 136
  Dutch tobacco imports from colonies of, 1: 221
  smuggling and contraband trade in, 2: 542–543, 546, 588, 589, 591
  tobacco factory in Seville, 1: 187; 2: 587, 588
  tobacco imports of, 2: 592
Spanish tobacco monopoly
  administration and organization of, 2: 587–589
  Caribbean tobacco trade, 1: 109–110; 2: 586
  Castile as first monopoly, 2: 587, 599
  Cuban cigar production control under, 1: 109, 150–151, 187

Cuban tobacco trade, 1: 187–189
  direct control by crown, 2: 588
  Inquisition and, 2: 587
  lack of complete control of, 2: 591–592
  lease holders *(arrendadores)* of, 2: 587
  mercantilism and, 2: 591–592
  Mexican tobacco trade, 1: 351–352
  Philippine tobacco trade, 2: 412–413, 591
  political crises and wars effects on, 2: 589
  Portuguese *conversos* in, 2: 587
  revenue from, 2: 588–589, 600, 601
  smuggling and contraband trading in, 2: 588, 589, 591
  South American tobacco trade, 2: 571
  in Spanish-American colonies, 2: 590–591
  See also Spanish Empire
  See also State tobacco monopolies
Spenser, Edmund, 1: 223, 303
Spinoza, Baruch, 2: 456
Sponsorship, **2: 593–596**
  of arts events, 1: 331–332, *333*
  Australian taxes for buy-out of, 2: 621
  of auto racing, 1: 333–334; 2: 594, *594*, 691
  banning of, 1: 335; 2: 596
  brand vs. corporate sponsorships, 2: 596
  definition of, 2: 593
  DOC organization "housecalls" at, 1: 55
  entertainment of troops as, 1: 330
  key objectives of, 2: 595–596
  patronage vs., 2: 593–594
  for public relations, 2: 478–479
  shift from broadcast advertising to, 2: 594–595, 691–693
  of sports events, 1: 332–334; 2: 594, *594–595*, 690–691
  television exposure of sponsored events, 2: 594, *594*
  television program sponsorship, 1: 332; 2: 408, 690, 690t
  See also Advertising
  See also Marketing
  See also Youth marketing
Sports, **2: 597–599**
  cigarette cards of athletes and, 2: 597
  tobacco advertising associated with, 2: 597–599, *598*, 671–672

tobacco industry sponsorship of, 1: 332–334; 2: 594, *594–595*, 690–691
  youth smoking and, 2: 597
Spuds, as first menthol cigarette brand, 1: 348
Squash, cultivation by Native Americans, 2: 379, *380*
Stallone, Sylvester, 1: 13; 2: 478, 692
Starting to smoke. *See* Initiation of smoking
State politics
  clean indoor air legislation and, 2: 431–432, *432*, 537
  large-scale tobacco control programs and, 2: 434–436
  overturn of tobacco control preemption laws in, 2: 433
  prohibition of tobacco sales to minors and, 2: 461
  taxes on tobacco and, 2: 433–434
  See also Politics
State tobacco monopolies, **2: 599–602**
  abolition of, 2: 601–602
  administration and organization of, 2: 587–589
  as alternative to prohibition, 2: 460–461
  Caribbean tobacco trade, 1: 109–110; 2: 586
  Castile as first monopoly, 2: 587, 599
  contraband trade associated with, 2: 588, 589, 591, 601
  Cuban tobacco trade, 1: 109, 150–151, 187–189
  direct vs. contracted management of, 2: 588, 600
  early monopolies, 2: 599, 600
  fiscal monopolies, 2: 599, 600, 601
  French colonial tobacco trade, 1: 241–242
  Iranian tobacco trade, 1: 272
  Japanese tobacco trade, 1: 20, 280–281
  lack of complete control of, 2: 591–592
  marginal members of society injured by, 2: 510–511
  mercantilism and, 2: 591–592
  Mexican tobacco trade, 1: 351–352
  Middle Eastern tobacco trade, 1: 359
  in Myanmar (Burma), 2: 582
  by papal states, 1: 141–142
  Philippine tobacco trade, 2: 412–413, 591

INDEX

State tobacco monopolies CONTINUED
  Portuguese Empire, 2: 444–445
  revenue from, 2: 588–589, 600, 601
  South American tobacco trade, 2: 571
  in Spanish-American colonies, 2: 590–591
  in Thailand, 2: 582
  types of income from, 2: 601
  in Vietnam, 2: 582
  See also Spanish tobacco monopoly
Statistical significance, in epidemiological studies, 2: 515
Steam power, in tobacco production, 1: 263
Steen, Jan, paintings of, 2: 666, 666
Steinfeld, Jesse, 1: 34, 54
Stemming of tobacco leaf
  by black women, 1: 297
  by women, 1: 296
Sternberg, Josef von, 1: 233, 236
Stogie (Conestoga) cigars, 1: 151
Stomach cancer, 1: 203t
Stone, Sharon, 1: 235
Strickman, Robert L., 1: 115
Stripping
  definition of, 1: 71
  in tobacco processing, 1: 71, 261; 2: 655
Stroke (cerebrovascular disease), 1: 202t
Strong, Josiah, 1: 47–48
Subsidiary, definition of, 1: 74
Substance abuse. *See* Addiction to nicotine; Drugs; Drugs, tobacco combined with
Sudden Infant Death Syndrome (SIDS), 1: 209, 210t
Suerdieck, August, 1: 89, 90
Sugar
  in alcohol, 1: 36
  in caffeinated beverages, 1: 36
  saccharine vs., in chewing tobacco, 1: 127
  as tobacco additive, 1: 7, 36
Sugar cane production, in non-Hispanic Caribbean islands, 1: 108
Sumptuary regulations, 2: 456
Sun-cured tobacco, 1: 113; 2: 454
Superman II (film), 1: 236; 2: 477, 692
Surgeon General reports.
  See United States Surgeon General reports
Survival. *See* Mortality
Sweat lodge ceremony, 2: 382

Sweden
  chewing tobacco usage, 1: 126, 128
  smokeless tobacco (Snus) produced in, 2: 633
  snuff use in, 2: 550, 560
Sweet Caporal cigarettes, 1: 332
Sxymanczyk, Michael E., 2: 477
Symposia, sponsored by tobacco industry, 2: 627–628
Synar Amendment, 2: 432, 500

T

Tabaco en Rama S.A. (Tersa), 1: 353
Tabaco Mexicanos, S.A. (Tabamex), 1: 353
Tabamex (*Tabaco Mexicanos, S.A.*), 1: 353
Talman, William, 2: 672
Tampers (stoppers), for pipes, 1: 157
Tar
  in bidi cigarettes, 2: 577
  cancerous lesions in animal studies of, 1: 206; 2: 506, 631, 631
  conversion to carcinogen, 1: 41
  definition of, 1: 5
  disclosure of levels in cigarettes, 2: 506, 619
  Federal Trade Commission report on levels of, 2: 495
  in light cigarettes, 1: 5, 174, 299, 300
  as particulate phase of smoke, 1: 112; 2: 632
  See also Carcinogens
  See also Chemistry of tobacco and tobacco smoke
Tariff, definition of, 1: 146
Taxation of tobacco products, **2: 603–608**
  as alternative to prohibition, 2: 460–461
  brief history of, 2: 603–604
  British taxes on Chesapeake tobacco, 1: 97
  Christian denominations supported by, 1: 141–142
  consumption linked to rate of, 2: 606–607
  Dutch tobacco imports, 1: 222
  duty-free vendors and, 2: 502
  European taxation of colonial tobacco, 2: 641–642
  fairness of, 2: 608
  as form of regulation, 2: 606–607
  French Empire, 1: 243
  Internet sales and, 2: 502
  Japan, 1: 280, 281
  Middle East, 1: 356–358

  Portuguese Empire, 2: 444–445
  problems related to, 2: 608
  retail stores and collection of taxes, 2: 501
  as revenue policy in past century, 2: 604–606
  smuggling associated with, 2: 608
  for sponsorship buy-outs, 2: 621
  United States taxes, 2: 433–434, 604–605, 604t–605t, 607, 607
  as user fees for public health costs, 2: 607–608
Technology. *See* Industrialization and technology
Teenagers. *See* Youth marketing; Youth tobacco use
Television
  ban on tobacco advertising on, 1: 19–20; 2: 463, 594, 672
  early tobacco advertising on, 1: 13–14
  exposure of advertising in sponsored events, 2: 594, 594
  product placement in film and television, 1: 13, 232; 2: 477–478, 691–692
  programs sponsored by tobacco industry, 1: 332; 2: 408, 690, 690t
  shift from broadcast advertising to sponsorship, 2: 594–595, 691–693
  smoking identified with "bad guys" on, 2: 672
  youth tobacco marketing on, 2: 690–691
Temburni leaf, definition of, 1: 195
Tenant farmers and farming
  after Civil War, 2: 524
  characteristics of, 2: 658
  definition of, 1: 71
  family labor in, 2: 658–659
  houses of, 1: 71
  labor and debts of, 1: 294
Tennis, sponsored by Virginia Slims, 1: 336; 2: 663–664, 672
Tennyson, Alfred Tennyson, 1st Baron, 1: 167
Terry, Luther L., 1: 53; 2: 676
Tersa (*Tabaco en Rama S.A.*), 1: 353
Thailand, 2: 582
Therapeutic use of tobacco, **2: 609–615**
  in Brazil, 2: 441–442
  in Chinese medicine, 1: 130–131; 2: 612–615
  early European concepts of, 1: 173; 2: 558, 610
  in English Renaissance literature, 1: 223–224

European debates over,
2: 611–612
humoral theory of medicine and,
1: 360; 2: 610–611
by Islamic peoples, 1: 274
Monardes' health claims for,
2: 558, 610–611
by Native Americans and
shamans, 2: 381, 519, 554,
555, 609
nicotian therapy as, 2: 611–612
in sixteenth century Europe, 1: 81
of snuff, 1: 132
unregulated health claims tobacco
products and, 2: 492, 493
See also Appetite suppression
See also Shamanism
See also Weight control
Throat cancer, 1: 203
TIRC (Tobacco Industry Research
Committee), 1: 206–207; 2: 429,
475–476
Tithes
definition of, 1: 187
late sixteenth century, 1: 187, 362
TMV (Tobacco Mosaic Virus),
2: 628–630
Tobacciana, 1: 156
See also Connoisseurship
"Tobacco" (Hemminger), 2: 672
"Tobacco" (Hume), 1: 226
Tobacco accessories. See Accessories
Tobacco as ornamental plant,
2: 616–618; 2: 617, 618
Tobacco combined with alcohol and
drugs. See Alcohol; Drugs,
tobacco combined with
Tobacco control
American Stop Smoking
Intervention Study (ASSIST)
program for, 1: 57; 2: 435–436
in Australia, 2: 619–621; 2: 620
in developing countries, 1: 197–198
Framework Convention on
Tobacco Control for, 1: 23;
2: 438–439, 439, 495, 677–678
in Japan, 1: 277, 280–281
in United Kingdom, 2: 622–624
See also Master Settlement
Agreement (MSA)
See also Prohibitions
See also Regulation of tobacco
products in United States
See also Smoking restrictions
Tobacco Industry Research
Committee (TIRC)
creation of, 2: 429
public relations as purpose of,
1: 206–207; 2: 429, 475–476
Tobacco industry science,
2: 625–628

funding of, 2: 625–626
Japanese Spousal Smoking Study
of, 2: 627
lawyers' role in, 2: 625–628
refutation of harmful effects of
tobacco as goal of, 2: 625, 628
secrecy of documents related to,
2: 626–627
See also Documents of tobacco
industry
See also Medical evidence (cause
and effect)
Tobacco Institute, 2: 429,
475–476
Tobacco jars, 1: 157–158, 157–158
Tobacco Monopoly Law (Japan),
1: 280–281
Tobacco Mosaic Virus (TMV),
2: 628–630
Tobacco plant
tobacco as ornamental plant,
2: 616–618, 617, 618
See also Botany (history)
See also Nicotiana
Tobacco presses
hydraulic, 1: 263
for plug making, 1: 262
Tobacco processing. See Processing
Tobacco production. See Production
of tobacco
Tobacco Products Export
Corporation, 1: 255
Tobacco shops, social gatherings at,
2: 497
Tobacco smoke. See Chemistry of
tobacco and tobacco smoke;
Secondhand smoke
Tobacco-specific nitrosamines
(TSNAs), 1: 113; 2: 632, 633
Tobacco substitutes, 2: 507
Tobacco tins, in World War I,
2: 568, 569
Tobacco use cessation. See Quitting
Tolerance of tobacco
dose strength and, 2: 389
nicotine addiction related to,
2: 391
Topsy-Turvy World (Steen), 2: 666,
666
Towns, Charles, 1: 38
Toxins, 2: 630–633
additives as, 1: 9–10
alkaloids as, 2: 632
in bidi cigarettes, 2: 577
chemistry of, in smoke, 1: 112,
112; 2: 390, 632
Chinese medicine and toxicity of
tobacco, 2: 612, 613–614
ciliatoxic compounds as, 1: 114
historical assessment of,
2: 630–631, 631

lethality of nicotine and
shamanistic use of, 2: 521
list of, produced by Dietrich
Hoffmann, 1: 114
measurement of, by smoking
machines, 2: 631, 632, 633
product design changes for
reduction of, 2: 632–633
as snuff additive, 2: 549
See also Carcinogens
See also Chemistry of tobacco and
tobacco smoke
Trade, 2: 634–648
Africa, 1: 25–26
Brazil, 1: 89–92
British Empire, 1: 96–101
Caribbean islands, 1: 107–111
changes after World War II,
2: 645–646
Chesapeake region, 1: 121
consignment system in, 2: 639
diffusion of tobacco to Europe
and, 2: 634–635
dissemination of tobacco vs.
coffee, 2: 635
Dutch Empire, 1: 221–222
Europe's expanding market for
tobacco, 2: 641–642
export of tobacco from New
World, 2: 635, 636, 638, 639
human costs of tobacco trade,
2: 642–643
import destinations for,
2: 645–646, 648
investment in New World tobacco
production, 2: 637
Mexican tobacco trade,
1: 351–352
Middle Eastern, 1: 356–358
Oceania, 2: 394–395
overdependency on tobacco in
Chesapeake, 2: 640
Philippines tobacco trade,
2: 412–413
Portuguese Empire, 2: 442–446
price of tobacco in Chesapeake
region, 1: 98t, 99–101; 2: 638,
639, 640, 641
re-export of tobacco to Europe,
2: 642, 642t
from Revolutionary War into
nineteenth century,
2: 643–644
shipping improvements in,
2: 639–640
Spanish Empire, 1: 107–111;
2: 542–543, 585–592
tobacco colonies and New World
settlement, 2: 635–640, 636
twenty-first century trade,
2: 647–648

INDEX

Trade CONTINUED
  worldwide trade in early twentieth century, 2: 644–645
  Zimbabwe tobacco trade, 2: 699–701
  See also Chesapeake region
  See also Globalization
  See also Smuggling and contraband
  See also Spanish tobacco monopoly
  See also State tobacco monopolies
Trade cards. See Cigarette cards
Tranquilizers, as quitting medication, 2: 484
Transgenic tobacco, 1: 245
Transportation, smoking restrictions on, 2: 537
Trinidad colony, contraband trade in, 2: 542–543
TSNAs (tobacco-specific nitrosamines), 1: 113; 2: 632, 633
Tubular smoking pipes, 1: 68–69
Turkey. See Middle East; Ottoman Empire
Turkish tobacco
  in American cigarette blend, 2: 644
  definition of, 1: 105
  Oriental leaf as, 2: 451
Turrett cigarettes, 1: 332
Twain, Mark, 1: 151, 270
Twist form of chewing tobacco, 1: 127, 261

U

Ubiquitous, definition of, 1: 132
Under Two Flags: A Story of the Household and the Desert (Ouida), 2: 533, 668, 681
United Kingdom
  advertising of tobacco in, 1: 13, 14, 15
  cigar production in, 1: 150–151
  classical music in, 1: 364–365
  English Renaissance literature of, 1: 223–226
  expansion of tobacco market in, 2: 641–642
  gardening of ornamental tobacco in, 2: 616–618, *617*
  literature about tobacco, 1: 223–226, 303–304
  re-export of tobacco to European countries, 2: 642, 642t
  resurgence of smoking in, 2: 561, 565
  safer cigarette research in, 2: 507
  smuggling of tobacco in, 2: 545, 636
  support for soldiers during World War I, 2: 568–569
  taxation of tobacco, 1: 97; 2: 641
  tobacco control in, **2: 622–624**
  as tobacco import destination, 2: 645, 648
  youth smoking in, 2: 694–695, 695
  See also British Empire
United States agriculture, **2: 653–660**
  Agricultural Adjustment Act of 1933 and, 1: 287; 2: 385, 524, 660
  Agricultural Adjustment Act of 1938 and, 2: 387
  Bright leaf tobacco production and, 2: 659
  cotton farming vs. tobacco farming, 2: 656–657
  cultivation of tobacco in, 2: 424–427, 654–657
  geographic expansion of, 2: 653–654, *654*, 660
  mechanization of, 2: 660
  New Deal policies and, 2: 385–387, *386*, 660
  plantations and, 2: 423–427, *425*, 657, 658
  post-slavery labor changes in, 2: 657–659
  sharecropping and, 1: 294; 2: 522–524, *523*, 532, 657–658
  tenant farming and, 1: 294; 2: 524, 658–659
  See also Chesapeake region
  See also Slavery and slave trade
United States Federal Trade Commission (FTC), 1: 5; 2: 675
United States Food and Drug Administration (FDA)
  attempt to regulate tobacco products, 2: 493–494, 500
  features of food and drug regulatory system, 2: 492
  Supreme Court decision against authority to regulation tobacco, 2: 494
  tobacco industry litigation against, 2: 494
United States Lobbying Disclosure Act (1995), 1: 314
United States Surgeon General reports
  on addictive role of nicotine, 2: 390–391
  The Changing Cigarette, 2: 507
  criteria for causal relationships invoked in, 1: 345, 346
  on diseases caused by smoking, 1: 201, 201t–203t
  epidemiologic studies as basis of, 1: 323, 346
  The Health Consequences of Involuntary Smoking, 1: 34; 2: 514
  Reducing Tobacco Use, 2: 677
  Smoking and Health: A Report of the Surgeon General's Advisory Committee (1964), 1: 53, 208; 2: 675
  on smoking as cause of cancer, 1: 53, 201, 201t–203t, 207, 208, 323; 2: 506
Universal Leaf Tobacco Company, 1: 255
Uptown cigarettes, 1: *228*

V

Vamps, in films, 1: 233–236
Vanillin, as tobacco additive, 1: 7
Vanitas (still-life) paintings, 2: 666
Vector Tobacco (company), 1: 247; 2: 493
Vegas, in Cuba, 1: 189
Ventilation holes in cigarette paper, 1: 300; 2: 455
Vice, tobacco use associated with, 1: 78–79
Vietnam, 2: 582
Virginia, 1: 117–119
  Civil War and, 1: 124
  contraband trade in, 2: 544–545
  early disasters in settlement of, 1: 118–119; 2: 637
  foundations for successful colonization of, 1: 119–120
  original colonization plans for, 1: 117–118
  population of, 1: 122
  See also Chesapeake region
Virginia Slims, **2: 663–665**
  advertising of, 1: 16, 63, 336; 2: 663, *664*, 664
  initiation of teenage smoking related to, 1: 173
  market share of, 2: 664–665
  success of, 2: 663, 685
  tennis sponsorship by, 1: 336; 2: 663–664, 672
  weight control associated with, 1: 63, 336
Virginia Slims Tennis Tour, 1: 336; 2: 663–664, 672
Virology, definition of, 2: 629
Visual arts, **2: 665–673**
  Cigarette Labeling and Advertising Act effects on, 2: 672–673
  exotic women in, 2: 669–670
  health issues and, 2: 671–672
  historical background of, 2: 665–667

impressionism and, 2: 669
new woman in, 2: 667–669, *668*
pipes in, 2: 422
recent developments in,
    2: 672–673
smoke as fluid element in,
    2: 665–666
twentieth-century images of
    glamour in, 2: 670–671
unglamorous portrayal of
    tobacco in, 2: 666, *666*
vanitas (still-life) paintings in,
    2: 666
wartime smoking and, 2: 667,
    670
See also Film
Voice, trauma to, in shamanistic
    initiations, 2: 520–521
Von Mises, Ludwig, 2: 458

W

Wahhabi movement, Islamic,
    1: 276
War
    cigarette smoking during Crimean
        War, 2: 670
    male virility associated with
        smoking in, 2: 667, 670
    smoking-related health of soldiers
        and, 2: 597, 694
    See also World War I
    See also World War II
Warao shamans of Orinoco Delta
    initiation ritual of, 2: 519–520
    tobacco as spirit food of, 2: 518
Warehouses, architecture of,
    1: 72–73
Warning labels, **2: 675–678**
    Australia, 2: 620, *620*
    Brazil, 1: 90
    Canada, 1: *211*; 2: 677
    Cigarette Labeling and Advertising
        Act, 2: 429
    developing nations, 2: 676–677
    European Union, 2: 677
    Framework Convention on
        Tobacco Control provisions,
        2: 677–678
    France, 1: *60*
    United States, 2: 675–676, 677,
        678
    wording of, 2: 675–676
Water pipes (hookahs)
    in *Alice in Wonderland*, 1: *304*
    in China, 1: 131
    definition of, 1: 131
    in developing countries, 1: 195
    invention of, 2: 575
    method of smoking, 1: 167
    Middle Eastern, 1: 355; 2: 420, *421*

Websites, of tobacco industry
    documents, 1: 218–219
Weight control
    body weight set-point and, 1: 65
    smoking for, 1: 61–65, 84, 131,
        336
    tobacco advertising related to,
        1: 62–63, *64*, 172, 336; 2: 686
    See also Appetite suppression
Weight gain, after smoking
    cessation, 1: 61, 65; 2: 484
Weissman, George, 1: 332
Westling, Håkan, 2: 488
White Burley tobacco, 1: 246
Wild at Heart (film), 1: 235
The Wild One (film), 1: 231
Wilde, Oscar, 2: 669
Wills (tobacco company), 1: 252;
    2: 452
The Wings of the Dove (film), 2: *680*
Withdrawal symptoms/syndrome,
    1: 3; 2: 391, 468, 486
Women
    as slaves, reproductive value of,
        2: 528
    tobacco personified as,
        2: 680–681
Women, advertising targeted to
    appetite and weight control
        strategy in, 1: 62–63, *64*, 172,
        336; 2: 686
    feminized versions of cigarettes
        and, 2: 453, 685
    gender-specific marketing
        strategies in, 1: 79; 2: 685–686
    increase of smoking associated
        with, 1: 17, 172, 173
    by Lucky Strike, 1: 318–319, *319*,
        335–336
    marketing strategies in,
        1: 335–336
    by Virginia Slims, 1: 16, 63, 173,
        336; 2: 663–665, *664*
    in women's magazines, 1: 13
    before World Wars I and II,
        1: 14–15, 17, 79; 2: 683
    See also Women, in
        advertisements
Women, in advertisements
    acceptability of women in, 1: 79
    on cigarette cards, 1: 11–12, *14*,
        32, 79; 2: 681
    Earhart, Amelia, in, 1: *319*
    exoticism of, 2: 669–670, 681
    McCall, June, as "Miss Perfecto,"
        2: *682*
    Mucha, Alphonse, Job cigarette
        poster of, 2: 667–668, *668*
    scantily clad women in, 1: 11–12,
        32, 78, *78*, 79; 2: 681, *682*,
        711

Women tobacco users, **2: 679–686**
    appetite suppression by smoking,
        1: 61–63, *62*, *64*, 65, 84, 172
    cigarettes as feminine method of
        smoking, 2: 452, 562–564
    cigarettes designed for women,
        2: 452, 453, 685
    colored cigarettes tips for, 2: 453,
        685
    criticism of, by men, 2: 682
    in developing countries, 1: 195
    in early twentieth century, 1: 32
    femme fatale role of, 1: 233–235,
        234; 2: 670, 680
    gender-based attitudes toward,
        2: 679–681
    global scale of cigarette smoking
        and, 2: 684–685
    in harems, 1: 356, *357*, 358;
        2: 669
    health risks of smoking and,
        2: 684–685
    in India, 2: 579
    initiation of smoking by, 1: 17,
        173, *175*, 175–176
    lung cancer death rates of,
        1: 320
    male control of women's body
        and, 2: 680–681
    male-female smoking rates,
        1: 177t, 178–181, 178t, 180t;
        2: 684
    in Middle East, 1: 355, 357, 358,
        359
    Nazi Germany's antismoking
        attitude and, 2: 383
    negritas, smoked by Filipinas,
        2: 583
    new (independent) women as,
        1: 32; 2: 667–669, *668*, 680,
        680
    in Oceania, 2: *394*, 395
    portrayal in English Renaissance
        literature, 1: 225
    in preindustrial Europe and North
        America, 2: 679
    respectability of smoking and,
        1: 77, 79, 148–149;
        2: 680–681, *682*
    secondhand smoke effects
        on, 2: 513–514, *515*,
        624, 627
    sexual attractiveness of, 2: 564
    sexual connotations of,
        2: 680–681, *682*
    vamp role, in films, 1: 233–236
    See also Sexual symbolism of
        smoking
    See also Women, advertising
        targeted to
    See also Women's liberation

Women tobacco workers,
  1: 295–297
 black women union leaders as,
  1: 297
 Carmen (opera character) as,
  1: 365; 2: 670, 681
 cheroot manufacturing by, 2: *581*
 cigar manufacturing by, 1: 266
 cigarreras, in Seville, 1: 295
 in Cuban factories, 1: 296
 in early industrial production,
  1: 263
 kretek manufacturing by, 1: *288*
 low pay of, 1: 263, 266
Women's Christian Temperance
  Union (WCTU), 1: 48, 141
Women's liberation
 as advertising theme, 1: 16, 335;
  2: 686
 cigarette smoking as symbol of,
  1: 148; 2: 563, 667–669
 Virginia Slims as image of, 1: 16,
  336; 2: 663, *664*
 of Western women, before 1950,
  2: 682–683
Wood, Edwin, 1: *112*
Wood pipes, 2: *415*, *418*, *419*
 briar pipes as, 1: 157, 167; 2: 418
 See also Briar pipes
Workplace, banning of smoking in
  from 1970 to present, 2: 463
 in airplanes and public
  transportation, 1: 34; 2: 537
 in bars, 2: 465, 540, *623*
 California's Smoke-Free
  Workplace Act, 2: 465
World War I
 gender patterns of tobacco use
  affected by, 2: 682
 patriotism linked to smoking
  during, 1: 148, 330; 2: 463, 569
 popularity of smoking during,
  1: 329
 "Smokes for Soldiers" campaigns,
  1: 49–50
 smoking by soldiers during,
  1: 32, 49–50, 79, 148, 172,
  329; 2: 568–569, *569*
 social connotations of smoking
  during, 1: 79
 tobacco advertising during,
  2: 670
 tobacco exports in pre-war
  period, 2: 645

World War II
 Japanese smoking consumption
  after, 1: *279*, *279*
 provision of cigarettes to soldiers
  during, 1: 330
 smoking by soldiers during,
  1: 172
 tobacco advertising during,
  2: 670
 tobacco trade after, 2: 645–646

X

X-Files (television program), 2: 672

Y

Yin and yang, in Chinese medicine,
  2: 612
Young Men's Christian Association
  (YMCA), 1: 48; 2: 485
Youth access laws
 "Action against Access" campaign,
  by tobacco industry, 2: 477,
  478
 retail sales of tobacco controlled
  by, 2: 500–501
 tobacco industry actions against,
  2: 431–432, 478
Youth marketing, **2: 689–693**
 advertising forms and strategies
  in, 1: 15, 334–335
 advertising restrictions for
  protecting youth, 1: 22
 on airways (radio and television),
  2: 690–691
 brand loyalty linked to age,
  2: 689
 importance to tobacco industry
  revenue, 2: 689–690
 Joe Camel advertising campaign
  in, 1: 22, 107, 173, 335;
  2: 692–693
 popular cigarettes brands and,
  2: 692t, 693
 product placement and,
  2: 691–692
 retail outlets advertising in,
  2: 499–500
 sports sponsorship and,
  2: 690–691
 by young women, 1: 173
 See also Children
Youth tobacco use, **2: 694–696**

alcohol and drug use associated
  with, 1: 38, 39–40
 in early twentieth century,
  1: 31–33; 2: 694–695
 in early twenty-first century,
  2: 694, 695–696
 in Great Britain, early twentieth
  century, 2: 694–695, *695*
 initiation of smoking and,
  2: 470–471, *471*, 689, 695–696
 in later twentieth century, 1: 33
 in Middle East, 1: 359
 by Native Americans, 1: 227
 by naughty boys before World
  War I, 1: 78
 prohibition of tobacco sales to
  minors, 2: 461
 as rebellious gesture, 1: 32, 235,
  235; 2: 464, 566
 smoothness and mildness of
  cigarettes as factors in, 1: 116
 sports participation and, 2: 597
 by young women, 1: 32, 173;
  2: 684
 See also Children
 See also Youth access laws
 See also Youth marketing

Z

Zambia, Africa, tobacco growing
  and processing, 1: 29
Zhang Jiebin, 1: 130; 2: 612
Zhang Lu, 2: 613–614
Zimbabwe, **2: 699–701**
 air-cured tobacco in, 2: 699
 colonial tobacco production in,
  2: 699–701
 flu-cured tobacco production in,
  2: 700
 inyoka tobacco of, 2: 699
 land redistribution and future
  tobacco trade in, 2: 701
 post-colonial tobacco production
  in, 2: 700, 701
 preferential purchase agreements
  with British tobacco
  manufacturers, 2: 700,
  700t–701t
 tobacco buyers in, 2: 647
 tobacco exports from, 2: 646, *646*
 tobacco workers in, 1: 294–295
 See also Rhodesia
Zimmermann, Detlef, 1: *93*

3 1333 03200 0232